Water
Quality & Quantity
Issues
for Turfgrasses in Urban Landscapes

Water Quality & Quantity Issues for Turfgrasses in Urban Landscapes

Water Quality and Quantity Issues for Turfgrasses in Urban Landscapes

CAST Special Publication 27

Proceedings of the Workshop on "Water Quality and Quantity Issues for Turfgrasses in Urban Landscapes"

Las Vegas, Nevada

January 2006

Water Quality and Quantity Issues for Turfgrasses in Urban Landscapes

Edited by
James B. Beard
Michael P. Kenna

Council for Agricultural Science and Technology
Ames, Iowa, USA

The Science Source for Food,
Agricultural, and Environmental Issues

© 2008 by The Council for Agricultural Science and Technology
All rights reserved
Printed in the United States of America
by PSI Printing Services, Inc., Belmond, Iowa
12 11 10 09 08   5 4 3 2 1

Library of Congress Cataloging-in-Publication
information is available.

ISBN 978-1-887383-29-5
ISSN 0194-407x

Contents

List of Figures.. xiii
List of Tables xv
Proceedings Contributors... xvii
Workshop Steering Committee.. xviii
Workshop Sponsors xix
Acknowledgments.. xix
Foreword... xx

Executive Summary 1

1   Urban Turfgrasses in Times of a Water Crisis: Benefits and Concerns
Douglas H. Fender 11
Introduction, 11
Landscape Plant Water Requirements, 12
Acreage and Economics of Turfgrass, 12
Water Availability for Humans, 13
A Crisis of Devastating Magnitude, 14
Multiple Factors Converging to a Crisis, 17
Unique "Water World" Words, 17
Peak Demand Requires Attention, 23
Conservation Solutions May Create a Crisis, 26
Reconciling Indoor versus Outdoor Water Uses, 27
Maximizing the Benefits of Turfgrass and Minimizing the Detrimental Impacts, 29
Literature Cited, 30

2   Integrated Multiple Factor Considerations in Low-Precipitation Landscape Approaches
James B. Beard... 33
Introduction, 33
Diseases and Airborne Dust, 33
Heat Stress Islands, 34
Wildfires, 35
Soil Erosion and Flooding, 36
Urban Pollutants, 36
Criminal Activity, 37
Human Disharmony, 37
Summary, 37
Literature Cited, 39

3   Regulatory Considerations for Water Quality Protection
Beth Hall.. 41
Summary, 41

Early State and Municipal Water Pollution Control Programs, 41
Early Federal Involvement with Water Resource Quality, 42
Post-Environmental Protection Agency Water Regulation, 43
    Clean Water Act, 44
    Evolution of Clean Water Act Implementation, 46
    Safe Drinking Water Act, 46
Water Security, 49
Clean Water Act/Safe Drinking Water Act Integration, 49
Source Water Protection, 50
Sustainable Development, 51
Conclusion, 52

4 Municipal Water Use Policies
Andrew W. Richardson 53
Introduction, 53
Historical and Current Municipal Water Use Policies, 54
    Institutional/Structural, 54
    Operational, 55
    Economic, 56
    Sociopolitical, 57
Water Industry Trends, 57
AWWA State of the Water Industry Results, 59
Trends' Impact on Future Municipal Water Use Policies, 60
    Institutional/Structural, 60
    Operational, 62
    Economic, 62
    Sociopolitical, 62
Summary, 63
Literature Cited, 63

5 Turfgrass and the Environment
Michael P. Kenna 65
Introduction, 65
Turfgrass Information File, 65
Turfgrass Biology and Distribution in the United States, 65
    U.S. Climate Zones, 67
    Cool-Season Turfgrasses, 67
    Warm-Season Turfgrasses, 68
    Native Grasses, 70
Turfgrass Water Use Rates, 71
    Drought Resistance, 71
    Water Quality and Turfgrass Culture, 74

Turfgrass Pest Management and Fertilization Effects on Water Quality, 75
    Volatilization, 75
    Water Solubility, 76
    Sorption, 77
    Plant Uptake, 77
    Degradation, 77
    Leaching, 78
    Runoff, 81
Biological Control, 83
Wildlife Programs, 84
Literature Cited, 85

6   Soil Water in Managed Turfgrass Landscapes
Ed McCoy 91
Introduction, 91
Soil Components and Their Associated Pores, 91
Soil Pores and Water, 93
The Fates of Water in Soil, 95
Soil Layering in the Urban Landscape, 98
Soil Compaction, 99
Root Zones for High-Traffic Areas, 99
Soil Profiles Found in High-Traffic Areas, 100
Slope Effects on Water within Layered Soils, 103
Impact of Layered Soils on Available Water, 104
Conclusion, 105
Literature Cited, 105

7   Leaching of Pesticides and Nitrate in Turfgrasses
Bruce Branham 107
Introduction, 107
Pesticide Leaching, 107
    Factors that Attenuate Leaching in Turf, 108
    Pesticide Leaching in Turf, 110
Nitrogen, 112
    Fate of Nitrogen in Turf, 113
    Clipping Removal, 114
Summary, 116
Literature Cited, 117

8 Nutrient and Pesticide Transport in Surface Runoff from Perennial Grasses in the Urban Landscape
K. W. King and J. C. Balogh 121

Introduction, 121
Hydrology and Runoff, 121
    Climate, 122
    Soil and Site Conditions, 123
    Turfgrass Management, 124
Runoff and Pollutant Discharges, 125
    Discharge Volume, 125
    Nutrients, 126
    Pesticides, 126
Turfgrass Management System Plans, 127
Directions for the Future, 127
Conclusions, 128
Literature Cited, 144

9 Pesticide and Nutrient Modeling
Stuart Z. Cohen, Qingli Ma, N. LaJan Barnes, and Scott Jackson 153

Introduction, 153
The Importance of Turfgrass Pesticides in the United States, 153
Mathematical Models for Turfgrass Chemicals: What are They?, 153
    Definition, 153
    Uses of Mathematical Models for Turfgrass Chemicals, 154
Unique Aspects of the Turfgrass System Relevant to Modeling Turfgrass, 155
Most Commonly Used Models, 155
    Use of Models in a Tiered Assessment, 156
    PRZM and PRZM/EXAMS, 157
    Comparison of PRZM/EXAMS Results with Monitoring Data, 158
    TurfPQ, 161
    RZWQM, 161
    Generic Estimated Environmental Concentration (GENEEC), 162
    Screening Concentration in Ground Water (SCI-GROW), 162
    Soil and Water Assessment Tool (SWAT), 162
    Nutrient Modeling, 163
Realistic Model Input Parameters, 163
    Critical Assumptions, 163
    Low-Probability Assumptions, 165
Summary, 165
Conclusions, 165

Recommendations, 166
Glossary, 167
Literature Cited, 167

10 Urban Landscape Water Conservation and the Species Effect
Dale A. Devitt and Robert L. Morris 171
Introduction, 171
Urban Landscapes, 172
Plant Water Use/Landscape Water Use, 173
  Is There a Species Effect on Landscape Water Use?, 176
  Turfgrass Water Use, 178
  Water Use of Trees and Ornamentals, 179
  Crop Coefficients, 182
Turfgrass and the Mixed Landscape, 183
Trees and the Mixed Landscape, 184
Conclusions, 185
Literature Cited, 186

11 Turfgrass Water Requirements and Factors Affecting Water Usage
Bingru Huang 193
Introduction, 193
Water Use Characteristics of Cool-Season and Warm-Season Turfgrass Species, 194
  Cool-Season Turfgrasses, 195
  Warm-Season Turfgrasses, 196
Plant Growth Characteristics Affecting Water Use, 196
  Canopy/Shoot Characteristics, 196
  Rooting Characteristics, 197
  Dormancy and Water Use, 197
Environmental Factors Influencing Turfgrass Water Use, 198
  Solar Radiation, 198
  Temperature, 199
  Relative Humidity, 199
  Wind, 199
  Soil Moisture, 200
  Soil Texture and Physical Properties, 200
Summary, 201
Literature Cited, 201

12 Turfgrass Cultural Practices for Water Conservation
Robert C. Shearman 205
Abstract, 205
Introduction, 205
Mowing Effects, 206
    Mowing Height, 206
    Mowing Frequency, 207
    Plant Growth Regulators, 208
    Mower Blade, 208
Turfgrass Nutrition, 209
Turfgrass Irrigation, 210
    Deficit Irrigation, 210
    Irrigation Scheduling, 212
Soil Cultivation, Topdressing, and Wetting Agents, 213
    Soil Cultivation, 213
    Topdressing, 214
    Wetting Agents, 214
Antitranspirants, Biostimulants, and Pesticides, 215
Summary, 216
Literature Cited, 216

13 Achieving High Efficiency in Water Application via Overhead Sprinkler Irrigation
Michael T. Huck and David F. Zoldoske 223
Introduction, 223
Water-Efficient Sprinkler Irrigation Systems, 223
Irrigation Design Guidelines for Water-Efficient Systems, 223
    Develop a Plan, 223
    Hydrozoning, 224
    Efficient Sprinkler Head Layouts, 224
    Hydraulics (Flow, Pressure, and Pipe Sizing), 224
    Other Requirements, 226
Sprinkler and Nozzle Technologies, 227
    Current Sprinkler Performance Limitations, 227
    Future Improvements, 228
    Role of Software in Irrigation Design, 229
    Modeling Sprinkler Coverage, 229
Development of Smart Water Application Technologies™ (SWAT™), 234
    Irrigation Controllers, 234
Soil Moisture Sensors, 236
    Future of SWAT™, 238

Irrigation System Installation, 238
Irrigation System Maintenance, 240
Summary, 240
Literature Cited, 241

14 Recycled, Gray, and Saline Water Irrigation for Turfgrass
M. Ali Harivandi, Kenneth B. Marcum, and Yaling Qian 243
Introduction, 243
Alternative Sources of Water, 243
    Recycled Water, 244
    Gray Water, 244
    Saline Water, 245
Agronomic Concerns, 245
    Salinity, 245
    Sodium, 246
    Bicarbonate and Carbonate, 247
    Hydrogen Ion Activity (pH), 247
    Chloride, 248
    Chlorine, 248
    Boron, 248
    Nutrients, 248
Irrigation System Issues, 249
Managing Alternative Sources of Water, 250
    Salt-Tolerant Turfgrasses, 251
    Blending Saline Water with Less Saline Water, 252
    Drainage and Leaching of Salts, 252
    Chemical Treatment (Amending) of Water and Soil, 253
Conclusions, 253
Literature Cited, 254

15 San Antonio Water Conservation Program Addresses Lawngrass/Landscapes
Calvin Finch 259
Introduction, 259
Why Is Turfgrass a Target of Water Purveyors Seeking to Decrease Water Use?, 260
The Beneficial Aspects of Lawns, 261
Other Sources of Savings, 262
    Infrastructure Improvements, 262
    Low-Flow Toilets, 263
Turfgrass-Related Programming, 263
    Conservation Pricing of Water, 263

General Education, 264
    The Media, 264
    Low-Water-Use Landscape, 265
    Irrigation Programming, 265
    Seasonal Irrigation Program, 266
    Water Saver Rebate, 267
    Water Conservation Ordinance, 268
    Drought Restrictions and Enforcement, 270
Conclusion, 271
Literature Cited, 271

16 Best Management Practices for Turfgrass Water Resources: Holistic-Systems Approach
Robert N. Carrow and Ronny R. Duncan 273
The Real Issue, 273
Best Management Practices, 274
    History, 274
    Characteristics, 275
    BMPs Applied to Other Water Issues, 276
BMPs for Irrigation Water Quality Management, 277
BMPs for Water-Use Efficiency/Conservation on Specific Sites, 279
    Site Assessment and Planning, 283
    Identify, Evaluate, and Select Water Conservation Options, 284
    Assess Benefits and Costs of Water Conservation Measures, 286
BMPs for Water-Use Efficiency/Conservation, 287
Integration of BMPs, 287
Conclusion, 289
Literature Cited, 290

Appendix A: Workshop Participant List 295

Appendix B: Workshop Agenda 297

Figures

1.1 Percentage change in U.S. population: 2000–2003, 15
1.2 Precipitation patterns in the United States, I-7
1.3 Estimated use of water in the United States in 2000, 16
1.4 Public-supplied and self-supplied populations in the United States, 22
1.5 Total freshwater withdrawals over a 50-year period compared to population growth, 23
1.6 Utah peak streamflows and municipal demand, 24
1.7 Weekly watering recommendations for Denver, Colorado, 25
2.1 Diagrammatic summary of benefits derived from turfgrass use, 38
5.1 Parts of the turfgrass plant and cross-section of the crown, 66
5.2 Major turfgrass climatic zones and geographic distribution of species in the United States, I-8
6.1 Soil water content as a function of height above a water table for three soils, I-9
6.2 Velocity of water flow and depth of water penetration within 2-m columns of three different soils, I-10
6.3 Principal paths of water flow in a soil containing a uniformly vegetated surface and with a slowly permeable layer at depth, 96
6.4 Soil water content within 1-m-long soil columns of three soils, I-11
6.5 Range of root zones and their respective depths as commonly found in high-traffic turfgrass soils, I-12
6.6 Impact of a 3% slope on lateral water flow in a layered soil profile, I-13
8.1 The hydrologic cycle, 122
9.1 Pesticide applications to turf, 154
9.2 Comparison of Tier I acute model predictions versus monitoring, 158
9.3 Comparison of Tier I with Tier II acute exposure predictions, 159

9.4   Relationship between total active applied and PRZM/EXAMS overprediction, 160

9.5   Implications of an erodibility parameter K=0.33 for the EPA/OPP cool-season turf scenario, 164

11.1 Evapotranspiration rate of creeping bentgrass, 195

13.1 Effect of pressure on sprinkler distribution profiles, 226

13.2 A nozzle with a modified orifice, 229

13.3 Examples of different-shaped single-leg profile data, 230

13.4 Two densograms represent coverage of two different nozzle combinations, I-14

13.5 Application uniformity versus sprinkler spacing distance for the nozzle and sprinkler combination and resultant profile presented in Figure 13.3, top, 232

13.6 Application uniformity versus sprinkler spacing distance for the nozzle and sprinkler combination and resultant profile presented in Figure 13.3, bottom, 233

15.1 San Antonio Water System water use by year in liters per person per day, 260

15.2 *San Antonio Landscape Care Guide*, I-15

15.3 Planting turfgrass test plots to determine which turfgrass species/cultivars will survive 60 d of drought, I-16

Tables

1.1 Relative amounts of various forms of water on the Earth, 14

1.2 Comparison of freshwater consumptive use in the United States for 1990, by category, 20

1.3 Fate of water in public water supplies of the United States, 1990, 20

1.4 Total per capita water withdrawals, deliveries, and leakage for 10 U.S. states, 21

1.5 Domestic indoor water use reported by the U.S. EPA and the city of Albuquerque, New Mexico, 27

1.6 Average national residential water use compared to California and Pennsylvania, 28

1.7 Indoor versus outdoor residential water use, 28

2.1 Temperature comparisons of four types of surfaces on August 20, in College Station, Texas, 35

5.1 Comparison of intensity of management practices between a golf course and a home lawn, 67

5.2 Summary of mean rates of turfgrass evapotranspiration (ET), 72

5.3 Turfgrass morphological, anatomical, and physical characteristics contributing to drought resistance, 73

5.4 Relative salt resistance of several turfgrass species used in the United States, 75

5.5 Percentage of nitrogen recovered from undisturbed soil columns with turfgrass cover, 79

5.6 Percentage of applied pesticide found in drainage water from experimental fairways, 80

5.7 Percentage of applied pesticide and concentration of pesticide transported from runoff plots during a storm event that occurred 24 hours after application, 82

7.1 Comparison of half-lives of pesticides applied to turf or to soil with turf removed, 109

8.1 Rainfall-runoff coefficients from turfgrass, residential, and forest land use classifications, 129

8.2 Selected studies identifying nutrient and sediment concentrations (mg L^{-1}) in surface waters from grassed and wooded catchments, 133

8.3 Selected studies identifying nutrient loads (kg/ha/yr) in surface waters from grassed, wooded, and agricultural catchments, 136

8.4 Concentrations (μgL^{-1}) of pesticide residues measured in the surface water in and around the urban landscape, 139

8.5 Selected list of best management practices (BMPs) tested under different turfgrass management scenarios, 142

9.1 Key model algorithms, 156

11.1 Relative maximum evapotranspiration rates of 24 turfgrass species, 194

13.1 Turfgrass and landscape sprinkler system field audit performance rankings by distribution uniformity and sprinkler type, 228

13.2 Current and potential soil moisture sensor technologies for landscapes, 237

14.1 Relative tolerances of turfgrass species to soil salinity (EC_e), 251

15.1 SAWS water saver rebate awards, 268

16.1 Criteria that can be used by water systems planners in selecting conservation measures for implementation on a community-wide or watershed basis, 280

16.2 Outline of the planning process and components of a golf course BMPs for water-use efficiency/conservation, 281

Proceedings Contributors

James B. Beard (Coeditor), Professor Emeritus, Texas A&M University, College Station, Texas

Michael P. Kenna (Coeditor), U.S. Golf Association, Green Section Research, Stillwater, Oklahoma

J. C. Balogh, Spectrum Research Inc., Duluth, Minnesota

N. LaJan Barnes, Environmental and Turf Services, Inc., Wheaton, Maryland

Bruce E. Branham, Department of Natural Resources and Environmental Sciences, University of Illinois, Urbana–Champaign

Robert N. Carrow, Crop and Soil Science Department, University of Georgia, Griffin

Stuart Cohen, Environmental and Turf Services, Inc., Wheaton, Maryland

Dale A. Devitt, Center for Urban Water Conservation, University of Nevada, Las Vegas

Ronny R. Duncan, Turf Ecosystems, LLC, San Antonio, Texas

Douglas H. Fender, D. Fender and Associates, Barrington, Illinois

Calvin R. Finch, San Antonio Water System, San Antonio, Texas

Beth Hall, Drinking Water Protection Division, U.S. Environmental Protection Agency, Washington, D.C.

M. Ali Harivandi, Environmental Horticulture, University of California Cooperative Extension, Alameda

Bingru Huang, Department of Plant Biology and Pathology, Rutgers University, New Brunswick, New Jersey

Michael T. Huck, Irrigation & Turfgrass Services, Dana Point, California

Scott Jackson, BASF Corporation, Research Triangle Park, North Carolina

Kevin W. King, USDA–ARS, Columbus, Ohio

Qingli Ma, Environmental and Turf Services, Inc., Wheaton, Maryland

Kenneth B. Marcum, Applied Bio Sciences, Arizona State University, Phoenix

Ed McCoy, School of Natural Resources, The Ohio State University, Wooster

Robert L. Morris, Cooperative Extension, The University of Nevada, Reno

Yaling Qian, Colorado State University, Fort Collins

Andrew W. Richardson, Greeley and Hansen LLC, Phoenix, Arizona

Robert C. Shearman, Department of Agronomy and Horticulture, University of Nebraska, Lincoln

David F. Zoldoske, The Center for Irrigation Technology, California State University, Fresno

Workshop Steering Committee

James B. Beard (Cochair), Professor Emeritus, Texas A&M University, College Station, Texas

Michael P. Kenna (Cochair), U.S. Golf Association, Green Section Research, Stillwater, Oklahoma

James L. Baker (Project Manager), Department of Agricultural and Biosystems Engineering, Iowa State University, Ames

Evert K. Byington, Rangeland, Pasture, and Forages, National Resources and Sustainable Agriculture

Robert N. Carrow, Crop and Soil Science Department, University of Georgia, Griffin

Dale A. Devitt, Center for Urban Water Conservation, University of Nevada, Las Vegas

Harry G. Fahnestock, Western Turf, Reno, Nevada

Douglas H. Fender, D. Fender and Associates, Barrington, Illinois

Calvin R. Finch, San Antonio Water System, San Antonio, Texas

Gary Grinnell, Las Vegas Valley Water District, Las Vegas, Nevada

Marvin H. Hall, Forage Management, Penn State University, University Park, Pennsylvania

M. Ali Harivandi, Environmental Horticulture, University of California Cooperative Extension, Alameda

T. Kirk Hunter, Turfgrass Producers International, East Dundee, Illinois

Clark Throssell, The Golf Course Superintendents Association of America, Billings, Montana

Brian Vinchesi, Irrigation Consulting, Inc., Pepperell, Massachusetts

Workshop Sponsors

United States Department of Agriculture–Agricultural Research Service
United States Golf Association
International Turf Producers Foundation
RISE–Responsible Industry for a Sound Environment
Irrigation Association

Acknowledgments

The Council for Agricultural Science and Technology (CAST) thanks the Steering Committee for planning and organizing this important workshop and for helping to increase understanding of the issues of water quality and conservation related to turfgrass use. CAST extends special appreciation to James Baker for serving as Project Manager and providing leadership in the organizational and planning phases of the project. CAST recognizes and thanks the Sponsors for providing financial support for the workshop and these proceedings. CAST also thanks all the speakers, moderators, and participants who made the workshop a success and the contributing authors whose work is represented in this Special Publication.

The Steering Committee and authors of this document offer their sincere thanks to the U.S. Department of Agriculture–Agricultural Research Service, the U.S. Golf Association, the International Turf Producers Foundation, Responsible Industry for a Sound Environment, and the Irrigation Association for their sponsorship of this workshop. They also thank CAST Past President Edward Runge, CAST Executive Vice President John Bonner, CAST Managing Scientific Editor Linda Chimenti, and the CAST staff who managed the workshop and reviewed and edited this document.

Foreword

Following a recommendation by the CAST National Concerns Committee, the CAST Board of Directors authorized development of a workshop to address issues regarding water quality and quantity as they relate to turfgrass use. An eminent group of experts was chosen as a steering committee, under the leadership of Dr. James L. Baker as project manager, to plan and conduct the workshop, which was held in Las Vegas, Nevada in January 2006. After that workshop, Drs. James Beard and Michael Kenna cochaired the task force for the development of this Special Publication.

The task force authors prepared initial drafts of their presentations and revised all subsequent drafts. The CAST Executive Committee and Editorial and Publications Committee reviewed the final draft, and the authors reviewed the proofs. The CAST staff provided editorial and structural suggestions and published the document. The task force authors are responsible for the document's scientific content.

On behalf of CAST, we thank the Steering Committee members, the cochairs, and the authors who gave of their time and expertise to conduct the workshop and prepare the Special Publication as a contribution by the scientific community for public understanding of the issues. We also thank the employers of the scientists, who permitted participation of these individuals at no cost to CAST. CAST thanks all members who made additional contributions to assist in the preparation of this document. The members of CAST deserve special recognition because their unrestricted contributions in support of CAST partially financed the preparation and distribution of this Special Publication.

Dr. Kassim Al-Khatib
President

Dr. John M. Bonner
Executive Vice President

Linda M. Chimenti
Managing Scientific Editor

Executive Summary

Turfgrasses used in urban areas impact Americans daily in many ways. There are an estimated 50 million acres of maintained turfgrass in the United States on home lawns, golf courses, sports fields, parks, playgrounds, cemeteries, and highway rights-of-way (see photos on Inset page I-1). The annual economic value of this turfgrass is estimated to be $40 billion. Scientists have documented an array of benefits to the environment and humans resulting from turfgrasses, but critics point out the excessive water requirements and pesticide use for turfgrass versus other landscape materials. It is important, however, to point out that plants do not conserve water, people do. Turfgrasses belong to the grass family, which evolved over millions of years without pesticides and irrigation systems. There are grasses adapted to the wettest and driest climates in the world. Academic and industry research on turfgrass can and will continue to provide quality turfgrass while reducing pesticide use and conserving water.

Water Crisis

Landscape plant water requirements, the acreage and economics in turfgrass, the amount of water available for humans, and the pending water crisis are discussed in Section 1. There is no longer a significant relationship between population distribution and water availability. The desert Southwest of the United States (Arizona, Nevada, and California) is among the fastest-growing areas, yet this is an area with undeniable water supply and distribution problems. According to the U.S. Geological Survey (USGS), total fresh water withdrawals during the last 45 years have declined as population has grown. The USGS concluded that more efficient industrial and agricultural water use accounted for the decrease in water withdrawals while population increased.

Urban water use can be divided into indoor and outdoor uses. Indoor water use remains fairly constant throughout the year; the peak demand for water during the summer, however, is the result of outdoor water use. Even in areas where water supplies are ample, an economic or investment concern exists whenever the peak demand becomes a driving force for water agencies' decision-making process. Flattening the peak demand is an objective of water agencies. Because the demand curve typically is highest during times of increased outdoor water use, conservation efforts target landscapes generally and turfgrasses specifically.

Clearly, water conservation can have positive benefits, such as extending the availability of water to more people or other uses and reducing the costs associated with developing new water resources. Outdoor water use estimations are complicated, however, and have many shortcomings. There is a need for more research and analysis to refine outdoor water use. There also is a need to clarify how much water is consumed by various landscape materials and how much is returned either through evaporation, runoff, or groundwater recharge.

Low-Precipitation Landscapes

Section 2 addresses the problems that can result from the loss of a turfgrass cover because of not allowing appropriate irrigation in low-precipitation regions. The seven categories of problems include diseases and airborne dust, heat stress islands, wildfires, soil erosion and flooding, urban pollutants, criminal activity, and human disharmony. There is a tendency to use a simplistic approach for eliminating certain water uses by enacting public laws. A single-issue approach of not permitting irrigation on all or a portion of the land area, such as grassed lawns, can lead to other potentially serious problems. Officials need to take these consequences into consideration when proposing legislation to exclude irrigation from all or part of the urban landscape. There are many other functional benefits attributed to the use of the turfgrass/soil ecosystem in urban landscapes that are summarized briefly. Certainly, the social and economic values of these benefits are substantial, but studies quantifying the economic aspects are needed.

Rather than eliminating certain water uses in low-precipitation landscapes, there are other substantial savings to be accomplished in furthering water conservation. These actions range from sustainable best management practices (BMPs) for irrigating turfgrass to repairing leaks in municipal water distribution systems. Incongruities in laws and "money-for-grass" approaches, which eliminate grassy areas but allow the use of ornamental shrubs and trees with higher water use rates, are not sound approaches. An integrated, holistic approach to water use in populated areas is essential. The elimination of turfgrasses from open areas in urban landscapes should be implemented only as a last resort in arid climates. Turfgrasses not only use water, but also collect, hold, and clean it while enhancing subsequent groundwater recharge and contributing to transpiration cooling.

Regulatory Considerations

The Environmental Protection Agency (EPA) is responsible for implementing the Clean Water Act and Safe Drinking Water Act, portions of the Coastal Zone Act, and several international agreements protecting our oceans and shores. The EPA's activities are targeted to prevent pollution wherever possible and to reduce risk for people and ecosystems in cost-effective ways. In recent years, water security also has become a more critical part of the EPA's mission. In Section 3, the legislative history and context of the Safe Drinking Water Act and Clean Water Act are discussed, along with how the goals of these two Acts are integrated through federal, state, and local implementation.

Municipal Policies

There are two fundamentally different legal systems that govern the allocation of water throughout the United States. Under the riparian system, which applies to 29 eastern states that were historically considered wet states, ownership of land along the waterway determines the right to use of the water. In times of shortage, all owners along a stream must reduce the use of water. Because of water scarcity in the West, it was impractical for water rights to depend on ownership of land along streams. This resulted in the prior appropriation system of water rights, which was originally developed by miners in California and adopted by nine arid western states. Under prior appropriation, a water right is obtained by diverting water and putting it to beneficial use. An entity whose appropriation is "first in time" has a right "senior" to one who later obtains a water right. In times of water shortages, senior rights must be fully satisfied before junior rights are met, sometimes resulting in juniors receiving no water at all. Section 4 further explains these systems and various other existing water policies.

In the United States, most water policy is at the state and local (municipal) level; the drinking water system is extremely decentralized and is structured in four basic ways: (1) owned by local governments, (2) independent government authorities, (3) privately owned companies, and (4) public-private partnerships. There are 53,000 community water systems in the United States, and they provide 90% of Americans with their tap water. Only 424 community water systems serve more than 100,000 people. In total, 80% of community water systems serve 82% of the U.S. population. Local governments or an independent government authority own 86% of the community water systems. Historically, pricing qualifies the costs of capture, treatment, and conveyance. Consequently, this method often obscures the larger, but less quantifiable, societal interest in preserving our water resources. In regard to water rates, there are well-established policies, primarily due to the efforts of the American Water Works Association (AWWA), whose members provide approximately 85% of the drinking water across the United States.

Turfgrass and the Environment

The first step toward water conservation is selecting the correct turfgrass for the climate in which it will be grown (see page I-2). A breakdown of climate zones in the United States and the differences between cool-season and warm-season turfgrasses are discussed briefly in Section 5. During the last 30 years, turfgrass scientists have determined the water use rates for major turfgrass species. Turfgrasses can survive on much lower amounts of water than most people realize; several turfgrass species have good drought resistance. A great deal of this information is available on the Internet through sources such as the Turfgrass Information File at Michigan State University (http://tic.msu.edu).

Agriculture chemicals registered with the EPA are applied to turfgrass, and through several processes, these chemicals break down into biologically inactive byproducts. Two concerns are whether pesticides and nutrients leach or run off from turfgrass areas. The downward movement of pesticides or nutrients through the soil system by water is called leaching. Runoff is the portion of precipitation or rainfall that leaves the area over the soil surface. There are several interacting processes that influence the fate of pesticides and fertilizers applied to turfgrass. Seven processes that influence the fate of pesticides and nutrients are introduced in Section 5 and include volatilization, water solubility, disruption, plant uptake, degradation, runoff, and leaching. Sections 7 and 8 further examine these processes and the likelihood that the pesticides will reach ground or surface water.

Soil Water

How water moves through soil under a growing turfgrass is reviewed in Section 6 (see page I-3). Water flow through soil is influenced partly by local weather conditions. Rainfall places water at the soil surface, and its intensity and duration dictate which portion will infiltrate or run off. Solar radiation, relative humidity, and wind control the rate of water evapotranspiration. Water flow through soil also is influenced by the characteristics and current growth stage of the turfgrass plant. The atmosphere's evaporative demand is tempered by the plant that draws water for transpiration from the soil. Consequently, intra- and inter-species differences in canopy resistance and variations in turfgrass cultural practices affect soil water uptake. Water flow through soil is controlled by retention and transmission capabilities of the soil pore space. Coarser-textured soils show greater transmission capabilities, and finer-textured soils show greater retention capabilities. Antecedent soil water content also affects the rate of water infiltration and flow through soil.

Groundwater

Turfgrasses and associated management practices reduce the potential for leaching of pesticides and nutrients to groundwater (see page I-3). Section 7 reviews the manner in which a healthy turfgrass protects groundwater. Turfgrass can provide considerable protection against leaching because of the high levels of organic matter and associated microbial activity that serve to immobilize and degrade applied pesticides and nitrates. Excessive irrigation or large rain events, which lead to preferential or macro-pore flow, can mitigate these advantages and push solutes below this zone of microbial activity. It is unwise to generalize when discussing pesticides because each pesticide has different characteristics that affect its distribution and fate; most pesticides currently used in turfgrass, however, present fairly low risks of producing significant groundwater contamination. A healthy turfgrass has a great capacity to use applied nutrients. Nitrate leaching may present problems, however, in some segments of the turfgrass industry where nitrogen fertilization rates have not been reduced to account for turfgrass age and clippings return.

Surface Water

Available knowledge about surface runoff quantity and chemistry from urban landscapes has increased over the last two decades; more information is required, however, before any overarching, widespread conclusions can be made (see page I-4). Section 8 discusses factors that affect surface runoff, such as climate, site and soil conditions, and management. The most significant climate factors are precipitation, evapotranspiration, and temperature. Site and soil conditions also affect potential off-site movement of sediment, nutrients, and pesticides. The most significant site and soil conditions are soil texture and organic matter content, bulk density, hydraulic conductivity, thatch layer, landscape slope, and proximity to water resources. The most critical factor affecting surface runoff is management, which includes irrigation, drainage, fertilizer and pesticide application, and cultural practices. Of the studies discussed in Section 8, a reasonable case could be made that runoff volume generally is small, and losses of pesticides and nutrients are less than those from agriculture. More geographically diverse, long-term data sets on both cool- and warm-season grasses and on well-defined catchments under natural conditions would further document this aspect.

Pesticide and Nutrient Modeling

Researchers who develop various approaches to turfgrass management, regulators and the regulated community concerned about off-site transport of pesticides and nutrients, and various scientists and engineers who designed the BMPs for managed turfgrass rely on mathematical models to predict the fate of turfgrass chemicals (see page I-4). Most of these models have not been designed for turfgrass, and the unique aspects of turfgrass relative to row crops should be incorporated into model algorithms and input guides. In addition, there can be fundamental questions about the overall model application scenarios regarding their ability to offer reliable predictions. Although models are useful tools, their content and application must be continually scrutinized and improved. Section 9 summarizes the key practices and research regarding techniques and applications of mathematical models that predict the off-site transport of turfgrass chemicals to water resources. These models are important tools for risk assessment and risk management of turfgrass chemicals, but they have potential to produce results that deviate significantly from reality. There are fundamental conceptual model and algorithm issues when evaluating chemical fate in turfgrass compared with row crop agricultural systems.

Plant Selection

Water use declines as the leaf area/leaf elongation rate decreases and the turfgrass density increases. Also, turfgrasses with deep, extensive root systems, coupled with decreased water use, are more drought resistant and have greater water conservation potential. Water usage rates vary with species and cultivars, as documented by extensive research, and are affected by external factors, especially environmental conditions. Selecting low water use and/or drought-resistant turfgrass species and cultivars is a primary means of decreasing water needs. Also, selection of turfgrass species and cultivars that are adapted to local climatic conditions can result in significant water savings. For example, in arid and semiarid climatic regions, warm-season turfgrasses use less water than cool-season turfgrasses. Section 10 addresses these plant selection factors as they relate to water conservation (see page I-5).

Currently, there is a lack of scientific data on the water use of trees, shrubs, and ground covers, as well as on how this water use is influenced by growing conditions and irrigation. Note that grassland-dominant plant communities occur in drier climates compared with forest lands. Emphasis should be placed on choosing functional landscapes and avoiding banning entire plant categories without justification. Turfgrasses that have lower water requirements should be used when possible.

Turfgrass Water Use

As water availability becomes increasingly limited and more costly, water conservation in turfgrass culture becomes extremely important. Without adequate water, turfgrass becomes brown and desiccated, and it may die in severe instances. Turfgrass growth characteristics that affect water use include differences in canopy configuration or leaf orientation, tiller or shoot density, growth habit, rooting depth, and root density. Water usage rates vary with species and cultivars and are affected by many external factors, especially environmental conditions. In Section 11, the water use characteristics of different turfgrasses and how environmental factors affect turfgrass water use are discussed.

Water use of turfgrasses is evaluated based on the total amount of water required for growth and transpiration (water lost from the leaf) plus the amount of water lost from the soil surface (evaporation). Transpiration water consumption accounts for more than 90% of the total amount of water transported into the plants, with 1 to 3% actually used for metabolic processes. Dormant turfgrass plants have limited or no transpiration water loss, and thus have low water usage. The leaves of dormant turfgrass turn brown in response to a water deficit, but the growing points in the stem are not dead. In general, turfgrasses, especially those with rhizomes (underground stems), can survive without water for several weeks or months with limited damage, depending on the air temperature. Allowing certain turfgrasses to go dormant in low maintenance areas can result in significant water savings without loss of turfgrass.

Water use of turfgrasses is influenced by environmental factors such as temperature, wind, solar radiation, relative humidity, soil texture, and soil moisture. These factors affect both plant transpiration and soil evaporation. Understanding the environmental factors influencing water use is important for developing efficient cultural strategies for turfgrass, especially in areas with limited water supply. Knowledge of critical plant physiological status and soil moisture content of different soil types is important for scheduling when to irrigate, how much water to apply by irrigation to replenish water loss through evapotranspiration, and how deep to irrigate the soil.

Cultural Practices

There is adequate research to substantiate specific cultural practices, or systems approaches, to decrease turfgrass water use, conserve water, and enhance drought resistance. Mowing height and frequency, nutrition, and irrigation are primary cultural practices that directly impact vertical elongation rate, leaf surface area, canopy resistance, rooting characteristics, and the resultant water use. These practices, explained in Section 12, can be used immediately to conserve water and maintain turfgrass quality and functional benefits. Secondary cultural practices, such as turfgrass cultivation, topdressing, wetting agents, plant growth regulators, and pest management, also influence turfgrass top and root growth and subsequently influence potential water conservation.

Achieving Efficient Irrigation

Section 13 discusses many elements of high water use efficiency in irrigation, beginning with proper system design and including installation, management, and maintenance of the irrigation system (see page I-6). One critical element is to apply the proper amount of water when the landscape needs the water to avoid both deep percolation and runoff. This practice may include cycling of control valves to minimize the surface movement of applied water.

A second important element to high water-use efficiency is to apply water as uniformly as possible. Innovative sprinkler designs for turfgrass and drip/micro irrigation for landscape plants have improved irrigation uniformity significantly in recent years, when properly designed and installed. Tools now exist for designers to model sprinkler application uniformity before the system is purchased and installed. Thus, it is reasonable to specify the irrigation application uniformity in a contract before purchasing an irrigation system. Auditing can be used to verify the system performance after installation.

Improved controllers for residential irrigation systems combined with highly uniform sprinkler and/or drip irrigation systems will produce high water-use efficiency, leading to significant water savings over conventional practices. This approach has been validated on extensive turfgrass areas and needs to be emphasized for home landscapes.

Recycled Water

In dry regions of the country, and in highly populated metropolitan areas where water is a limited natural resource, irrigation of landscapes with municipal recycled water, untreated household gray water, or other low-quality (saline) water is a viable means of coping with potable water shortages. Section 14 explains these methods and the associated benefits and concerns of their use. Many years of practice and field observation on extensive turfgrass areas confirm that recycled or brackish water can be used successfully to irrigate turfgrasses. Water conservation resulting from this practice far outweighs the potential negative impacts.

Nonetheless, recycled or brackish water quality must be evaluated thoroughly before developing appropriate plant cultural strategies for its use. Irrigation water quality, which is a function of the volume and type of dissolved salts present in the water, affects the chemical and physical properties of soil, and therefore plant-soil-water relations. The interrelationships can be monitored by regular chemical analysis, and in many situations managed. Currently, the use of household gray water for irrigating home landscapes is not widely practiced. More research is needed to determine the most effective, least expensive, and safest (vis-à-vis human health) methods for using such water.

Public Policy Approach

A water conservation program can be very effective. It can be based on science, and it can be embraced by the citizens of the community (see page I-6). The water conservation program in San Antonio, Texas, fits that description. San Antonio is a community in a semiarid climate that has decreased per capita water use by more than 40% since the early 1980s and has avoided conflict over landscape watering. Success has been achieved because the San Antonio Water System recognized the value of lawns to its citizens and worked with them to develop a comprehensive water conservation program that addressed infrastructure improvements, inefficient plumbing, industrial technology, and other water-saving opportunities along with savings in landscape watering. The landscape watering savings were based on opportunities identified in outside research and local studies, resulting in changes in turfgrass management, variety or cultivar selection, and irrigation technology, without attempting to eliminate lawns. Every community's situation is different, and the formula for decreasing water use may be different. The example provided by San Antonio, summarized in Section 15, shows that water use can be decreased in a manner that takes advantage of turfgrass benefits and is consistent with local positive attitudes toward turfgrass use.

Comprehensive Assessment

Section 16 reviews various approaches for comprehensive water quality and environmental management. The Best Management Practices approach developed over the past 35 years by the EPA for protection of surface and subsurface waters from sediment, nutrients, and pesticides has a long track record for being successfully implemented because of certain critical characteristics. It is science-based; incorporates all strategies in the ecosystem (holistic); embodies all stakeholders and their social, economic, and environmental concerns; values education and communication outreach; allows integration of new technologies; has been applied at the regulatory, watershed, community, and site-specific levels, as well as in educational realms; and maintains flexibility to adjust to new situations. Thus, this BMP model is the template for dealing with other complex environmental issues, such as water conservation. An Environmental Management System (EMS) approach brings under one umbrella all environmental issues and consequences on a site. When a single issue (e.g., water conservation) is targeted by a group toward the turfgrass industry or a single facility, it is not uncommon for the only determination of success to be the decrease in water use, without any consideration for economic/job or unintended environmental consequences. Within an EMS, all environmental issues are addressed, including potential adverse effects.

Summary

There is a pending water crisis due to population growth in areas with inadequate water supplies. Even in areas where water supplies are ample, an economic or investment concern exists whenever peak demand becomes a driving force in decisions about providing water to the public. There is a tendency to use a simplistic approach for eliminating certain water uses by enacting public laws. A single-issue approach of not permitting irrigation on all or a portion of the land area, such as grassed lawns, can lead to other potentially serious problems. Officials need to take these consequences into consideration when proposing legislation to exclude irrigation from all or part of the urban landscape.

In the United States, there is currently no national water policy, partly because of the history of the country and partly because most water issues have been treated as local issues, resulting in an extremely decentralized water delivery system. The nation's water issues need to be addressed in an integrated manner, focusing on programs at the watershed and basin levels. There is a need to reconcile the myriad laws, executive orders, and congressional guidance that have created a disjointed, ad hoc national water policy. The fiscal realities facing the nation need to be recognized to effectively coordinate the actions of federal, state, tribal, and local governments dealing with water.

For grassed landscapes, the first step toward water conservation is selecting the correct turfgrass for the climate in which it will be grown. There is adequate research to substantiate the use of specific cultural practices, or systems approaches, to decrease turfgrass water use, conserve water, and enhance drought resistance. These practices could be used immediately to conserve water and maintain turfgrass quality and functional benefits.

Recycled or brackish water can be used successfully to irrigate turfgrasses. Water conservation resulting from this practice far outweighs the potential negative impacts. Nonetheless, recycled or brackish water quality must be evaluated thoroughly before developing appropriate plant cultural strategies for its use. If irrigation systems are employed, proper design, installation, management, and maintenance are very important. One critical element is to apply the proper amount of water when the landscape needs the water to avoid both deep percolation and runoff.

Other concerns include potential pesticide and nutrient leaching and runoff from turfgrass areas. The legislative history and context of the Safe Drinking Water Act and the Clean Water Act demonstrate that the federal, state, and local governments provide a clean and safe drinking water supply. It is important to understand that a healthy turfgrass has a great capacity to use applied nutrients, break down pesticides, help recharge groundwater, and reduce surface runoff. The critical aspect is management, which includes irrigation, drainage, fertilizer and pesticide application, and cultural practices. Based on turfgrass landscape research, runoff volume generally is small and losses of pesticides and nutrients are less than those from agriculture. This information is being used to develop models for risk assessment and risk management of turfgrass chemicals.

The BMPs approach developed by the EPA has a long track record of being implemented successfully. A water conservation program using a similar approach could be very effective. It can be based on science, and it can be embraced by the citizens of the community. The ultimate goal is to provide quality urban areas for daily activities and recreation while conserving and protecting the water supply.

1

Urban Turfgrasses in Times of a Water Crisis: Benefits and Concerns

Douglas H. Fender

Introduction

Turfgrasses used in urban areas impact Americans in many ways on a daily basis. Millions of acres of turfgrass are found on home lawns, commercial property, roadsides, parks, athletic fields, and golf courses. Turfgrass improves our quality of life by providing open space, recreational and business opportunities, enhanced property values, and by conserving natural resources. Turfgrass to some people, however, is nothing more than a water-wasting, pesticide-addicted, fertilizer-dependent, landfill-clogging, energy-consuming insult to mankind and the environment. These critics conclude turfgrass use in the landscape is unimaginative at best, dangerous at worst.

Critics of grass maintain that it wastes time, money, and resources, and, even worse, that efforts to grow grass result in environmental pollution, the poisoning of pets, and human cancers. Critics find no redeeming values for grass and recommend its total replacement with "native plants."

Before advancing to specific scientific presentations, it is essential to consider whether or not properly managed turfgrass has benefits to society. If this question cannot be answered positively, there is no practical reason to proceed.

Although individuals or groups may debate the relative merits of any single landscape material, the efforts of Drs. James B. Beard and Robert L. Green have scientifically documented the multiple and wide-ranging benefits of turfgrass (Beard and Green 1994). Additionally, numerous psychologists, sociologists, and environmentalists have identified multiple benefits associated with maintained turfgrass and landscape areas. In Section 2 of this report, Dr. Beard reviews these benefits thoroughly.

In addition to the currently documented benefits of turfgrass, future research projects that would be extremely helpful for policymakers include an economic benefits analysis resulting from the presence of turfgrass. Topics of these standardized studies could include the potential energy-cost savings resulting from reducing or eliminating heat-islands; the groundwater recharge capacity of turfgrass- versus asphalt-covered areas; the sequestration of atmospheric carbon; the fire-buffering capabilities of nondormant turfgrass; the long- and short-term psychological and physical fitness relationships of turfgrass areas in a community; and the local economic impacts resulting from the production, installation, maintenance, and use of turfgrass.

Although scientists have documented vast and diverse benefits to the environment and humans resulting from turfgrasses, there remains the criticism of excessive water requirements for turfgrass versus other landscape materials.

Landscape Plant Water Requirements

When discussing the water requirements for landscape plants, two important issues need to be addressed. First, the water requirements of most turfgrasses have been established by scientific study. The application of water to turfgrass in amounts exceeding its requirements can be attributed to human factors, not plant needs (Beard and Green 1994).

Second, very few scientific studies have been undertaken to determine the water requirements of trees, shrubs, or plants termed "native" or "low water use" (Beard and Green 1994). Recommendations favoring one plant, or plant group, over another should be based on scientific information, not anecdotal information, wishful thinking, or even marketing efforts.

In addition, two landscape water use studies found that xeric style landscapes alone do not necessarily equate to decreased amounts of applied water. A broad-based study of homeowners who installed low water use landscaping versus those who did not reports, "a comparison of average annual outdoor consumption between these groups resulted in the finding that the low water use landscape group actually used slightly more water outdoors annually than the standard landscape group" (Mayer et al. 1999).

Dr. Chris Martin, based on his own work in Phoenix, Arizona, reported, "On the whole, residents with xeric designs and programmable irrigation systems do not adjust their water applications to seasonal changes in evapotranspiration and plant water needs. In contrast, residents with oasis or mesic landscapes tend to follow rates of monthly evapotranspiration and water more in summer and less in winter" (Martin 2001). Dr. Martin's conclusion: "Plants do not conserve water, people do."

Although it will ultimately be people and not plants that conserve water, it is clear that people have an appreciation of turfgrass when measured by the area of the United States blanketed by turfgrass and the annual investments people make in their landscapes.

Acreage and Economics of Turfgrass

There are an estimated 50 million acres of maintained turfgrass in the United States on home lawns, golf courses, sports fields, parks, playgrounds, cemeteries, and highway rights-of-way. The 50 million acres of grass would blanket the nine smallest states (Rhode Island, Delaware, Connecticut, New Jersey, New Hampshire, Vermont, Massachusetts, Hawaii, and Maryland), but would leave the remaining 41 states (totaling 127,098,800 acres) devoid of any turfgrass. The annual economic value of this turfgrass is estimated to be $40 billion (Morris 2003). Additional measures of the significance Americans place on yards and gardens were reported in the 2004 National Gardening Survey *Fact Sheet* (Butterfield 2004):

- 78% of all U.S. households (84 million households) participated in one or more types of do-it-yourself lawn and garden activities last year.

- Consumers spent a total of $38.4 billion on do-it-yourself lawn and garden products.

- They also spent $31.3 billion to hire professional services.

To establish, maintain, and enjoy the benefits of turfgrass, some amount of water—either naturally occurring or applied—is required. The question, or in some locales the debate, centers on the balance between competing demands for this valuable, limited resource and the benefits provided by its use.

Water Availability for Humans

Living on a planet where water covers 71% of the Earth's surface may beg the question for some of why water availability for turfgrass is even an issue. But the fact is, only a small fraction of the Earth's water is available to humans.

The U.S. Geological Survey (USGS) estimates a total of 332.5 million cubic miles of water on Earth (USGS 2005a). Ninety-seven percent of the total is saline, mostly in the oceans, seas, and bays. The remaining 3% is fresh, but of that amount nearly two-thirds (68.7%) is contained in ice caps, glaciers, and permanent snow. Groundwater represents 30.1%, soil moisture or ground ice and permafrost is 0.9%, and only 0.3 of 1% is surface water. That 0.3 of 1% of the world's freshwater is divided as follows: lakes 87%, swamps 11%, and rivers 2%.

To appreciate what these numbers mean on a relative basis, think of placing all water on Earth, in all its forms, into a single one-gallon (3.785 liters) container. Table 1.1 gives the various equivalents in fluid ounces, teaspoons, and milliliters or drops.

The liquid freshwater in lakes, swamps, and rivers would not be a drop in the ocean—it would be just over one-tenth of one drop—if all the Earth's water were contained in a gallon jug.

Relying on groundwater as a source of freshwater is limiting because the practice of "over-drafting" extracts greater volumes than are being replenished. Groundwater supplies ultimately may be exhausted and will take eons to restore. Freshwater also is not naturally distributed uniformly across the planet or equally distributed across populations or areas of use. Therefore, one must conclude that although water is a global matter, the real concern is much more localized. Demand exceeding supply at the local level causes concern and debate about the use of a limited and precious resource to produce a landscape plant that some people see as having no particular value.

Unless we can find practical, affordable, and environmentally satisfactory means of drilling deeply or melting immense amounts of glacial ice into freshwater and then distributing either or both, we currently have all the accessible freshwater we will ever have. This is true regardless of how the population expands or its uses for water increase. There is an urgent need to address the likelihood that a devastating, worldwide water crisis is developing.

Table 1.1. Relative amounts of various forms of water on the Earth

		Earth's total water as one gallon		
Water type	% by type	Fluid ounces	Teaspoons	Milliliters/Drops
One gallon	100.00	133.4400	800.6400	3,945.8208
Saline – Oceans	97.00	129.4368	776.6208	3,827.4462
Freshwater	3.00	4.0032	24.0192	118.3746
Total water	**100.00**			
Icecaps, glaciers	68.70	2.7502	16.5012	81.3234
Groundwater	30.10	1.2050	7.2298	35.6308
Other	0.90	0.0360	0.2162	1.0654
Surface water	0.30	0.0120	0.0721	0.3551
Total freshwater	**100.00**			
Lakes	87.00	0.0026	0.0157	0.0772
Swamps	11.00	0.0013	0.0079	0.0391
Rivers	2.00	0.0002	0.0014	0.0071
Total fresh surface water	**100.00**			

A Crisis of Devastating Magnitude

In October 1991, a highly unusual series of meteorological events came together to create what is now called "The Perfect Storm." The event, which caused the deaths of several New England fishermen and property damage in the billions of dollars, has been described by Dr. Shirley Ann Jackson, president of Rensselaer Polytechnic Institute, as "the unlikely confluence of conditions…in which multiple factors converged to bring about an event of devastating magnitude" (Jackson 2004).

Today, when we study population growth, demographics, economics, and the environment, as well as social and cultural factors as they relate to water availability and use, we may be looking at a quite different, but even more serious and widespread "unlikely confluence of conditions…in which multiple factors converge to bring about an event of devastating magnitude." This may seem to be an overly broad and highly anxious statement for this publication. Yet, as a nation, we need to consider an unlikely confluence of conditions so that we might lessen or eliminate any resulting devastating consequences. Limiting the scope of this paper to the United States, we can document the growth of our population and, with that growth, an increased water demand. More people equates to more water consumption, for both their daily personal needs and the production of the goods and services they require or desire.

There also is a significant relationship between population distribution and water availability. Historically, people have settled near a water source, but we are seeing a U.S. population growth trend into areas where water supplies are limited or unreliable. The desert southwest of the United States (Arizona, Nevada, and California) is among the fastest growing areas, yet this is an area with undeniable water supply and distribution problems (Figure 1.1).

When we compare U.S. population growth with precipitation maps (see Figure 1.2, page I-7), we see that the population is indeed expanding in areas with limited natural water resources (Negative 2005).

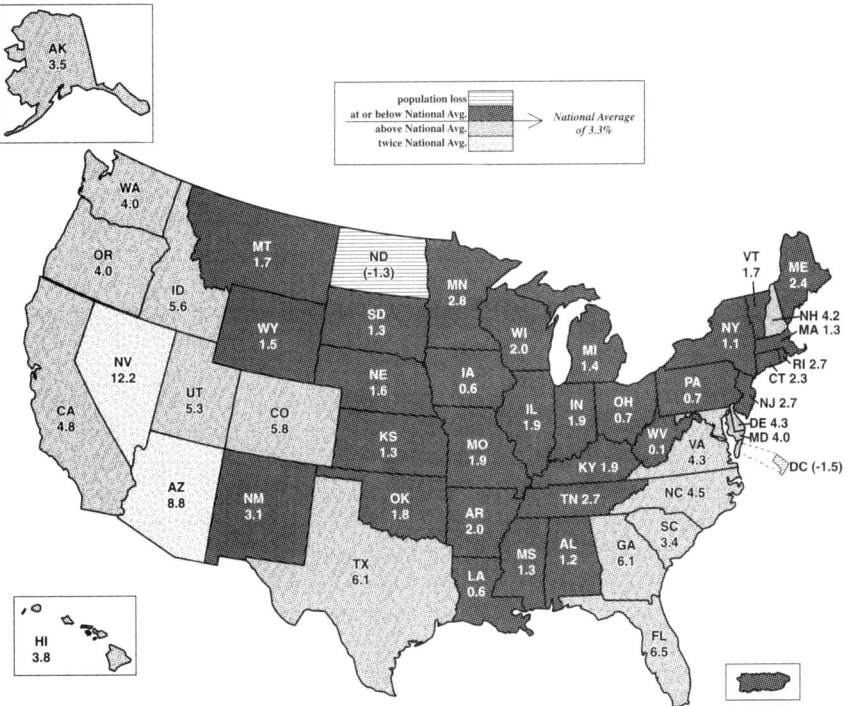

Figure 1.1. Percentage change in U.S. population: 2000–2003. Scale based on U.S. national average.

Nevada and Arizona populations are growing at more than twice the national average. Texas, Colorado, Utah, California, Idaho, Oregon, and Washington are well above the national average of 3.3% growth (National Atlas 2005). Yet, except for the western halves of Washington and Oregon, these are areas of exceptionally low precipitation and limited surface and groundwater resources. Worse still, these same areas are at the upper end of the water use scale (Figure 1.3). Rapidly expanding population in areas of limited water resources and high water-use rates must be viewed as an unlikely confluence of unfavorable conditions.

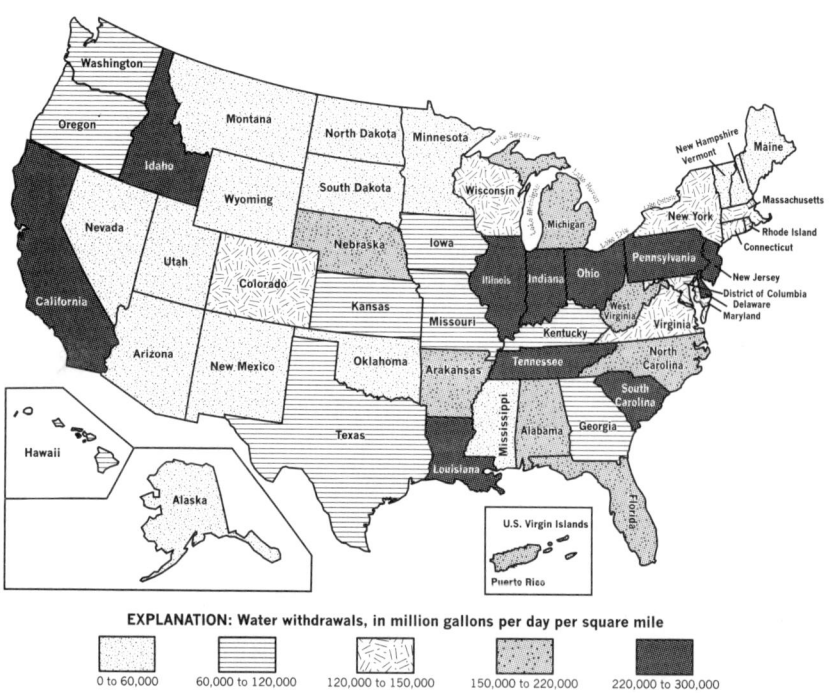

Figure 1.3. Estimated use of water in the United States in 2000 (Source USDOI-USGS 2004).

Multiple Factors Converging to a Crisis

The economy is another of the possible converging factors. We can understand the benefits of an economy that expands with new home construction, new business development, and higher rates of employment. Mortgage bankers, builders, lumberyards, brick and block people, bureaucrats and tax collectors, and even green industry companies have traditionally pushed for more development, more houses, more construction, more sales, and more profits. But have we considered the resulting demands for and availability of water resources in those areas? How expansive can the economy be if water conservation is the limiting factor? How severely and how willingly would current residents and businesses restrict their traditional water use patterns so the economy of their community could expand?

Environmental damage is yet another potential factor for concern. As water is withdrawn from any source for human use, we are becoming more aware of a potential shift of environmental conditions for plants and animals that are reliant on those bodies of water for survival. Governmental and nongovernmental organizations have been vigorous in their efforts to reserve water use for nature's purposes, while at the same time eliminating or severely restricting the ability of the government to create new water supplies and storage reservoirs. It will not be easy or simple to choose between water for drinking and water for wildlife habitats if multiple factors converge and require a choice.

Water is essential to all life forms, regardless of species, but there also are significant social and cultural aspects associated with the availability of clean, fresh water. These aspects include not only food preparation, sanitation, and safety considerations, but also recreational and emotional matters. These factors add to the list of unlikely conditions that are coming together in a potentially devastating way. Given what scientists have already documented regarding the multiple benefits of maintained turfgrass, we also must ask ourselves what the health and environmental consequences would be of removing grass and managed landscapes from our country. Would such an act, taken in the name of water conservation, contribute unlikely or unrecognized factors that could ultimately converge to bring about an event of devastating magnitude? This scenario, too, is an area that requires further scientific study on multiple levels.

There are any number of other unlikely conditions we have yet to identify that could converge to bring about an event of devastating magnitude, but only by starting to consider the ones we can identify might we begin to avert the disastrous consequences of flawed or fogged thinking.

Unique "Water World" Words

Clearly this is a complex issue, and in order for all parties concerned to communicate properly and understand it fully, it is imperative that all parties speak the same language. In many respects, the lexicon of water is not always what it may first seem to the nonprofessional. There also are wide ranges of water use categories or definitions that can lead to misunderstanding or miscommunication. Therefore, care must be taken to address some of the more commonly confronted word/definition issues and their significance to turfgrass or landscape water usage.

Water use usually is defined and measured in terms of withdrawal or consumption, that which is taken and that which is used (USEPA 2005).

Withdrawal refers to water extracted from surface or groundwater sources, with **consumption** being that part of a withdrawal that is ultimately used and removed from the immediate water environment by evaporation, transpiration, incorporation into crops or a product, or other consumption.

Return flow is the portion of a withdrawal that is actually not consumed, but is instead returned to a surface or groundwater source from a point of use and becomes available for further use.

The United States Environmental Protection Agency (EPA) notes, "Water can also be divided into **offstream** and **instream** uses."

Offstream water involves the withdrawal or diversion of water from a surface or groundwater source for domestic, residential, industrial, agricultural, and energy development uses.

Instream water uses are those that do not require a diversion or withdrawal from the surface or groundwater sources, such as water quality and habitat improvement, recreation, navigation, fish propagation, or hydroelectric power production.

Another set of water use terms that requires clarification is **public use, domestic use, and commercial use.**

Public use and losses include water for fire protection, street cleaning, municipal parks and pools, as well as water considered as **Unaccounted-for loss,** which is the amount of treated water placed into a distribution system that is lost to leaks, meter malfunctions, or other nonrecorded means.

Domestic use typically defines water delivered to a single- or multi-family residence by a municipality or water service company or agency and includes everyday uses that take place in a residential setting. It is only one component of public supply.

Domestic indoor use is composed of toilet flushing, bathing/showering, cleaning clothes or the home, drinking, and kitchen uses that can include food preparation and clean-up with either automatic or hand dishwashing.

Domestic outdoor use is composed of landscape irrigation (lawns, trees, shrubs, and flower and vegetable gardens), swimming pools, fountains and water features, vehicle washing, siding or sidewalk wash-downs, and children's yard/water toys.

Commercial water uses are those that take place in office buildings, hotels, restaurants, civilian and military institutions, public and private golf courses, and other nonindustrial commercial facilities.

Gallons per capita per day (GPCD) or **Gallons per person per day (GPD)** are also terms requiring additional clarification. Typically, this is a measure of the water system's total daily production divided by the population and includes all water-receiving categories such as industrial, commercial, and domestic customers. It is not necessarily a measure of the average individual's domestic indoor and outdoor water consumption on a personal daily basis—although it may seem that way in some usages.

A clear understanding of the definition of GPCD or GPD in use can become significant when historic or projected totals are reported. For example, actual industrial/ commercial and domestic water use could shift in opposite directions but not be recognizable because the combined GPCD/GPD would be unchanged. The success of a residential conservation program could be unrecognized if industrial or commercial uses expand by a greater amount.

Peak instantaneous water use is a term significant to water system managers with resultant impact on water end-users. According to the 1999 American Water Works Association (AWWA) report, peak demand "is the high flow rate observed during a given time interval." As will be shown, peak demand is one of the reasons summer outdoor water use is a driving factor for conservation.

The purpose of considering these terms is to assist in understanding that although there may seem to be a straight pipeline of facts and logic between the water source and a lawn sprinkler, there are multiple kinks in the system that need to be examined. Before policymakers and the general citizenry embark on any water conservation program, they need to appreciate that some solutions might create their own unfortunate confluence of factors that could lead to a devastating, though unintended, consequence.

Although a considerable amount of water conservation attention focuses on domestic withdrawals and consumption, that category accounts for only 7.5% of all freshwater withdrawals and only 6% of all freshwater consumptive uses (Table 1.2). When measured as a percentage of "public supplied water," however, domestic usage jumps to 57% of the total, clearly a significant amount of water going to a single receiving category (Table 1.3).

Equally noteworthy is the increase in the category "Public use and losses," which is only 1.6% of all freshwater use (Table 1.2), but jumps to 14% of the public water supply (Table 1.3). Although there are important benefits associated with public use, the portion of that amount that is considered "unaccounted-for losses," or more specifically system leaks, deserves examination.

Table 1.2. Comparison of freshwater consumptive use in the United States for 1990, by category (USEPA 2005)

	Total freshwater withdrawals Millions of gallons/day	Percentage of total withdrawals	Consumed percentage of total withdrawal	Consumptive use Millions of gallons/day	Percentage of total consumed
Irrigation	137,000	40.0	56	76,200	81
Thermo-electric	131,000	39.0	3	3,500	4
Industrial and mining	27,800	8.2	16	4,500	5
Domestic	25,300	7.5	23	5,900	6
Commercial	8,300	2.4	11	900	1
Public use and losses	5,500	1.6	N/A	N/A	N/A
Livestock	4,500	1.3	67	3,000	3
Total	**339,000**	**100.0**	**28**	**94,000**	**100**

Table 1.3. Fate of water in public water supplies of the United States, 1990 (USEPA 2005)

Receiving category	Volume (Mgal/day)	Percentage of total
Domestic	21,900	57
Commercial	5,900	15
Public use and losses	5,460	14
Industrial	5,190	13
Thermoelectric power	80	<1
Total	**38,530**	**100**

For the sake of brevity and illustration, Table 1.4 itemizes the per capita water withdrawals and deliveries for 10 U.S. states, as well as the per capita leakage and the leakage as a percentage of total withdrawals. Based on the leakage percentage, a system efficiency rating is given. Utah ranks as the second highest state for withdrawals and delivery but has the most efficient system because its leakage rate is relatively small at 5.2%. At the other end of the scale, Pennsylvania is the most inefficient system, with a leakage rate of 36.9%. In other words, one of every three gallons of water put into the system in Pennsylvania is lost before it can be used.

Table 1.4. Total per capita water withdrawals, deliveries, and leakage for 10 U.S. states (Houston 2002)

State	Rank		Per capita		Per capita leakage	Leakage as % of withdrawals	Rank: Efficiency of system
	Withdrawals	Deliveries	Withdrawals	Deliveries			
Utah	2	2	269	255	14	5.20	1
Oklahoma	17	12	194	182	12	6.20	2
Montana	9	20	222	161	61	27.40	48
Kentucky	42	29	148	135	13	8.70	5
Maryland	14	30	200	135	65	32.50	49
South Dakota	44	32	147	134	13	8.60	4
New Hampshire	47	38	140	131	9	6.40	3
Wisconsin	28	42	169	127	42	24.90	46
Pennsylvania	27	48	171	108	63	36.90	50
West Virginia	48	50	134	100	34	25.40	47

As noted in the Utah Foundation *Research Report* (Houston 2002), "Those states with older, less efficient water systems will be larger water consumers in the withdrawals category than they will be in the delivery category. This is of particular importance in the discussion of conservation and efficient systems as part of an overall state conservation plan."

System-wide leakage and conservation are not mutually exclusive factors. Therefore, any consideration of conditions that could converge into a devastating event must include not only the end-uses of available water, but also the cost to repair leaks, the available technology, and the low price or perceived value of water.

This is a public policy matter because the trend is clearly toward less reliance on self-supplied water and greater dependence on public-supplied sources (Figure 1.4). As reported by the USGS, the 45-year-trend (1955–2000) is certainly heavily weighted in that direction.

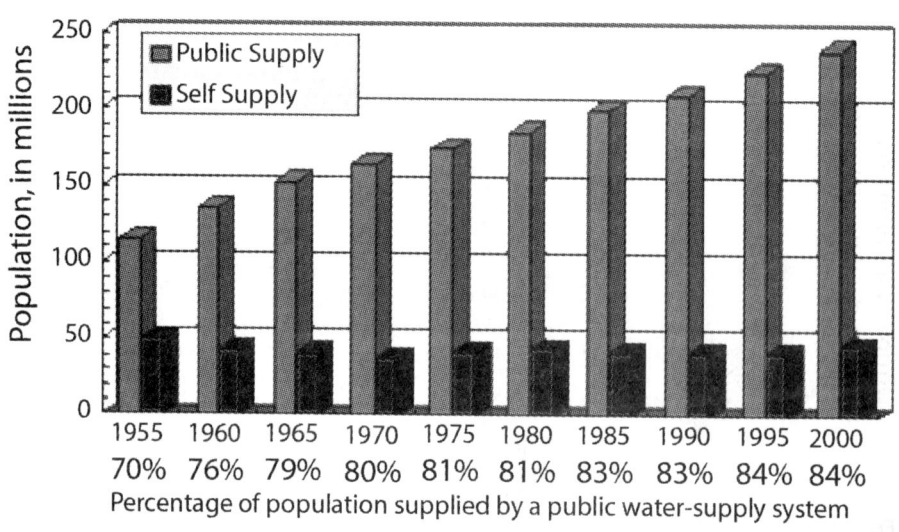

Figure 1.4. Public-supplied and self-supplied populations in the United States, 1955–2000 (USGS 2005b).

During the same 45-year period, according to the USGS, total freshwater withdrawals have declined as population has grown (Figure 1.5). It has been concluded generally that industrial and agricultural water use efficiencies account for the decrease in water withdrawals at a time when the population has continued to increase. The question is whether or not there are additional industrial and agricultural water use efficiencies still available, or if an increase in withdrawals should be anticipated as the population grows and domestic use increases. Additionally, there is the question of how much more efficient or effective domestic water users might become.

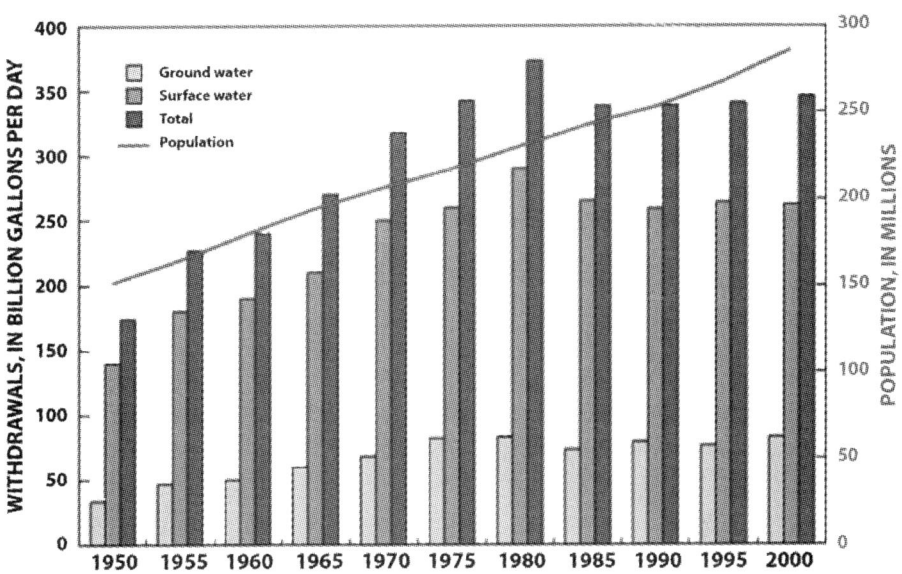

Figure 1.5. Total freshwater withdrawals over a 50-year period (1950–2000) compared to population growth (USGS 2005c).

Peak Demand Requires Attention

Before examining the specifics of domestic water use more closely, the system-wide impact of the "peak demand" phenomenon should be considered. In bridging between the topics of peak demand and domestic water use, keep in mind the EPA has noted, "When divided into indoor uses and outdoor uses, the amount of indoor water use remains fairly constant throughout the year." It therefore seems reasonable to conclude that peak demand would be the result of increased outdoor water use and could be expected to have a seasonal relationship.

Figure 1.6 provides a good example of peak demand: Utah's water storage and release shows that the July peak demand is roughly 50% higher than the remainder of the year. This phenomenon is critically important to water resource agencies because it impacts not only the availability of water supply but also the design of the water storage and distribution system in areas such as pumping and water treatment capacities, as well as the size of piping and meters. Even in areas where water supplies are ample, an economic or investment concern exists whenever the peak demand becomes a driving force for the water agencies' decision-making processes.

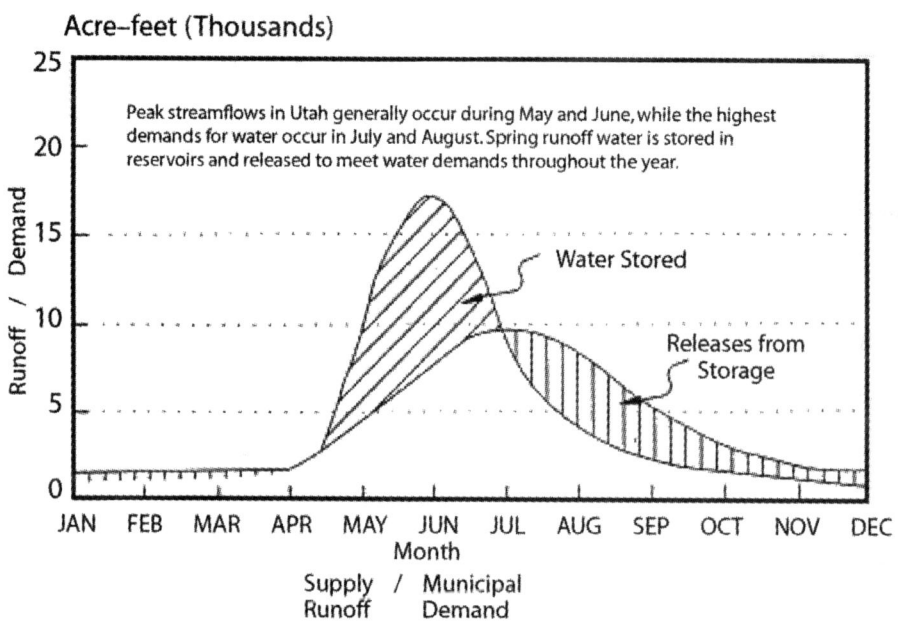

Figure 1.6. Utah peak streamflows and municipal demand (Utah 2005a).

In areas where water supplies are not ample, storage is limited, or water use demands come at a time when rainfall is low or supply restoration is not taking place, the problems created by demand peaks are all the more significant.

The Denver Water Board's Website provides a graphic explanation for this seasonal demand peak (Figure 1.7). There is a close correlation between the Utah graph of water demand (Figure 1.6) and the Denver Water Board's efficient turfgrass watering recommendations (Figure 1.7).

Three Options for Efficient Watering

1. Weekly Watering Amounts for Bluegrass Turf

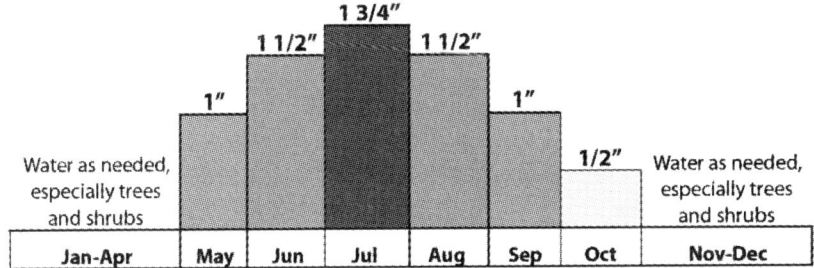

Figure 1.7. Weekly watering recommendations for Denver, Colorado, indicate turfgrass needs water during peak demand in the summer months (Denver 2005).

Flattening the peak demand is an objective of all water agencies. Because the demand curve is typically highest during times of increased outdoor water use, conservation efforts target landscapes generally and turfgrass specifically.

Keying into conservation, and therefore decreasing the demand peak, has multiple benefits for the water agency and the area it serves. Utah provides a vivid example of multiple factors converging and being addressed on the state's Website (Utah 2005b):

- Utah has some of the lowest prices for water in the western United States.

- 63% of water used in Utah homes is used for outdoor purposes, including watering the lawn.

- Currently, Utah residents consume approximately 293 gallons per person per day, second only to Nevada.

  ○ If Utah residents can decrease per capita consumption 25% by 2050, they will conserve the equivalent of 400,000 acre-feet of water per year.

  ○ 400,000 acre-feet of water is more than can be held in the Jordanelle Reservoir and more than any water project Utah has developed.

- Without water conservation, Utah will have to develop a large volume of new water.

- With conservation, some water development projects can be postponed or delayed by several years.

Clearly, water conservation can have multiple positive benefits. These benefits would include extending the availability of water to more people and more uses while lessening the costs (financial and environmental) associated with developing new water resources. At the same time, care must be taken to ensure that a proposed solution does not create its own "unlikely confluence of conditions…in which multiple factors are converging to bring about an event of devastating magnitude." One real world example may help to illustrate this point.

Conservation Solutions May Create a Crisis

For at least two decades, Denver, Colorado, experienced remarkable growth rates. The population expanded, and along with that expansion came an economic boom. During the same time, however, there was virtually no expansion of the city's water resources. In 2001 and 2002, severe drought struck, leaving the area with tremendous water shortages. Throughout much of Colorado, but particularly in the area served by the Denver Water Board, all outdoor watering was banned. No new landscapes could be installed, and existing landscapes suffered.

A November 25, 2002, headline from the *Denver Post* stated, "Spruces done in by bugs and drought." The article reported that at least 300 dead or dying Colorado blue spruce trees (the state tree of Colorado and a plant native to the area) would have to be removed from Denver public parks because the drought depleted the trees' defense against beetles (Borwsky 2002).

The population and industrial expansions, an extended drought, insufficient water storage and distribution capacity, and bans on landscape water use resulted in the death of a large number of native trees. It seems doubtful that the cause of death for the trees will be noted as "insufficient planning, leading to an unlikely confluence of conditions which resulted in death," because, at least for now, we are discussing only trees.

It is interesting to observe that the death of the Colorado blue spruce trees resulted in meaningful news coverage; the death or decline of turfgrass areas in parks, public places, and homes did not receive so much as a word of recognition. It is debatable whether this lack of media interest is because of the lesser value assigned to turfgrass as opposed to trees, or the ability of turfgrass to survive drought or be economically and efficiently replaced.

Whatever the situation is in Denver, the city is not alone in facing the problem of water demand exceeding supply. We need only to read the news or watch television to be aware of other places facing similar water shortage problems. To assess fully and accurately the potential for a water crisis of any magnitude, and to reach meaningful and workable solutions that will decrease and not increase converging negative factors, there is a need to understand how and where water is being used. Yet this is not the simple, straightforward task it might first seem to be.

Reconciling Indoor versus Outdoor Water Uses

As reports indicate, indoor water use seems to remain constant throughout the year. But there seem to be significant inconsistencies in both indoor and outdoor water usage levels among locations when reported in terms of either total volume or percentage of use (Table 1.5).

Table 1.5. Domestic indoor water use reported by the U.S. EPA and the city of Albuquerque, New Mexico (City of Albuquerque 2002; USEPA 2005)

Domestic indoor use	EPA	Albuquerque
Showers and bathing	33%	18.5%
Toilet flushing	41%	26.7%
Kitchen	5%	1.4%
Cleaning/Clothes washing	21%	21.7%
Faucet	--	15.7%
Leaks	--	13.7%
Other	--	2.2%

When the two breakouts are studied, the only matches are for the EPA's "Cleaning" and Albuquerque's "Clothes washing" at roughly 21% each. One must question whether or not it is reasonable that residents of Albuquerque use or flush their toilets only half as often as the national average (26.7% versus 41%), or that they bathe or shower less (18.5% versus 33%), and still clean their clothes an equal amount (21.7% versus 21%). These data illustrate a serious shortcoming of attempting to draw conclusions or make projections related to water use on limited data sets, and endeavoring to apply those conclusions or projections across larger or different areas.

Outdoor water use estimations present similar, if not more complicated, shortcomings. Estimating the amount of water applied to turfgrass in the United States is a daunting and, ultimately, seriously error-prone exercise for several reasons, not the least of which is determining the total area of turfgrass coverage and what, within that total, is receiving applied water. Additionally, measurement or reporting methods for turfgrass areas and the amount of applied water are not always comparable.

This question is confounded further by issues of what percentage of the applied water intended for turfgrass actually is used by trees, bushes, or other landscape plants and what percentage of the applied water is in excess of the plant's needs or is misapplied and becomes runoff.

The EPA states that outdoor residential water use "varies greatly depending on geographic location and season" (USEPA 2005). The AWWA reports other factors influencing outdoor water use are lot size and the percentage of the lot that is irrigable landscape (Mayer et al. 1999). At its Website, the EPA compares the average national residential outdoor water use with that of California and Pennsylvania (Table 1.6) as a percentage of all household water use (USEPA 2005).

Table 1.6. Average national residential water use compared to California and Pennsylvania (USEPA 2005)

Water use	CA	PA	U.S. avg.
Indoor	56%	93%	68%
Outdoor	44%	7%	32%

The 1999 AWWA study on residential water use across the United States and Canada found wide ranges of data (Table 1.7). Residents of Waterloo and Cambridge, Ontario, used roughly 10% of their relatively small total water usage outdoors, whereas Californians used on average 64% of a much larger total of water outdoors. Of all measured water, 56.8% was used outdoors across the various cities included in the study. Which source is correct for estimating the amount or percentage of water that is used for outdoor purposes? The EPA and its estimate of 32%, or the AWWA with nearly double the EPA's rate — a total of 56.8%? Both rates frequently are cited by a variety of individuals, companies, and causes.

Table 1.7. Indoor versus outdoor residential water use (Mayer et al. 1999)

Area	City	Outdoor annual (kgal/home)	Indoor annual (kgal/home)	Total annual use	Outdoor % of total	Indoor % of total
Ontario	Waterloo	7.8	67.7	75.5	10.3	89.7
Ontario	Cambridge	7.8	71.2	79	9.9	90.1
Arizona	Phoenix	161.9	70.8	232.7	69.6	30.4
Arizona	Scottsdale	156.6	60.1	216.6	72.3	27.7
Arizona	Tempe	100.3	65.2	165.5	60.6	39.4
California	Las Virgenes	213.3	70.9	284.1	75.1	25.0
California	Lompoc	43.5	62.1	105.6	41.2	58.8
California	San Diego	99.3	55.3	154.6	64.2	35.8
California	Walnut Valley	114.8	76.3	191.1	60.1	39.9
Colorado	Boulder	73.6	54.4	128	57.5	42.5
Colorado	Denver	104.7	61.9	166.6	62.8	37.2
Florida	Tampa	30.5	56.1	86.6	35.2	64.8
Oregon	Eugene	48.8	65.1	113.9	42.8	57.2
Washington	Seattle	21.7	54.1	86.6	25.1	62.5
	Average	84.61	63.7	149.0	56.8	42.7
	Avg w/o hi/lo	80.3	63.8	143.6	55.7	44.3

Clearly, research and analysis are needed to clarify the amount of outdoor water that is used. Eliminating differences in protocols, standardizing measurement methods and basic terms, and using greater delineations would yield more accurate and meaningful data. In addition to how much water is applied, there is an equally important need for research on how much water is required by the landscape.

There also is a need to clarify with documented research how much water is consumed by various landscape materials and how much is returned either through evaporation, runoff, or groundwater recharge. A determination should be made concerning whether these rates are beneficial or harmful to the hydrologic cycle, the decrease of heat-islands and energy consumption, and even the temperature comfort index as influenced by higher or lower percentages of relative humidity.

To concentrate on a single-focus, one-size-fits-all solution, such as mandating the significant reduction or elimination of landscape or lawn water use in the name of conservation, may result in greater problems than was first apparent. Without scientifically obtained documentation on all these issues, the course of action may simply be adding another converging factor that leads to a devastating result.

Maximizing the Benefits of Turfgrass and Minimizing the Detrimental Impacts

In keeping with a scientific and fact-finding approach, the developments made during the last 20 years in integrated management practices and technologies that will enhance our ability to take advantage of the environmental, cultural, and economic benefits that turfgrass systems provide to society, while simultaneously scientifically examining ways to eliminate or mitigate any detrimental impacts, are documented throughout this publication.

There is an obvious, widespread, and increasingly dire need for water conservation to become a way of life, as well as a need to educate people about the environmental benefits of landscapes and the related return on various investments that this requires. For too long, landscapes generally, and turfgrass specifically, have been sold almost exclusively on the basis of beauty. Often, landscapes are poorly designed, installed, and maintained, with practically no regard for water requirements or water conservation.

In many respects, the green industry has failed to educate its professionals and the consumers on proper landscape water usage. This failure has resulted in over-watering, even in instances where Xeriscape™ designs have been used.

As a society, we are realizing that beauty is not only in the eye of the beholder, it also is a transitory or fleeting objective with little perceived long-term value or benefit. Beauty alone is an easy target for elimination, particularly when considered in terms of a beautiful landscape's freshwater investment requirements and the overall return on that investment. Conversely, the environmental and societal benefits of properly established and maintained landscapes, including grassed areas, have a significant value, but only to the extent that these values are fully understood and the proper means are practiced. When landscape water is withheld, clean air and clean water, natural cooling, noise reduction, good human health and healing, and wildlife—all significant environmental benefits—are jeopardized.

From a sociological and psychological point of view, studies have shown that as the population expands and densities increase, there will be a greater need to increase plants in the human habitat. This publication attempts to provide the best management strategies and technologies to enhance the environmental quality of urban grass systems, including the appropriateness of water use by "Urban Turfgrasses in Times of a Water Crisis."

I believe that the United States has the capacity to identify, address, and avoid an "unlikely confluence of conditions…in which multiple factors could converge to bring about an event of devastating magnitude." Is there any other viable choice?

Literature Cited

Beard, J. B. and R. L. Green. 1994. The role of turfgrasses in environmental protection and their benefits to humans. *J Environ Qual* 23(9):452–460.

Borwsky, C. 2002. "Spruces done in by bugs, drought; At least 300 will have to be axed." *The Denver Post*, November 22, p. B1.

Butterfield, B. 2004. *National Gardening Survey Fact Sheet*. National Gardening Association, South Burlington, Vermont.

City of Albuquerque. 2002. *Save Water Indoors,* http://www.cabq.gov/waterconservation/indoor.html (7 December 2005)

Denver Water. 2005. *Three Options for Efficient Watering,* http://www.water.denver.co.gov/ (7 September 2005)

Houston, J. 2002. *Research Report No. 650*. Utah Foundation, Salt Lake City, Utah. 11 pp. [Original citation in U.S. Geological Survey, *Estimated Use of Water in the United States, 1985, 1990 & 1995*.]

Jackson, S. A. 2004. Leadership to sustain our national capacity for innovation. Paper presented at the John F. Kennedy School of Government, Center for Public Leadership, May 3, at Harvard University in Cambridge, Massachusetts.

Martin, C. A. 2001. Landscape water use in Phoenix, Arizona. *Desert Plants* 17(2): 26–31.

Mayer, P. W., W. B. DeOreo, E. M. Optiz, J. C. Kiefer, W. Y. Davis, B. Dziegielewski, and J. O. Nelson. 1999. *Residential End Uses of Water.* American Water Works Association Research Foundation and American Water Works Association, Denver.

Morris, K. 2003. *National Turfgrass Research Initiative.* National Turfgrass Federation, Inc., and National Turfgrass Evaluation Program, Beltsville, Maryland. 7 pp.

National Atlas. 2005. *U. S. Annual Precipitation Map,* http://nationalatlas.gov/natlas/Natlasstart.asp (6 September 2005)

Negative Population Growth. 2005. *U. S. Resident Population Change 1990–2000,* http://www.npg.org/popfacts.htm (6 September 2005)

U.S. Department of the Interior–U.S. Geological Survey (USDOI–USGS). 2004. *Estimated Use of Water in the United States in 2000.* USDOI–USGS, Washington. D.C. 13 pp.

U.S. Environmental Protection Agency (USEPA). 2005. *How We Use Water in These United States,* http://www.epa.gov/OW/you/chap1.html (8 September 2005)

U.S. Geological Survey (USGS). 2005a. *Water Cycle: Summary, USGA Water Science,* http://ga.water.usgs.gov/edu/watercyclesummary.html (1 September 2005)

U.S. Geological Survey (USGS). 2005b. *Trends in Domestic Water Use,* http://ga.water.usgs.gov/edu/dopsss.html (22 August 2005)

U.S. Geological Survey (USGS). 2005c. Estimated Use of Water in the United States in 2000, Figure 12, Trends in population and freshwater withdrawals by source, 1950–2000, http://water.usge.gov/pubs/circ/2004/circ1268/htdocs/figure13.html (22 August 2005)

Utah Division of Water Resources. 2005a. *Utah's Water Supply,* http://www.water.utah.gov/brochures/UWS_BROC.HTM (7 December 2005)

Utah Division of Water Resources. 2005b. *Utah State Water Plan, Planning for the Future,* http://water.utah.gov/waterplan (7 December 2005)

2

Integrated Multiple Factor Considerations in Low-Precipitation Landscape Approaches

James B. Beard

Introduction

Water has multiple impacts on the urban environment as expressed through the living landscape. There is a tendency to use a simplistic approach for eliminating certain water uses by means of enacting public laws. A single-issue approach of not permitting irrigation on all or a portion of the land area, such as grassed lawns, can lead to other potentially serious problems (Beard 1993). For example, the original Xeriscape theory used a single-issue approach, but sound, science-based research and review by knowledgeable scholars caused major changes to be made in the Xeriscape guidelines. Significant problems that can result from the loss of a turfgrass cover because of not allowing appropriate irrigation in low-precipitation regions are addressed in this section.

Diseases and Airborne Dust

Dust by the hundreds of millions of tons circles the earth annually (Garrison et al. 2003). Airborne dust is an atmospheric pollutant that is unfavorable to human habitats and activities. One million tons of dust may contain up to 10 quadrillion (10^{16}) microbes (USGS 2003). Microbes in airborne dust known to be pathogenic to humans include anthrax, aspergillosis, coccidioidomycosis, hantavirus pulmonary syndrome, influenza, meningococcal meningitis, and tuberculosis (Weinhold 2004).

Airborne dust is a carrier of disease-causing organisms that can adversely affect human health. The risk of respiratory disease increases with the outdoor concentration of total suspended particles (Liu and Diamond 2005).

During the Cultural Revolution of the late 1960s and early 1970s in China, the grassed lawns, trees, and shrubs were removed in major urban areas because they were viewed negatively by the government officials as symbols of a capitalistic society. As a consequence, the dust problem became so severe that at times car traffic on roads had to be closed because of a lack of visibility for safe travel. Eventually, the rate of human diseases associated with the aerial dust pollution in these Chinese cities was much greater than for comparable cities in other parts of the world. Initially, government officials attempted to solve the problem by employing large groups of workers to regularly sweep up the dust from barren ground. The program was unsuccessful. This author was then contacted by the Chinese Botanical Society about revegetation strategies. The officials decided to reintroduce trees, but the problem was not significantly lessened until after an official policy of reintroducing turfgrasses onto the bare areas was implemented. It should be noted that windborne dust from rural areas remains a problem.

This is an extreme, real-world example illustrating that the spread of disease-causing pathogens attached to airborne dust will increase the exposure and potential for disease occurrence in humans, especially in urban areas. Another example of an increasing bare soil–dust–human disease problem is one that occurs in the urban areas of Arizona. Known as valley fever, or coccidioidomycosis, this disease is caused by a soil-inhabiting fungus (*Coccidioides immitis.*). These problems emphasize the importance of dust stabilization by turfgrass use in highly populated urban areas. Studies are needed to determine the extent to which these dust-human disease problems are occurring in other cities around the world. The turfgrass ecosystem not only stabilizes the dust, but also restores it back to a functional soil resource.

Lyme disease is caused by the bacterium *Borrelis burgdorferi* and is transmitted by ticks on certain animals, such as mice and shrews, that commonly occur in unmowed, tall grass and woodland-shrub habitats. A similar threat exists from West Nile Virus that can be transmitted by mosquitoes. Mowed grass lawns around residences offer a less favorable habitat, or a buffer zone, for disease vectors as well as for nuisance insects such as chiggers (Clopton and Gold 1993).

Dust also transports nonbiological material that can adversely affect the health of humans, plants, and animals. An example is the 110 square (sq.) mile (42.5 sq. kilometer) Owens Dry Lake in California. This former lake dried up because of water diversion for agricultural uses and, especially, for a water supply to Los Angeles. The result has been major dust emissions that have exceeded the National Ambient Air Quality Standards for particulate matter (PM) by 25-fold, with peak 24-hour PM-10 values of 8,346 tons per day (USEPA 1999). This dust contains arsenic, nickel, and cadmium, plus very high salt levels. Such wind-borne contaminants place humans at risk through inhalation.

Heat Stress Islands

Bare soils, stones, and other hard surfaces are heat sinks (Table 2.1) that can result in a buildup of maximum daytime temperatures ranging to more than 70°F (39°C) higher than a transpiring green grass cover (Beard 1993). Temperatures in urban areas can average 10°F (5.5°C) higher than nearby rural environments. In contrast, the transpiration of water from grass leaves has a substantial cooling impact, especially in urban areas (Table 2.1).

Table 2.1. Temperature comparisons of four types of surfaces on August 20, in College Station, Texas

Type of surface	Maximum daytime surface temperature °F (°C)	Minimum nocturnal surface temperature °F (°C)
Green, growing turfgrass*	88 (31)	78 (24)
Dry, bare soil	102 (39)	79 (26)
Brown, summer-drought turfgrass*	126 (52)	81 (27)
Dry, synthetic turf	158 (70)	84 (29)

*bermudagrass (*Cynodon dactylon*)

In the 1960s, the national news media featured stories about turfed lawns in southern California being removed, replaced with concrete, and painted green. The resultant heat levels, however, restricted outdoor activities because of human discomfort and substantially increased air conditioning costs and allied energy use. Most of these green-painted, "hard surface" lawns were replaced with transpiring grass lawns within a few years, with an absence of news media fanfare. Now, similar promotions are occurring for green, artificial turfs, which also can act as heat stress sinks.

Wildfires

A green grass cover retards the spread of wildfires because of its low fuel value, and it provides a defendable space around structures where firefighters can work effectively (Youngner 1970). Devastating urban fires have occurred, such as one in San Diego County, California, in October 2003, in which 273,246 acres (110,626 hectares [ha]) burned, 2,232 homes and 588 other structures were destroyed, and 14 lives were lost (California Department of Forestry and Fire Protection 2004). The dry biomass of grass, shrubs, and trees facilitated rapid spread of the firewall from house to house. The wildfire did not reach buildings shielded by an elongated, open, irrigated, grass area of 107 acres (43.3 ha). Post-forest wildfire hazards include accelerated loss of surface soils by wind and water erosion.

Certainly, a buffer area of green, low-growing, perennial turfgrasses functions as a valuable firebreak that decreases the fire threat to human life and property in hazardous fire-prone regions with combustible shrubs and trees (McKell et al. 1966). Also, a firebreak area for a specified distance from a building can lower insurance costs in some localities.

Soil Erosion and Flooding

Surface soils are a vital natural resource that must be protected through proper stewardship. A cost-effective means of controlling soil erosion is a live, functioning grass cover, including urban lawns. The superior aboveground shoot density of mowed turfgrasses—with from 185 million to 49 billion shoots per acre (75 million to >20 billion ha^{-1}), depending on the species, plus 890 to 26,785 pounds per acre (1,000 to 30,000 kilograms [kg] ha^{-1}) of leaf/stem biomass (Lush 1990)—provides substantial resistance to lateral surface water movement that slows otherwise erosive water velocities. Consequently, there is a major decrease in eroded soil sediments entering rivers, lakes, and seas (Gross et al. 1991).

The turfgrass biomass functions essentially as a sponge that traps water and increases vital groundwater recharge. Turfgrass areas can be designed with contours to temporarily hold water, thereby decreasing storm-water runoff. These water-retaining properties of turfgrasses contribute to decreased storm flow via grass waterways and associated flooding, as well as minimize the need for costly mechanical-concrete, water-control structures in urban areas.

The belowground characteristics of grasses also are important. A grass root system is one of the most effective in soil stabilization because of the fibrous, dense character of the roots. For example, Kentucky bluegrass (*Poa pratensis* L.) has a root biomass of 9,790 to 14,329 pounds per acre (11,000 to 16,100 kg ha^{-1}) (Boeker 1974; Falk 1976).

Urban Pollutants

Runoff water from intense rains is much greater from the impervious surfaces of urban areas compared with turfgrass lawns. This runoff carries organic pollutants such as oils, greases, fuels, paint thinners, organic preservatives, and solvents (Schuyler 1987). As previously discussed, turfgrasses are effective in decreasing runoff water and thereby can trap or filter out significant quantities of associated organic pollutants. Runoff waters also can contain potentially toxic metals such as lead, cadmium, copper, and zinc. These metals can be trapped in the turfgrass ecosystem, thereby further decreasing the pollution threat to rivers, lakes, and seas in the associated watershed.

The large, diverse microbial population in the turfgrass/soil ecosystem is one of the most active biological systems for the decomposition of organic pollutants washed from hard-surface, urban areas. The average microbial biomass pool for arable, forest, and grassland systems are 625, 759, and 973 pounds of carbon (C) per acre (700, 850, and 1,090 kg C ha^{-1}), respectively (Smith and Paul 1990). This turfgrass ecosystem microbial activity serves a valuable function in the decomposition of trapped organic pollutants from urban areas.

Also, recent research showed that the soil microbial biomass C and nitrogen (N) were approximately six times greater for 95-year-old turfgrass on a golf course than for an adjacent native pine forest (Shi, Yao, and Bowman 2006). The results indicate that soil microbial properties are not negatively affected by long-term cultural practices in turfgrass/soil ecosystems. A tight coupling between N mineralization and immobilization can be sustained in mature turfgrass systems because of their increased microbial C and N use efficiency, thus decreasing the potential for N loss from soils.

Criminal Activity

Security is important to minimize harm to people and to protect property. Police and military officials have indicated that low-growing, turfgrass lawns provide key lines-of-sight around homes and buildings, thus aiding crime prevention and facilitating searches for criminals. In contrast, trees and shrubs block lines-of-sight needed by law enforcement officials to function effectively in urban areas.

Human Disharmony

The lack of a green, vegetative surface cover in the urban landscape contributes to a decline in human social harmony and productivity. When the visual content responses of humans to a golf course turfgrass landscape were compared with the responses from viewing a structured urban building site and a forest setting, the turfgrass landscape lowered the average blood pressure level and the skin conductance level, with both returning to baseline levels more rapidly (Parsons et al. 1998). Furthermore, the viewer subsequently performed mental arithmetic tasks more rapidly. A visually aesthetic external environment of grass, trees, and shrubs improves mental health (Ulrich 1984, 1986) and enhances the quality of life in urban communities (Kaplan and Kaplan 1989).

Summary

The seven categories described here represent significant potential problems if officials do not take them into consideration when proposing legislation to exclude irrigation from all or part of the urban landscape. There are other functional benefits attributed to turfgrass/soil ecosystem use in urban landscapes; these benefits are summarized in Figure 2.1 (Beard and Green 1994). Certainly, the social and economic values of these benefits are substantial, but studies quantifying the economic aspects are needed.

Functional	Recreational	Quality of life
Soil erosion control	Low cost surfaces for:	Beauty contributing to:
Dust prevention	Physical health	Mental health
Disease prevention	Mental health	Social harmony
Natural filtering system	Decreased injury risk	Community pride
Buffer areas	Family-lawn activities	Human productivity
Flood control	Community recreation	Property values
Organic pollutant decomposition	Community sports	Compliments trees and shrubs in landscape
Bioremediation	Spectator entertainment	
Soil restoration		
Carbon sequestration		
Groundwater recharge		
Heat dissipation		
Air pollution control		
Fire barrier		
Noise abatement		
Glare reduction		
Roadside safety		
Crime control		
Nuisance animal reduction		
Wildlife habitat		
Pollen/Weed control		

Figure 2.1. Diagrammatic summary of benefits derived from turfgrass use.

To summarize, rather than eliminating certain water uses in low-precipitation landscapes, there are other substantial savings to be accomplished in furthering water conservation. These actions range from sustainable best management practices for irrigating urban, grassed landscapes to repairing leaks in municipal water distribution systems. Incongruities in laws and "money-for-grass" approaches, which eliminate grassy areas but allow the use of ornamental shrubs and trees with higher water use rates, are not sound approaches (Park et al. 2005). An integrated, holistic approach to water use in populated areas is essential. The elimination of turfgrasses from open areas in urban landscapes should be implemented only as a last resort in arid climates. Turfgrasses not only use water, but also collect, hold, and clean it while enhancing subsequent groundwater recharge and contributing to transpiration cooling.

Literature Cited

Beard, J. B. 1993. The Xeriscaping concept: What about turfgrasses. *Intl Turfgrass Soc Res J* 7:87–98.

Beard, J. B. and R. L. Green. 1994. The role of turfgrasses in environmental protection and their benefits to humans. *J Environ Qual* 23(3):452–460.

Boeker, P. 1974. Root development of selected turfgrass species and cultivars. Pp. 55–61. In E. C. Roberts (ed.). *Proceedings of the Second International Turfgrass Conference*. International Turfgrass Society, American Society of Agronomy, and Crop Science Society of America, Madison, Wisconsin.

California Department of Forestry and Fire Protection. 2004. *California Fire Siege 2003—The Story*, http://www.fire.ca.gov/fire_er_siege.php (24 January 2007)

Clopton, R. E. and R. E. Gold. 1993. Distribution, seasonal, and diurnal activity patterns of *Eutrombicula alfreeddugesi* (Acari: Trombiculidae) in a forest edge ecosystem. *J Med Entomol* 30:47–53.

Falk, J. H. 1976. Energetics of a suburban lawn ecosystem. *Ecol* 57:141–150.

Garrison, V. H., E. A. Shinn, W. T. Foreman, D. W. Griffin, C. W. Holmes, C. A. Kellogg, M. S. Majewski, L. L. Richardson, K .B. Ritchie, and G. W. Smith. 2003. African and Asian dust: From desert soils to coral reefs. *BioSci* 53(5):469–480.

Gross, C. M., J. S. Angle, R. L. Hill, and M. S. Welterlen. 1991. Runoff and sediment losses from tall fescue under simulated rainfall. *J Environ Qual* 20:604–607.

Kaplan, R. and S. Kaplan. 1989. *The Experience of Nature*. Cambridge Univ. Press, New York. 352 pp.

Liu, J. and J. Diamond. 2005. China's environment in a globalizing world. *Nature* 435:1179–1186.

Lush, W. M. 1990. Turf growth and performance evaluation based on turf biomass and tiller density. *Agron J* 82:505–511.

McKell, C. M., V. Stoutemyer, C. Perry, L. Pyeatt, and J. R. Goodin. 1966. Hillside clearing and revegetation of fire hazard areas. *California Agric* 20(12):8–9.

Park, D. M., J. L. Cisar, G. H. Snyder, J. E. Erickson, S. H. Daroub, and K. E. Williams. 2005. Comparison of actual and predicted water budgets from two contrasting residential landscapes in South Florida. *Intl Turfgrass Soc Res J* 10:885–890.

Parson, R., L. G. Tassinary, R. S. Ulrich, M. R. Hebl, and M. Brossman-Alexander. 1998. The view from the road: Implications for stress recovery and immunization. *J Environ Psych* 18(2):113–140.

Schuyler, T. 1987. *Controlling Urban Runoff: A Practical Manual for Planning and Designing Urban BMPs.* Metropolitan Washington Council of Governments, Washington, D.C.

Shi, W., H. Yao, and D. Bowman. 2006. Soil microbial biomass, activity and nitrogen transformations in a turfgrass chronosequence. *Soil Biol and Biochem* 38:311–319.

Smith, J. L. and E. A. Paul. 1990. The significance of soil microbial biomass estimations. Pp. 357–396. In J. M. Bollag and G. Stotzky (eds.). *Soil Biochemistry, Vol. 6.* Marcel Dekker, New York.

Ulrich, R. S. 1984. View through a window may influence recovery from surgery. *Science* 224:420–421.

Ulrich, R. S. 1986. Human responses to vegetation and landscapes. *Landscape Urban Planning* 13:29–44.

United States Geological Survey (USGS). 2003. African dust carries microbes across the ocean: Are they affecting humans and ecosystem health? USGS Open-File Report 03-028. 4 pp.

U.S. Environmental Protection Agency (USEPA). 1999. Approval and promulgation of implementation plans; California-Owens Valley nonattainment area; PM-10. *Federal Register* 64(122). 20 pp.

Weinhold, B. 2004. Infectious disease: The human costs of our environmental errors. *Environ Health Perspectives* 112(1):1–5.

Youngner, V. B. 1970. Landscaping to protect homes from wildfires. *California Turfgrass Culture* 20(4):28–32.

3

Regulatory Considerations for Water Quality Protection

Beth Hall

Summary

The Environmental Protection Agency's (EPA) Office of Water (OW) is responsible for implementing the Clean Water Act (CWA) and Safe Drinking Water Act (SDWA), as well as portions of the Coastal Zone Act Reauthorization Amendments of 1990; Resource Conservation and Recovery Act; Ocean Dumping Ban Act; Marine Protection, Research, and Sanctuaries Act; Shore Protection Act; Marine Plastics Pollution Research and Control Act; London Dumping Convention; International Convention for the Prevention of Pollution from Ships; and several other statutes. Its activities are targeted to prevent pollution wherever possible and to decrease the risk for people and ecosystems in the most cost-effective ways possible. In the last few years, water security also has become a more critical part of the EPA's mission. The Homeland Security and Bioterrorism Act of 2003 specifically denotes the responsibilities of the EPA and the water sector.

This paper focuses on the legislative history and context of the SDWA and the CWA and concludes with a discussion of how the goals of the two Acts can be integrated through federal, state, and local implementation.

Early State and Municipal Water Pollution Control Programs

People have long recognized the relationship between contaminated water supplies and disease outbreaks. For example, in the fourth century B.C., Hippocrates advised citizens to boil and strain water before drinking it, to prevent hoarseness.

In the mid-1800s, authorities in the United States began to recognize and address public health concerns related to drinking water. In the late 1800s, cities recognized the relationship between typhoid fever outbreaks and the use of untreated surface water as drinking water. It was not until the germ theory of disease was broadly accepted in the early 1900s, however, that treatment of water (to mitigate disease spread through untreated water) began on a significant level. The earliest treatment provided disinfection and sometimes filtration of surface water sources.

In the early 1900s, reacting to the large number of typhoid and other disease outbreaks, states and local governments began establishing public health programs to protect water supplies. The first were water pollution control programs, which focused on keeping surface water supplies safe by identifying and limiting sources of contamination. Early water pollution control programs concentrated on keeping raw sewage out of surface waters used for drinking water.

Early drinking water programs were aimed at providing safe and adequate drinking water to a community. At first, these programs were not separate from the water pollution control programs because they also focused on identifying and maintaining safe sources of drinking water. For example, efforts were made to site intakes used to collect drinking water upstream from sewage discharges. By the mid-1900s, state public health departments were well-established regulatory agencies. The primary contaminants of concern were microbes, and states used the following "multiple barrier approach" to prevent microbial contamination of drinking water:

- Selection and protection of an appropriate source. For surface water sources, this meant locating and constructing water intakes to ensure little or no contamination from fecal bacteria. For groundwater sources, this meant constructing wells in appropriate locations, at appropriate depths, and with approved construction methods (e.g., casing and grouting).

- Treatment to be appropriate to the quality of the source water. Treatment was designed to eliminate all contaminants of concern identified during testing of source water. Under the umbrella of treatment, there were multiple barriers. For example, settling, filtration, and disinfection may all be used to treat the same water for different constituents.

- Well-engineered distribution systems to promote full circulation and avoid stagnant water conditions that might facilitate microbial contamination. The integrity of distribution systems was checked periodically to avoid any cross-connection whereby untreated or contaminated water might enter the system. State agencies insisted on well-engineered and constructed storage facilities that reliably protected finished water from contamination.

States used several regulatory methods to implement the multiple barrier approach. Most required that plans and specifications for new water systems (or major alterations to existing systems) be approved before construction. Some states also required a postconstruction inspection to ensure that "as-built" systems conformed to the approved plans and specifications. In addition, routine sanitary surveys were conducted by a state sanitarian or engineer who checked all components of the system from source to tap. Operator training and certification also were important components.

Early Federal Involvement with Water Resource Quality

The origins of the Public Health Service (PHS) are traced to the passage of an Act in 1798 that provided for the care and relief of sick and injured merchant seamen. After the Civil War, the PHS began studying illnesses associated with contaminated drinking water. But early federal laws were limited to activities that state laws could not address—primarily interstate commerce that included the following legislation:

- The Rivers and Harbors Act of 1899, which applied primarily to discharges such as mine tailings, rocks, or other objects that would interfere with navigation.

- The Interstate Quarantine Act, which provided federal authority to establish drinking water regulations to prevent the spread of disease from foreign countries to the states or from state to state. This resulted in promulgation of the first interstate quarantine regulations in 1894.

- The first water-related regulation, adopted in 1912, which prohibited the use of the "common cup" on carriers of interstate commerce, such as trains.

- The first federal drinking water standards, which were established in 1914 by the PHS. The standards applied to water supplied to interstate carriers—primarily passenger trains—and included a 100-cc (100 organisms/cubic centimeter) limit for total bacterial plate count. Further, they stipulated not more than one of five 10-cc portions of each sample examined could contain *B. coli* (now called *E. coli*). The standards were legally binding only on water supplies used by interstate carriers, but many state and local governments adopted them as guidelines.

Post-Environmental Protection Agency Water Regulation

In 1970, the EPA was established as an independent agency. A major factor in its establishment was an implicit understanding of the need for federal enforcement authority. The drinking water, air pollution control, and solid waste programs were moved from the PHS to the EPA. Water pollution control was moved from the Department of the Interior to the EPA.

In 1972, Congress enacted the Federal Water Pollution Control Act Amendment (PL 92-500). On the basis of the Act's new provisions, the newly created EPA assumed the dominant role in directing and defining water pollution control programs across the country. This Act became the basis of the Clean Water Act in effect today. Congress was able to use this experience in crafting the next major piece of legislation—the Safe Drinking Water Act. In the late 1960s and early 1970s, several surveys of drinking water quality were conducted including a national survey that showed drinking water to be widely contaminated on a national scale, particularly with synthetic organic contaminants. Because of the concerns about drinking water contamination and nascent water pollution control programs, Congress established national health-based standards for finished drinking water. Congress also required that suppliers routinely monitor water to ensure that the established standards were achieved.

Clean Water Act

Seeking a way to respond to public concern about water pollution, the Nixon Administration attempted to bring back the Rivers and Harbors Act, empowering the U.S. Army Corps of Engineers to issue discharge permits from the national level. The law, however, had no provision for decision criteria or standards on which to base the permits, and by reviving the Act without Congressional authorization, a clear legal basis, a legislative record, or consultation with key policymakers in Congress, the Administration gave the House and Senate strong reasons to override their traditional differences and fashion a program of their own.

The Congressional response was the enactment, in 1972, of the Federal Water Pollution Control Act Amendment (PL 92-500) For the past 35 years, U.S. water quality policy has been based on objectives stated in the 1972 statute, one of which was "...to restore and maintain the chemical, physical, and biological integrity of the Nation's waters." The second national goal expressed in the statute was "Water quality which provides for the protection and propagation of fish, shellfish, and wildlife and provides for recreation in and on the water." In addition, the Act presented several other precepts that still remain:

- States have the primary responsibility for implementing programs to meet the above goals.

- There is no right to pollute the navigable waters of the United States. Anyone wishing to discharge pollutants from a "point source" must obtain a permit to do so.

- All point sources must meet the best controls that technology can produce at a reasonable cost, regardless of the receiving water's ability to purify itself naturally.

Since its creation, the CWA has emphasized two dramatically different strategies for achieving and maintaining these goals: the water quality-based approach and the technology-based approach. The first strategy is based on the risk-based approach that states were using before the CWA was enacted. This approach starts by looking at the condition and the uses that the state or tribe wants a particular water body to support, such as protecting aquatic life and drinking water and establishing water quality standards, which include the water quality criteria that support those uses. If it is determined through monitoring that the water is impaired (not meeting water quality standards), the next step is determining the rate at which a pollutant can enter the water body without exceeding the water quality standards (i.e., establishment of a Total Maximum Daily Load [TMDL]).

For the first decade after CWA passage, the focus changed almost entirely to implementation of the second strategy—a new technology-based approach. This was intended to prevent pollution of water bodies by implementation of technologically and economically achievable controls on major categories of pollution sources, such as publicly owned treatment works (POTW) and industrial facilities. Each of these facilities then is required to meet end-of-the-pipe discharge limits established through regulation as effluent guidelines. These programs and standards are now well established.

Implementing programs that achieve needed reductions in pollutants is more problematic. For point sources, the EPA can use enforceable regulatory and permitting authorities, i.e., the National Pollutant Discharge Elimination System (NPDES) permits, which combine technology-based and water quality-based effluent limits. In contrast, the CWA provides funding through the Section 319 Nonpoint Source National Monitoring Program, but there is no regulatory authority for achieving nonpoint source reductions.

In 1977, the Act was amended and became formally known for the first time as the Clean Water Act. The 1977 Amendments clarified and expanded the concept of controls based on best available technology to include toxic pollutants. Congress established both schedules for the EPA to set limits and deadlines for industry to meet them. Section 404 of the 1977 Act required the EPA to develop a program to control discharges of dredged and fill materials into wetlands and other waters of the United States. The Agency is required to monitor the protection of these water areas in coordination with other federal agencies and the states through a permit program.

The Water Quality Act of 1987 addressed a number of issues on which Congress deemed progress to be unsatisfactory. These issues included toxics, nonpoint sources, storm water, coastal pollution, and the use and disposal of domestic sewage sludge (biosolids). The Act extended the construction grants program only through FY 1990; the Amendments phased out the construction grants program in favor of a state revolving fund (SRF).

Congress responded to the lack of numeric criteria for toxic pollutants within state ambient water quality standards by mandating state adoption of such criteria. In addition, the EPA was required to establish concentration limits for toxics in sludge and to develop regulations for sludge use and disposal and state permit programs. New provisions required the EPA (or states authorized for the NPDES program) to issue permits for storm water from separate storm sewers and industrial sources of storm water. These statutory provisions are the result of lawsuits. The Natural Resources Defense Council (NRDC) won cases in which the courts said that CWA already provided this authority, so Congress made it explicit in the law.

The Act also explicitly recognized the EPA's antidegradation policy for the first time. The intent of this policy was to preserve the level of water quality necessary to protect existing uses and to provide a means for assessing activities that may lower water quality. The Act also provided federal funding for state nonpoint source programs. It required each state to identify nonpoint sources of pollution that contribute to water quality problems and waters unlikely to meet the water quality standards without nonpoint source controls. States also had to adopt management programs to control nonpoint source pollution and then implement the management programs.

The 1987 statute extended participation in the CWA programs to Indian tribes. The Act directed the EPA to establish procedures by which a tribe could qualify for "treatment as a state," at its option, for purposes of administering CWA programs and receiving grant funds.

Over the years, many other laws have changed parts of the Clean Water Act. Title I of the Great Lakes Critical Programs Act of 1990, for example, put into place parts of the Great Lakes Water Quality Agreement of 1978, signed by the United States and Canada, in which the two nations agreed to decrease certain toxic pollutants in the Great Lakes. That law required the EPA to establish water quality criteria for the Great Lakes, addressing 29 toxic pollutants with maximum levels that are safe for humans, wildlife, and aquatic life. It also required the EPA to help the states implement the criteria on a specific schedule.

Evolution of Clean Water Act Implementation

For many years following the passage of the CWA in 1972, the EPA, states, and Indian tribes focused mainly on the chemical aspects of the "integrity" goal. During the last decade, however, more attention has been given to physical and biological integrity. Also, in the early decades of the Act's implementation, efforts focused on regulating discharges from traditional "point source" facilities such as municipal sewage plants and industrial facilities, with little attention paid to runoff from streets, construction sites, farms, and other "wet-weather" sources.

Starting in the late 1980s, efforts to address polluted runoff have increased significantly. For "nonpoint" runoff, voluntary programs, including cost-sharing with landowners, provide the key tool. For "wet-weather point sources" such as urban storm sewer systems and construction sites, a regulatory approach is being employed.

Evolution of the CWA programs during the last decade also has included somewhat of a shift from a program-by-program, source-by-source, pollutant-by-pollutant approach to more holistic watershed-based strategies. Under the watershed approach, equal emphasis is placed on protecting healthy waters and restoring impaired ones. A full array of issues is addressed, not only issues subject to CWA regulatory authority. Involvement of stakeholder groups in the development and implementation of strategies for achieving and maintaining state water quality and other environmental goals is another hallmark of this approach.

Safe Drinking Water Act

1974 SDWA

Increased concern and awareness of contamination of drinking water supplies prompted Congress to enact the SDWA in 1974. Concerns raised in the late 1960s and early 1970s about drinking water quality prompted the EPA to conduct a national survey to detail the quality of drinking water. The survey showed that drinking water was widely contaminated on a national scale, particularly with synthetic organic chemicals. Contamination was especially alarming in large cities. This survey raised concerns about drinking water in both the public health community and the general public.

The 1974 SDWA expanded the focus from water system planning and prevention of contamination to include developing standards, monitoring for contaminants, and taking enforcement action. It required the EPA to establish national enforceable standards for drinking water quality and to guarantee that water suppliers would monitor water to ensure that it met national standards. Further, the following three new programs were established:

- The Public Water System Supervision Program (PWSS), which set up a higher level of responsibility for regulating drinking water systems than established state programs.

- The Under Ground Injection Control (UIC) Program, which regulated injections of fluids that could endanger sources of drinking water.

- The Sole Source Aquifer Program, which provided special status to those designated aquifers that represented a primary source of drinking water. Such designation gave the EPA the ability to review and comment on the groundwater impacts of federally funded projects in those areas.

The law acknowledged that protection of drinking water was still primarily a state responsibility. The 1974 SDWA established a major focus on delegating primary responsibility for program implementation (i.e., primacy) to the states.

1986 Amendments

From 1974 to 1986 when the SDWA was amended, state regulations varied in many respects. For example, states differed in requirements for groundwater disinfection, mandated filtration, monitoring of organic chemicals, and operator certification requirements. Data management and the requirements for contaminant monitoring were relatively simple. The EPA conducted the first inventory of community water supply systems in 1976. The survey revealed that the majority of systems were small, privately owned groundwater systems, but most people were customers of large, publicly owned systems using surface water.

By 1986, Congress was concerned about the EPA's lack of progress in developing drinking water regulations. Congress also was concerned about the lack of regulation for microbial contamination, synthetic organic chemicals, and other industrial wastes. In reaction, the 1986 Amendments were prescriptive and required the EPA to regulate 83 contaminants within 3 years after enactment. The Amendments declared the interim standards promulgated in 1975 to be final and required the EPA to require disinfection of all public water supplies and filtration for surface water systems. Further, the EPA was required to regulate an additional 25 contaminants (to be specified by the EPA) every 3 years and to designate the best available treatment technology for each contaminant regulated. States with primacy were required to adopt regulations and begin enforcement within 18 months of the EPA's promulgation.

Along with increased treatment requirements for surface water systems, some groundwater supplies were recognized as providing water of essentially surface water quality and were required to be regulated as surface water systems.

New public notification requirements increased the communication between water systems and consumers, further increasing awareness of drinking water contamination. Public notification requirements were prescribed strictly and included broadcast and printed notices, depending on the severity of the contamination problem.

More stringent coliform monitoring requirements in the 1986 Amendments increased the frequency of coliform detection. Increased requirements for follow-up monitoring after initial detection revealed even more problems and led to greater awareness of the inadequacy of some water sources, even after treatment.

The lead and copper requirements affected systems of all sizes, making implementation an enormous undertaking. The lead and copper requirements also were difficult to implement because the need for relatively high pH water to prevent corrosion seemed to contradict microbial treatment needs of a lower pH for effective coagulation and disinfection practices. Balancing water chemistry, treatment needs, and compliance with several regulations became an increasing challenge.

The 1986 Amendments also initiated the groundwater protection program, including the Wellhead Protection Program. The law specified that certain program activities, such as delineation, contaminant source inventory, and source management, be incorporated into state Wellhead Protection Programs, which are approved by the EPA before implementation.

1996 Amendments

The implementation of the 1986 Amendments raised many issues. Unlike the CWA programs, in which many of these issues were addressed through guidance and policy, the reauthorization of the SDWA was able to directly implement change in the focus of the program. First, the Amendments addressed concerns about the existence of an overly burdensome regulatory structure. Congress eliminated the 1986 requirement that the EPA regulate an additional 25 contaminants every 3 years. Instead, the EPA was allowed to establish a process for selecting the contaminants to regulate based on scientific merit. The EPA now has the flexibility to decide whether or not to regulate a contaminant after completing a required review of at least five contaminants every 5 years. The EPA also is required to conduct cost-benefit analyses of new regulations and analyze the likely effect of the regulation on the viability of public water systems. The Act added new and stronger prevention approaches. The comprehensive, preventive approach of the 1996 SDWA Amendments introduced the nonregulatory source water assessment and protection program.

The 1996 Amendments also addressed concerns about funding needs for PWS infrastructure and state program management by establishing the Drinking Water State Revolving Fund (DWSRF) modeled after the Clean Water State Revolving Fund. They also strengthened the EPA's enforcement authority and included provisions to help increase the ability of small systems to comply with the regulations. The SDWA Section 1420 mandates that the EPA assist states in developing water systems' financial, managerial, and technical capacity. Consumer awareness and public information requirements were increased significantly through a requirement that water systems issue an annual Consumer Quality/Water Quality report.

Water Security

Recently, the government has promulgated legislation and directives in recognition of the increased need to protect the nation's water supply and utilities from terrorist attacks. The Homeland Security Presidential Directives (HSPDs) and the Public Health Security and Bioterrorism Preparedness and Response Act (Bioterrorism Act) of 2002 specifically denote the responsibilities of the EPA and the water sector in

- assessing vulnerabilities of water utilities,
- developing strategies for responding to and preparing for emergencies and incidents,
- promoting information exchange among stakeholders, and
- developing and using technological advances in water security.

These recent directives and laws supplement existing legislation, such as the SDWA and the CWA, which have always had the goals of promoting a clean and safe supply of water for the nation's population and protecting the integrity of the nation's waterways. These directives and laws affect the actions and obligations of the EPA, the Water Security Division, and water utilities.

Clean Water Act / Safe Drinking Water Act Integration

Since the EPA was established in 1970, the Agency and country have made great progress in improving surface water quality and ensuring safe drinking water. Under the provisions of the CWA, the nation invested more than $75 billion to construct municipal sewage treatment facilities, nearly doubling the number of people served with secondary treatment to almost 150 million.

Through federal and state actions that issued permits, the EPA has controlled more than 48,000 individual industrial facilities, and it has controlled thousands more through general permits. By establishing nationwide discharge standards for more than 50 industrial categories, the EPA has helped to decrease industrial loadings by as much as 90%. Industrial waste and sewage sludge—which, at their peaks, produced 5.9 million tons and 8.7 million tons, respectively—are no longer dumped into U.S. coastal waters. Based on current water quality standards, more than 70% of rivers, 68% of estuaries, and 60% of lakes now meet legislatively mandated goals. Fish are coming back, the rate at which wetland habitats are lost is slowing, and many miles of formerly contaminated beaches are now safe for swimmers.

Since 1986, the EPA has more than tripled the number of contaminants tested for drinking water standards, bringing the total to 94. Once implemented, the Surface Water Treatment Rule is expected to prevent 83,000 cases of illness due to waterborne diseases. The EPA expects to prevent more than 600,000 children from having dangerously elevated levels of lead in their blood by implementation of the Lead and Copper Rule of 1991. Through the SDWA and parts of other laws, the EPA is regulating many high risk sources of groundwater contamination including pesticides, underground storage tanks, underground injection wells, and landfills, helping to ensure the safety of drinking water supplies.

But as both programs—the CWA and the SDWA—have been able to evolve, this legislative framework makes comprehensive solutions and their implementation problematic and complicates protection of ecosystems and habitat. The traditional command and control approach, combined with single media laws, precludes flexibility and deflects attention from developing and applying alternative solutions that include market mechanisms, economic incentives, voluntary approaches, alternative enforcement penalties, prevention, negotiation, education, and land use planning. In spite of those challenges, significant and deliberate efforts are being made to meet new challenges.

Source Water Protection

The 1996 SDWA Amendments required and funded a new critical tool for the local and state protection of drinking water: source water assessments. For the first time, in a comprehensive way, water utilities and community members could get the information they needed to decide how to protect their drinking water sources. The SDWA required that the states develop the EPA-approved programs to carry out assessments of all source waters in the state. The assessment is a study that defines the land area contributing water to each public water system, Identifies the major potential sources of contamination that could affect the drinking water supply, and determines how susceptible the public water supply is to this potential contamination. Public utilities and citizens can then use the publicly available study results to the take actions to decrease potential sources of contamination and protect drinking water.

These assessments can be of invaluable assistance to the many programs and organizations that bear some responsibility for water quality and land use planning. These can range from a town's conservation commission or local county extension agent to state agencies, nonprofit organizations, and federal agencies such as the Forest Service. Some programs work specifically with small communities and water systems. For example, funded through grants from the EPA and the Department of Agriculture, source water protection specialists are working with small communities in 48 states on wellhead and surface water protection planning. Agricultural specialists are working in 32 states to address agricultural sources of contamination.

Protecting sources of drinking water also can help various federal programs, states, and communities meet other environmental and social goals such as green space conservation, storm water planning, management of nonpoint source pollution (such as runoff from agricultural lands), and brownfields redevelopment. Protection of drinking water quality is a high priority for the public and can serve as a driver for more comprehensive water quality efforts.

The EPA's Office of Ground Water and Drinking Water is working with a number of other stakeholders and offices within the EPA to encourage those programs to better integrate drinking water source quality considerations into their priority setting. The EPA currently is coordinating with our Pesticides Office on development of §304(a) water quality criteria for Atrazine, Alachlor, and Cryptosporidium. The EPA's UIC Class V program and Underground Storage Tank Programs are working with states to prioritize inspection and enforcement sites based on the source water assessments.

For surface water, the CWA is the federal and state regulatory complement to local source water protection. The EPA is completing a baseline of state water quality standards that apply to CWS intake. The SDWA program is working with our CWA counterparts to track all state waters that are listed or should be listed as impaired for the public water supply use and to track their restoration. On the SDWA side, the EPA recently promulgated a new drinking water protection rule, the Long Term 2 Enhanced Surface Water Treatment Rule (LT2), which lowers the risk of disease-causing microorganisms from entering water supplies.

Sustainable Development

One of the EPA's newest initiatives is to work with drinking and wastewater utility managers and associations to move the management paradigm beyond compliance to sustainability. New technologies, research and development, and a market-based approach can help ensure that the nation's water infrastructure can leverage needed resources through better utility management and operations, improvements in water efficiency, full-cost pricing of water supply and wastewater treatment, and watershed-based approaches to solving water quality and water quantity problems. Again, from the perspective of protection of drinking water sources, the goal is to take advantage of the opportunities within watershed-based approaches to minimize infrastructure and operating costs for drinking water utilities. The EPA also is encouraging water quality protection through the promotion of cost-effective approaches for future development, such as low-impact development and smart growth, that will help predict and mitigate the impact of development on water resources.

Conclusion

Through integrated federal, state and local implementation, the SDWA and the CWA have had a positive impact on the protection and conservation of water in the United States. The EPA's Office of Water is responsible for implementing the CWA and SDWA, as well as portions of other important legislation dealing with ocean and coastal waters. The EPA's activities are targeted to prevent pollution wherever possible and to decrease risk for people and ecosystems in the most cost-effective ways possible. In the last few years, water security also has become a more critical part of the EPA's mission.

4

Municipal Water Use Policies

Andrew W. Richardson

Introduction

The story of man could be told in terms of his struggle for water and his use of it. The first great civilizations arose in the valleys of great rivers—the Nile Valley of Egypt, the Tigris-Euphrates Valley of Mesopotamia, the Indus Valley of India, and the Hwang Ho Valley of China. All these civilizations built large irrigation systems, made the land productive, and prospered.

Civilizations crumbled when water supplies failed or were poorly managed. Many historians believe the Sumerian civilization of Mesopotamia fell because of poor irrigation practices. Salt in irrigation water is left behind during evaporation and tends to build up in the soil. The ancient Sumerians failed to achieve a balance between salt accumulation and drainage.

The ancient Romans built aqueducts, canals, and reservoirs throughout their empire. They turned regions along the coast of northern Africa into prosperous civilizations. But after the Romans left, their water projects were abandoned. Now, these places are deserts.

The challenge today, as in ancient times, is for man to make the best use of water. But the challenge is even greater than before because man needs more and more water as industry and populations grow. Although the Earth has enough water to meet the growing demand, that water is distributed unevenly. To meet this challenge of effective water management, policies often are developed. Sometimes they are based on the lessons of the past, sometimes not.

What is a policy? Webster defines a policy as "Prudence or wisdom in the management of affairs; management or procedure based primarily on material interest, rather than on higher principles; hence worldly wisdom. A settled course adopted and followed by a government, institution, body, or individual."

In the United States, there is currently no national "water policy." This is due in part to the history of the country and in part to the understanding that most water issues have been historically treated as local issues. Three Congressional actions—(1) the establishment of the Federal Bureau of Reclamation in the early twentieth century, (2) the passage of the Clean Water Act (CWA) in 1972, and (3) the passage of the Safe Drinking Water Act (SDWA) in 1974—are the closest the country has come to a national water policy. These actions together have had some impact on municipal water policies through the years. Even today, however, most municipal water use policies are developed based on local water conditions.

To understand how municipal water use policies have been developed and what we might expect in the future, this section will examine historical and current municipal water use policies, current water industry trends, and the impact those trends will have on future municipal water use policies.

Historical and Current Municipal Water Use Policies

Institutional/Structural

The historical, institutional, and structural aspects of water policies in the United States are similar to those of the early civilizations of mankind. All major population centers were developed in the vicinity of abundant water resources. As the eastern part of the country developed and population centers grew, it became apparent that the West needed to be developed and populated. To do this, water was key. As the country developed, so did two fundamentally different legal systems that govern the allocation of water throughout the United States. Under the "riparian system," which applies in 29 eastern states that were historically considered "wet" states, ownership of land along a waterway determines the right to use the water. In times of shortage, all owners along a stream must decrease their water use. Because of water scarcity in the West, it was impractical for water rights to depend on ownership of land along streams.

The "prior appropriation system" of water rights, originally developed by miners in California, was adopted by nine arid western states. Under prior appropriation, a water right is obtained by diverting water and putting it to beneficial use. "Beneficial use" was historically interpreted to mean domestic, municipal, agricultural, or industrial uses, but more recently has been expanded to include recreational and wildlife uses. An entity whose appropriation is "first in time" has a right that is senior to one who later obtains a water right. In times of water shortages, "senior" rights must be fully satisfied before "junior" rights are met, sometimes resulting in juniors receiving no water at all. Water rights can be bought and sold; the priority date of the right remains unchanged after the sale, making senior rights highly desirable. These two legal concepts, plus the establishment of the Federal Bureau of Reclamation and the passage of the CWA and the SDWA, form the basis for all water policy in the United States. It is from this basis that most municipal water use policies are developed.

To encourage growth in the western United States, the Federal Bureau of Reclamation was created to develop and manage water resources. For the early twentieth century, this was a major policy decision. The Bureau's projects are located in all 17 western states and are in virtually every major river basin. The sheer volume of water controlled in those projects— the dams and reservoirs—has had a significant impact on western water policy, often contributing to local (municipal) water policies as well.

In the eastern United States, as metropolitan areas developed, so did water policy. In essence, most water policy is really at the state and local level. A major characteristic contributing to this fact is the structural decentralization of our nation's water systems. Accordingly, across the United States, local water use policy will be influenced by the physical area covered by its water system, as well as its institutional structure.

The drinking water system in the United States is extremely decentralized. Community or municipal water systems are structured in four basic ways that will influence water use policies: (1) owned by local governments, (2) independent government authorities, (3) privately owned companies, and (4) public-private partnerships. Community water systems are regulated by the states, except in Wyoming, the District of Columbia, and the territories. The states implement the United States Environmental Protection Agency (EPA) regulations through "primacy" agreements. There are 53,000 community water systems in the United States, and they provide 90% of Americans with their tap water. Thirty-one thousand community water systems serve 500 people or fewer. Only 424 community water systems serve more than 100,000 people. In total, 8% of community water systems serve 82% of the United States' population. In the United States, 86% of community water systems are owned by local governments or independent government authorities; 14% are owned privately.

Operational

The structure of a community or municipal water system will determine how it is operated. The operations associated with each of the four types of structures are discussed below.

Local Government-Owned Community Water Systems

Local government-owned community water systems are owned and controlled by a local government entity and are usually a department or entity of the governance structure. Local government-owned community water systems are usually part of the local government budget. Rates charged for drinking water are set through the political process of the local elected governing body.

Independent Government Authorities

Independent authorities are established by local governments to own and manage a community water system separate from a local government structure. Independent authorities are controlled by an independent board that determines the budget and sets rates. Independent authorities are not subject to the direct political process of the local government and operate as an enterprise.

Privately Owned Community Water Systems

Privately owned community water systems may be investor-owned stock corporations, privately held corporations, or not-for-profit corporations. Privately owned community water systems include private mobile home communities, not-for-profit cooperative utilities owned by the citizens, and homeowner associations. Private ownership is heavily concentrated in small- and medium-sized community water systems. Forty-two percent of community water systems serving 500 or fewer people are privately owned. Privately owned community water systems are regulated by state public utility commissions that approve the rates the privately owned community water systems may charge their customers. Private ownership is a means of providing capital investment in community water systems in those instances when a local government may not have the resources for investment, or when it is more efficient for a private company to obtain capital and manage the community water system. Private ownership may provide increased flexibility in management.

Public-Private Partnerships

In public-private partnerships, the local government usually retains ownership of the community water system assets but contracts with a private company to manage the day-to-day operations of all, or part, of a community water system. Community water systems in a public-private partnership operate as an enterprise, and the management is not subject to direct political control by the local government. Public-private partnership provides increased flexibility in management.

Economic

The importance of water to our survival renders it, literally, priceless. But this intrinsic value of water is frequently left out of traditional pricing. Historically, pricing quantifies the costs of capture, treatment, and conveyance. Consequently, this method often obscures the larger, but less quantifiable, societal interests in preserving our water resources.

Supplementing historical pricing methods with incentives for consumers to manage demand is a combination that serves both financial and environmental goals. This practice is known as demand management pricing. The economic water use policy may vary greatly depending on whether one is in the eastern or western United States. Historically, in the East, most utilities have priced water to sell to meet their revenue goals, whereas in the West, there always has been a struggle to meet revenue goals while not overtaxing sources of supply.

Water demand can be manipulated by price to some degree. Water for necessities (sanitation, cleaning, and cooking) is far less responsive to price than water for more discretionary use (landscape watering, car washing, and swimming pools). Clearly, water demand is inelastic, meaning that when price increases, consumption decreases but at a lower rate than the price increase. To foster conservation, some municipal water use policies may be developed to influence the demand of consumers. This tactic would be accomplished through water rates, as well as water conservation policies, such as plumbing codes and landscaping requirements or incentives.

In regard to water rates, through the years there have been well-established policies on how water rates should be determined. A leader in this area is the American Water Works Association (AWWA), whose members provide approximately 85% of the drinking water across the United States. In 1965, the AWWA Board of Directors adopted a policy in regard to the financing, accounting, and rates associated with municipal water systems. The policy states, "AWWA believes that the public can best be provided water service by self-sustained enterprises adequately financed with rates and charges based on sound accounting, engineering, financial, and economical principles." In essence, AWWA is advocating a municipal water use policy of "full cost of service" rate-making, with those rates (prices) being passed on to the consumer.

Sociopolitical

Until recently, the water industry prided itself on being the "silent servant." Both urban and massive water projects were built with little public involvement. There was a general understanding from a political perspective that water projects needed to be built to support economic growth and development; they were a necessity. The policy was, it could be suggested, that the water project had priority over all else. To understand this fully, one must consider the volume of dams, reservoirs, canals, pumps, and levees built from the beginning of the twentieth century until the 1960s. Since the 1960s, when there was movement on several social issues including the environment, the universal support for large massive water projects has declined steadily.

In the early twentieth century, when the United States was investing large sums of private and public money to adapt to the vicissitudes of the hydrologic cycle, the general public seemed to support this investment. As time went on, however, there was a shift in the value society attached to water projects versus the environment. At first, society seemed to value water projects, but now it questions, what are the environmental costs of storing and diverting water? What are the environmental costs on the decline of free-flowing streams? What are the environmental costs that impact-flow diversions have on wetlands and wildlife habitat? The sociopolitical water use policy may have shifted 180 degrees as we entered the twenty-first century.

Water Industry Trends

The AWWA recently conducted research to determine the trends that could shape the future of the water industry and their impact on future water policy. That research identified nine major changes:

1. **Infrastructure management** will be a critical issue resulting in an increased capital investment with an impact on ratepayers in the future. The AWWA has indicated that we are at the dawn of the "replacement era," and, based on demographics and the age of our infrastructure, critical funding issues will need to be addressed in the immediate future. In May 2001, AWWA's Water Utility Council published a report titled *Dawn of the Replacement*

Era: Reinvesting in Drinking Water Infrastructure (Cromwell 2001). That report indicated that the drinking water infrastructure needs over the next 30 years would be approximately $300 billion. One of the report's main conclusions was "Overall, the findings confirm that replacement needs are large and on the way....Ultimately, the rate-paying public will have to finance the replacement of the nation's drinking water infrastructure either through rates or taxes. The AWWA expects local funds to cover the great majority of the nation's water infrastructure needs and remains committed to the principle of full-cost recovery through rates" (Cromwell 2001).

Since that time, numerous reports have been published concerning the projected "gap" between available resources and the needs of local water utilities to address regulatory and infrastructure challenges. Recent reports from the EPA and the Congressional Budget Office (CBO), as well as competing national priorities and the struggling economy, offer little hope of significant federal assistance any time soon (Beider and Tawil 2002). That same CBO report indicated that by 2019, average annual expenditures (capital and operations and maintenance) will be $71 to $98 billion for both drinking water and wastewater.

2. **Environmental regulations** will become more stringent for water, with eight major regulations being implemented in the next few years that will impact smaller utilities the most. This era is what the industry is calling the "Big Wave" of regulations. At the turn of the twentieth century, the average life expectancy in the United States was 47 years. The major cause of death was waterborne diseases—typhoid and cholera. Today, the average life expectancy is 77 years, and the major cause of death is heart disease. Over time, regulations have moved the water industry from concerns of acute disease and death-causing agents to chronic long-term effects of exposure to micropollutants, as well as organic and inorganic chemicals.

3. **Water utilities** will continue to reorganize and evaluate the best available management practices to be as efficient as possible, particularly in light of the need to fill the impending funding gap and provide more with less. In addition, it is expected that there will be some consolidation. Larger utilities may be looking to offer assistance to or may even be running smaller utilities as consolidation continues to occur across the North American water industry.

4. **Good customer relations** will be the key to success when dealing with the impending funding shortfall that is perceived to meet future regulations, as well as to rehabilitate older systems. As water scarcity and quality problems become more serious and apparent, the public will become more informed. Utilities already are required to inform the public through their Consumer Confidence Reports. As the general public becomes more aware and concerned about water, their perceptions and demands will be a key driver in the industry.

5. **The overall work environment** within the water industry, as in other industries, will continue to change with diversity, age, gender, and ethnicity being important aspects of the new work force. Technology will require entities to provide two to three times more training than was necessary in the past. Over the next few years, many water utilities will experience the "brain drain" with their most experienced people retiring. For water, wastewater, and combined utilities, 9% of employees are eligible to retire now, 22% of employees are eligible to retire in less than 5 years, and 35% of employees are eligible to retire in fewer than 10 years. The attrition rate for the last 3 years has been 6%, and the attrition rate expected for the next 3 years is 8% (Oldstein et al. 2004).

6. **Management, training, and financial planning** will be key to the future survival of water utilities. The understanding of rates and asset management, along with creating innovative financing to minimize rate impacts to consumers, will become a basic requirement in the future of all utilities' business operations. Utilities also may bring in key decision makers from the financial and management fields, compared to the past reliance on "water people" to manage a water utility.

7. **Total quality management** will be critical to doing more with less and looking at trends on how existing utilities can provide the same service to customers at less cost. Utilities will examine internal procedures for potential cross-training and to be as efficient as possible.

8. **Total water resources management** is the key to the future. Many water utilities are looking at the impacts of watershed management and source water protection, while also building a partnership with the agricultural community and its impacts on regulations. Total water resource management is expected to lead to "portfolio" management of water resources, as well as the idea of "appropriate quality for appropriate use." This practice will lead to more "water reuse."

9. **Rising water rates** will be needed to pay for the capital improvements and the rehabilitation of the water industry's existing infrastructure. The other eight trends all will have some type of impact on water prices. In some locations, political will, as well as a common sense approach to educating the public, will be required. As prices continue to increase, decisions about water usage will begin to take on greater significance in the overall economy.

AWWA State of the Water Industry Results

To monitor the trends previously mentioned and determine if they are accurate, the AWWA in 2004 and 2005 conducted "State of the Industry" surveys. These surveys, which asked members what they saw as the important issues in the next 3 to 5 years, gave the AWWA an idea of what would consume the water industry's time and resources. The reports were published in *Journal*

AWWA, December 2004 and October 2005, respectively. The reports show that AWWA's previous studies correctly predicted some of the trends. The results of the 2004 and 2005 "State of the Industry" reports indicate that utilities are concerned about the following factors (Murphy 2004; Runge and Mann 2005):

1. **New complex regulations.** It is difficult for smaller utilities with few resources to (a) understand regulations and (b) comply with them. The regulations they are concerned about include the Stage 2 Disinfectant Byproducts Rule, the Long-Term 2 Enhanced Surface Water Treatment Rule, the Arsenic Rule, and the Groundwater Rule.

2. **Infrastructure management.** Utilities are very concerned that we are just at the start of the "Dawn of the Replacement Era" as reported by the AWWA in 2001 (Cromwell 2001).

3. **Future water supply.** Utilities are concerned with where water will come from in the future. Communities are expanding into arid and semiarid climates, and groundwater sources are being depleted in some places. Droughts have magnified the limitations of surface water. Because of the concern for future water supply, utilities indicated they will be focusing on source water protection (the better source waters are protected, the less expensive it is to treat drinking water), water reuse (particularly for nondrinking purposes, such as watering urban landscapes), and desalination (promising but expensive, and raises the issue of what to do with the resulting waste).

4. **Security.** Utilities are feeling more prepared to deal with security issues after their vulnerability assessments and resulting investments. Yet security registers as a large concern, particularly in the area of developing better online contaminant monitoring systems. Since September 2001, utilities have made security a priority and have made resource investments to back public confidence.

5. **Overarching concern: How do we pay for all this?** Ultimately, utilities believe water customers will have to pay for the costs of these growing expenses, and that means utilities have to think seriously about their rate structures and implementation of full-cost recovery rates.

Trends' Impact on Future Municipal Water Use Policies

Institutional/Structural

The trends previously discussed and the results of the two "State of the Industry" surveys will have an impact on future municipal water use policies from an institutional/structural perspective. In 2002, and again in 2005, more than 250 of the nation's experts on water resources met to discuss a national water resources policy (AWRA 2005). That group's findings were as follows:

1. The nation's water issues need to be addressed in an integrated manner, focusing not on single projects, but on programs and watershed- and basin-level issues. The successful cooperative and holistic efforts evidenced in evolving programs to restore the Everglades, manage the California Bay Delta, and protect Coastal Louisiana need to be replicated across the country.

2. There is a need to reconcile the myriad laws, executive orders, and Congressional guidance that have created a disjointed, ad-hoc national water policy and to define clearly our twenty-first century goals. Many important laws were passed early in the last century when national objectives and physical conditions were far different than they are today. Many of these laws are in conflict, placing the federal, state, and tribal agencies that execute those laws in tenuous and sometimes adversarial situations. Reexamination of these laws would eliminate contradiction and confusion, leading to far more effective water policies and policy implementation.

3. Recognizing the fiscal realities facing the nation, there is a need to more effectively coordinate the actions of federal, state, tribal, and local governments in dealing with water. Collaboration instead of competition will provide better and more fiscally efficient use of scarce resources and will assist in overcoming decision gridlock on key water programs.

4. The nation is blessed with access to superb scientific capabilities and cutting-edge technologies that can support water-related decision making. These capabilities and technologies need to be focused clearly on supporting water policy decision makers as they carry out their challenging responsibilities.

Dialogue participants also noted that much of the general public, as well as many public officials, lack a full understanding of the extent and complexity of our water challenges. Education about water must parallel efforts to solve water problems. Furthermore, the participants noted that funding to support our water resources infrastructure has not kept pace with needed repairs, replacements, and modernization. Because of this, we see a greater discussion on the value of water as a result of trends in the industry.

Water is our most precious natural resource. And while most of us do not give it a second thought when we grab a glass and head to the faucet, drinking water should not be taken for granted. A safe, reliable drinking water supply creates jobs, attracts industry and investment, and most importantly, provides for the health and welfare of our citizens in ways ranging from disease prevention to fire suppression.

It is hard to say what policy change will develop from this dialogue. It is important to keep in mind that most municipal water use policy does follow national input, but is most often based on local needs. In recent years, however, we have seen some major institutional/structural changes at a regional watershed level. An example of this change was the creation of the Southern Nevada Water Authority (SNWA) in the early 1990s. The SNWA was formed by southern Nevada water utilities to give them a common voice on water issues impacting their communities. Thus far, the SNWA has been effective in managing water resources at a regional level.

Operational

The impact the trends will have on the operational policies of municipal water agencies will be related to having to do more with less. The four basic operational models mentioned earlier will stay the same. But there will continue to be consolidation of utilities, causing many large regional entities to deal with watershed and source water protection issues.

We also may see small utilities being taken over by medium to large utilities for economy of scale operations. Overall, regionalization also may occur as utilities and states strive for cooperation on issues associated with next source of supply.

Economic

As indicated by the trends, water prices will rise. This rise will be a result of municipal water use policies associated with full-cost recovery. The real question is, what impact will this have on economic development in certain areas of the country and, more importantly, on "wet" industries or commercial enterprise? Water is still cheap—ridiculously cheap in many ways. In fact, water is so cheap that its price has not really become a significant driver of change. In terms of "Economics 101," we are still on the inelastic portion of the demand curve for water, where increases in price have relatively little impact on usage. As prices continue to increase, decisions about water usage will begin to take on greater significance in the overall economy. Focus will begin to intensify on more efficient water usage and demand management techniques, and technologies and supply sources such as desalination and water recycling will be more and more common. In fact, this is already happening. There is clear evidence that water prices are increasing faster than inflation, much faster in many instances.

Sociopolitical

With each passing day, people are demanding more choice in their water service as in other areas of their lives. Citizens also want water that meets increasingly stringent standards of quality, but they do not necessarily expect to pay more for it. More stringent standards and more choice mean that all across the United States, water customers have had, and will continue to have, a larger bill to pay for water service.

Although government control generally is viewed as undesirable, citizens are demanding government action to preserve the environment and agricultural lands, as well as to ensure a healthy economy, the sustainability of water supplies, and better quality of life. Some past policies governing the level of water service must be adjusted to enable the utility to satisfy its water customers in this atmosphere of rising expectations and fiscal constraints.

Summary

We need to remember that the amount of fresh water on this Earth is relatively fixed, and we need to become much smarter and much more efficient in the usage of this scarce, precious resource. The facts are very straightforward: water is life. To sustain and improve our modern industrial economy and maintain our quality of living, we need to manage our water wisely. Will we learn from the civilizations that have gone before us and develop water use policies that will stand the test of time? That question will be answered based on how we collectively respond to the trends in the water industry.

Literature Cited

American Water Resources Association (AWRA). 2005. *Second Annual National Water Resources Policy Dialogue*, Tucson, Arizona, February 14–15, 2005.

Beider, P. and N. Tawil. 2002. *Future Investment in Drinking Water and Wastewater Infrastructure*. Congressional Budget Office, Washington, D.C.

Cromwell, J. 2001. *Dawn of the Replacement Era: Reinvesting in Drinking Water Infrastructure*. American Water Works Association, Denver, Colorado.

Murphy, M. 2004. State of the Industry Report 2004. *J AWWA* 96(12):60–62, 64, 66–72.

Oldstein, M., D. L. Martin, J. G. Voeller, J. D. Jennings, P. M. Hannan, and D. Brinkman. 2004. *Succession Planning for a Vital Workforce in the Information Age* (Draft). American Water Works Association Research Foundation, Denver, Colorado.

Runge, J. and J. Mann. 2005. State of the Industry Report 2005–A Guide for Good Health. *J AWWA* 97(10):58–67.

5

Turfgrass and the Environment

Michael P. Kenna

Introduction

Population pressures along with depletion and contamination of traditional water supplies are straining the world's water resources. The growth in urban population has resulted in an increased demand for low-maintenance green space and recreational areas that can withstand increased use. At the same time, turfgrass managers have experienced increased maintenance costs and tougher government regulation with regard to water use and environmental issues. This has placed increased emphasis on developing turfgrasses that can tolerate environmental stress and drought. Knowledge on how to produce, establish, and maintain the turfgrasses used in urban and recreational settings has grown tremendously in the past 50 years.

Turfgrass Information File

The amount of information on turfgrass science and management was growing faster than researchers, golf course superintendents, and many others could possibly keep pace. Even the university libraries across the country could not adequately maintain a complete record of all the important turfgrass literature. The Turfgrass Information File[1] at Michigan State University Library is the most well-organized and complete collection on turfgrass science and management in the world. This collection is available to anyone interested by calling, writing, or accessing directly by computer.

Turfgrass Biology and Distribution in the United States

This section provides background on general turfgrass biology, the major climate zones in the United States, several of the turfgrasses commonly used, and some of the turfgrass breeding efforts underway. All grasses have three major vegetative organs: stem, leaf, and root. In grasses, the apical meristem arises near the soil level and consists of a series of compressed nodes. This region of the plant, the interface between roots and shoots, is commonly referred to as the crown of the plant. The bulk of the crown is just below or at the soil surface (Figure 5.1). A grass stem (or culm) is a tube that typically is hollow and round, but may be elliptical or flattened. The tube is interrupted by thickened sections called nodes. A node is the point of attachment for the base of each leaf sheath. Each node gives rise to one leaf. The stem region between nodes

---

1 Visit the Turfgrass Information File at Turfgrass Information Center, W212 Main Library, Michigan State University, East Lansing, MI 48824-1048, (800) 446-8443, http://www.lib.msu.edu/tgif

is called the internode. A tiller consists of a hollow whorl of leaves, protecting the crown (growing point) from which each leaf originates (Figure 5.1).

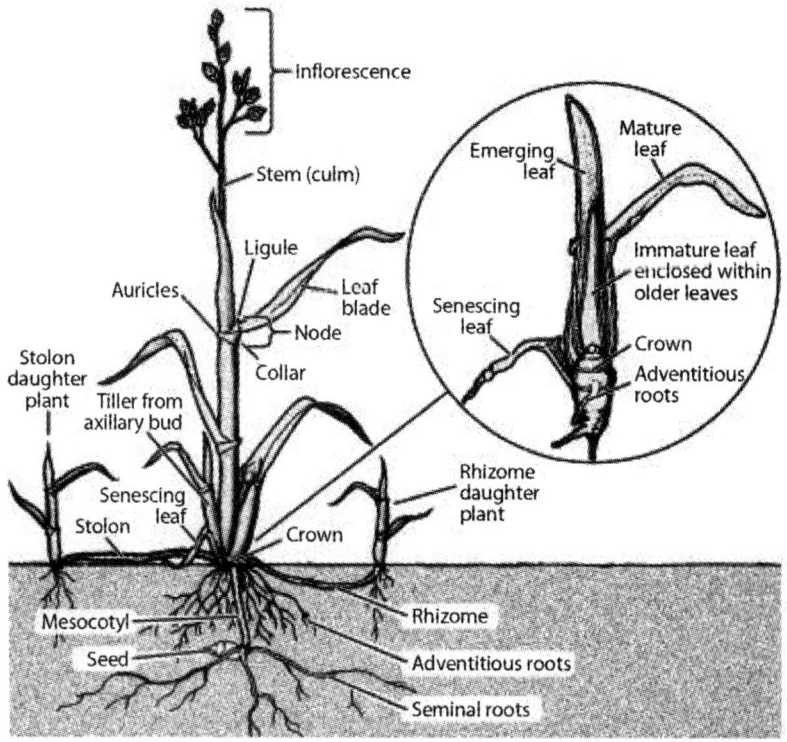

Figure 5.1. Parts of the turfgrass plant and cross-section of the crown with the organization of leaves (adapted from Turgeon 1996).

Each grass leaf consists of two structures: a sheath and a blade (or lamina). The sheath and blade are separated by a collar that may have additional structures, such as a ligule (a hairy or membranous appendage at the junction of the sheath and blade) or auricles (a pair of appendages on either side of the collar).

Grass root systems are fibrous, highly branched, and consist almost entirely of secondary and adventitious roots (secondary roots originating from the crown or nodes). Roots may originate from lower nodes of the stem or from stem nodes (with axillary buds) that come in contact with soil. Roots function as organs for nutrient and water uptake, storage of carbohydrate reserves, and an anchor for the grass plant.

The management of turfgrasses involves five basic practices: mowing, fertilization, irrigation, pest control, and cultivation. The extent or frequency of each of these practices is determined by how the turfgrass is used. For example, managing a golf course is considerably different from managing a residential lawn. Table 5.1 compares the various management practices for two uses of turfgrasses.

Table 5.1. Comparison of intensity of management practices between a golf course and a home lawn

Practice	Golf course	Home lawn
Fertilization	Biweekly (light amounts)	Once or twice per year
Mowing	Daily (putting greens)	Weekly
Irrigation	Daily	Occasionally
Pest control	Often	Occasionally
Cultivation	Many times per year	Rarely

U.S. Climate Zones

Climate is the combination of light, temperature, precipitation, and wind that influences the growth and development of turfgrasses. Temperature extremes and precipitation patterns are the most important environmental factors that influence the range of turfgrass adaptation (Beard 2002). Cool-season turfgrasses grow best at soil temperatures between 16 and 24°Celsius (C) [60 to 75°Fahrenheit (F)]. In contrast, warm-season turfgrasses grow best in soil temperatures between 27 and 35°C (80 to 95°F). Figure 5.2 (see page I-8) depicts the geographical distribution of turfgrass species in relation to the major climatic zones.

Cool-Season Turfgrasses

Kentucky bluegrass is a general purpose turfgrass commonly used on lawns, parks, athletic fields, cemeteries, and golf course roughs and fairways. Kentucky bluegrass is a long-lived perennial that is widely adapted throughout the cool-season growing areas. It also can be used in the cool semiarid and arid regions if irrigated. Kentucky bluegrass is capable of surviving extended drought periods and can initiate new shoot growth when moisture conditions improve. Summer dormancy may occur with the above ground foliage becoming brown.

Perennial ryegrass is generally considered to be a short-lived perennial, but it can persist indefinitely if not subjected to extremes in high or low temperature. Typically, perennial ryegrass persists under cold winter conditions where it is protected by consistent snow cover. Perennial ryegrass has two primary uses. In the cool-season zone, it is frequently used to compliment Kentucky bluegrass in sunny lawn mixes. In the South, perennial ryegrass is the primary overseeding grass. Seeded in late August and early September, it remains green until late spring when it dies as the underlying bermudagrass breaks dormancy.

Fine fescues are long-lived perennials widely distributed through the cool-season area of the United States. Fine fescue is limited in its geographic distribution compared with Kentucky bluegrass because of lower heat tolerance. Creeping Red Fescue is distinguished from other fine fescues because it spreads through a creeping growth habit, although somewhat less vigorously than Kentucky bluegrass. Chewings and hard fescue have a bunch-type growth habit. All the fine fescues are distinguished by their narrow, upright leaves. Fine fescues are superior to other cool-season grasses in their shade adaptation, and their water use rate is lower than Kentucky bluegrass and perennial ryegrass.

Tall fescue is a long-lived perennial when grown in the transition zone between the cool humid and warm humid regions. It persists as far north as the Great Lakes and as far south as Atlanta and Dallas. Tall fescue is fairly heat tolerant compared with other cool-season species. Farther north, tall fescue is prone to winter injury. It is a deep-rooted cool-season grass that remains green during summer heat and drought stress when Kentucky bluegrass, ryegrass, and fine fescue go dormant. It is the most drought and wear tolerant cool-season species. Tall fescue is superior to bluegrass and perennial ryegrass in shade tolerance, but it is inferior to fine fescue in the shade.

Creeping bentgrass is primarily known as a grass for golf course putting greens and fairways. The name creeping bentgrass is derived from the vigorous, creeping stolons that develop at the surface of the ground. When closely mowed, it forms a fine-textured turf with superior shoot density, uniformity, and turfgrass quality. Colonial bentgrass differs from creeping bentgrass because it has less spreading capability through rhizomes or stolons. Velvet bentgrass is extremely fine textured, forming a very dense turf. Its rate of spread by stolons is greater than colonial bentgrass but less than creeping bentgrass. Bentgrasses are seldom used on home lawns because of poor stress tolerance and high maintenance requirements. Bentgrass requires lower mowing than is practical with most rotary mowers.

Several bentgrass cultivars now have superior heat tolerance for both high soil and air temperatures. These conditions impair the transpirational cooling process. Most bentgrasses exhibit a definite degeneration of root tissue and shortening of roots under high soil temperatures, close frequent mowing, and heavy traffic (Beard and Daniel 1965; Huang, Liu, and Fry 1998a, b). Screening techniques were developed that examine leaf and shoot water content as it relates to bentgrass plants grown in high ambient and soil temperatures (Lehman and Engelke 1993). In addition, root-screening procedures helped identify individual plants with superior root growth (Lehman and Engelke 1991).

Warm-Season Turfgrasses

Bermudagrass is an aggressive, warm-season turfgrass species that spreads rapidly by stolons and rhizomes. It has excellent drought tolerance in the summer. Extremely heat tolerant, but very intolerant of shade, bermudagrass is the dominant sunny lawn grass in the South and hot summer climates of the Far West. Once established, bermudagrass is hard to kill. It is one of the few warm-season grasses that can be taken north. Bermuda is recommended for use in parts of Tennessee, North Carolina, Arkansas, and Oklahoma, as well as the Central Valley of California.

Bermudagrass varieties have been developed by two different methods. The seeded varieties of bermudagrass are improvements in common bermudagrass achieved by plant breeders using conventional plant selection techniques. Vegetatively planted bermudagrasses are produced from sterile hybrids, which initially were crosses between common bermudagrass and African bermudagrass. Because sterile hybrids do not produce seed, these lawns must be established using sprigs, stolons, or sod at considerable time and expense.

Improvements in vegetative and seed-propagated bermudagrasses used for golf courses and sports fields have been successful (Taliaferro 2003). Improvements in seed establishment as well as cold tolerance will help provide bermudagrass cultivars for the transition zone climates of the United States. Decreases in water use and maintenance costs will be met by providing better adapted warm-season turfgrasses for the transition climate zone where poorly adapted cool-season species require excessive cultural inputs.

Zoysiagrass is a perennial, sod-forming species that is widely adapted across the warm-season growing area of the United States. Most zoysiagrass lawns discolor with the advent of 10 to 13°C (50 to 55°F) temperatures and remain dormant throughout the winter and early spring. It forms a uniform, dense, low-growing, high-quality turf that has a slow rate of growth. Zoysiagrass spreads by thick stolons and rhizomes that form a very tight, vigorous, tough, prostrate growing turf. Thatch management is vital for a Zoysia lawn because it is more thatch-prone than other warm-season grasses.

Although zoysiagrass does best in full sun, its primary advantage is moderate shade tolerance. With heat tolerance equal to bermudagrass and better shade tolerance, it is one of the best choices for southern lawns with partial shade. Zoysia traditionally has been planted vegetatively by sprigs, plugs, or sod, but new seeded varieties are now available.

Zoysiagrasses with better sod production characteristics have been developed, and they are better adapted to a broader range of environmental conditions (Engelke and Anderson 2003). These entries range in texture from rather broad-leaved, aggressive *Zoysia japonica* types to fine-textured, highly rhizomatous *Z. matrella* types. New cultivars were selected for a combination of characters related to survival and turfgrass quality under natural environmental conditions. Specific emphasis was placed on low water use, competitive ability against weed invasion, recovery from injury, low fertility, and sod production characters.

Centipedegrass is a creeping perennial that is well adapted to sandy, acidic soils. It tolerates low fertility and requires little maintenance. Centipedegrass spreads by stolons and has a coarse texture with short, upright stems that grow to a height of 3 to 5 inches. It requires infrequent mowing and will survive mild cold temperatures. Centipedegrass has moderate shade tolerance. It is primarily used where low-maintenance turf is desired. Where higher-maintenance lawn is planned, bermudagrass is superior in the sun, and zoysiagrass is recommended in the shade.

Buffalograss is a native prairie grass that can be used for low-maintenance lawns and other turf areas. Buffalograss is an example of how a species native to North America can be used for low-maintenance lawns and golf course roughs (Riordan and Browning 2003). It greens up earlier than bermudagrass in the spring and turns brown after the first fall freeze. Buffalograss grows best in full sun, requiring at least 6 to 8 hours of direct sun daily. Good soil drainage is essential. Although buffalograss will persist in clay soils, it will not survive in sandy soils. Continued research efforts will hopefully allow this species to be more widely used where traditional grasses are not well adapted. Buffalograss is by no means a panacea, but represents a major step in the recognition of a valuable natural resource for the turfgrass industry to use.

St. Augustinegrass is a warm-season grass of tropical origin that does best in sun or partial shade conditions in the Southern Coastal Plain, Florida, and the California Central Valley. It grows vigorously during the warm (80 to 95°F) months of spring, summer, and early fall. Like other warm-season grasses, it goes dormant and turns brown in the winter. It is very susceptible to winter injury and cannot be grown as far north as bermudagrass and zoysiagrass. St. Augustinegrass has large, flat stems and broad-coarse leaves somewhat similar to centipedegrass. With attractive blue-green color, it spreads by stolons and forms a deep, dense turf. St. Augustinegrass is shade tolerant and is superior to zoysiagrass in the shade in the warmest regions of the south.

Seashore *Paspalums* can survive high levels of salt. The focus has been on collecting interesting *Paspalums* from around the world. Breeding efforts to improve cold tolerance, color, density, and other turfgrass characteristics are well under way. Vegetative cultivars such as Sea Isle I, Seal Isle 2000, Seadwarf, or the seeded variety Sea Spray, have been developed (Duncan 2000).

Native Grasses

Native grasses have the greatest potential in regions where water, poor soils, or climate are the limiting factors in providing quality turfgrass. Taking advantage of the natural selection that has occurred over millions of years may be more successful than a 10-year breeding program, but the domestication of native species is not a simple task. Alkaligrass, blue grama, fairway wheatgrass, and inland saltgrass have been evaluated as potential turfgrasses for the arid west or irrigation with saline water.

Alkaligrass accessions from Eurasia were screened for turfgrass color variation because of cool- and warm-weather response and susceptibility to biennial seed habit causing die-off (Cuany 1992). Blue grama clones derived from collections originating from the Great Plains of the United States were screened for seed productivity, caryopsis weight, and plant type over two generations. Fairway crested wheatgrass was surveyed from 17 Eurasian sources, and parents were selected in two successive generations for rhizome development tendency, disease resistance, leaf width, and seed yield (Cuany 1992). Inland saltgrass collections were surveyed for growth pattern (Christensen and Qian 2004). Many accessions were strongly rhizomatous and produced a dense, short canopy requiring little maintenance. Management studies, including variable mowing heights and fertility rates, are being used to evaluate alkaligrass, blue grama, and fairway crested wheatgrass selections.

Turfgrass Water Use Rates

Research completed over the last 30 years provides a much clearer understanding of water use rates among the most important turfgrasses used throughout the United States. Water use rates are summarized using the evapotranspiration (ET) rates of turfgrasses under well-watered conditions (Table 5.2). As a group, warm-season species had lower ET rates than cool-season species. The range of ET rates for the warm-season turfgrasses was 3 to 9 millimeters day^{-1} (mm d^{-1}) as compared with 3.6 to 12.6 mm d^{-1} for cool-season species. High-density, low-growing turfgrasses, such as hybrid bermudagrass, zoysiagrass, buffalograss, and centipedegrass, exhibited the lowest water use rates. For cool-season species, the fine-leafed fescues ranked medium, whereas Kentucky bluegrass, annual bluegrass, and creeping bentgrass exhibited very high water use rates.

Most of the turfgrass water use research was conducted in arid parts of the United States and the world. These estimates of ET are not necessarily valid for humid regions. Therefore, ET ranges were determined for three warm-season grasses under moderate stress irrigation in large field plots in Georgia (Carrow 1995). Water use varied, but the relative rankings for the warm-season grasses were similar to estimates from arid areas. For a well-watered irrigation regime common in the Southeast, hybrid bermudagrass (Tifway) used the least water compared with zoysiagrass (Meyer) and centipedegrass (common type). Water use rates were 39% and 11% greater than hybrid bermudagrass during August for zoysiagrass and centipedegrass, respectively. Under severe moisture stress, such as for rough areas, zoysiagrass was severely wilted and centipedegrass used 43% more water than hybrid bermudagrass.

Drought Resistance

Drought resistance is a term that encompasses a range of mechanisms, which allow plants to withstand periods of drought. The categories of drought resistance are avoidance and tolerance (Table 5.3). The drought resistance of 11 warm-season turfgrasses was compared for a drought stress period of 48 days without irrigation (Kim and Beard 1988). After this period, irrigation resumed and the plants' ability to recover after the stress was observed. Significant differences in leaf firing and shoot recovery were observed during and after the period of induced drought. In general, those species that turned yellow or brown earlier tended to have poorer post-drought stress shoot recovery, or in other words, poor drought resistance.

Table 5.2. Summary of mean rates of turfgrass evapotranspiration (ET)

Turfgrass species[1]		Mean summer ET rate (mm d^{-1})[2]	Relative ranking
Cool season	Warm season		
	Buffalograss	5 – 7	Very low
	Bermudagrass hybrids	3.1 – 7	Low
	Centipedegrass	3.8 – 9	
	Bermudagrass	3 – 9	
	Zoysiagrass[1]	3.5 – 8	
Hard fescue		7 – 8.5	Medium
Chewings fescue		7 – 8.5	
Red fescue		7 – 8.5	
	Bahiagrass	6 – 8.5	
	Seashore paspalum	6 – 8.5	
	St. Augustinegrass	3.3 – 6.9	
Perennial ryegrass		6.6 – 11.2	High
	Carpetgrass	8.8 – 10	
	Kikuyugrass	8.5 – 10	
Tall fascue		3.6 – 12.6	
Creeping bentgrass		5 – 10	
Annual bluegrass		>10	
Kentucky bluegrass		4 – >10	
Italian ryegrass		>10	

[1] Based on the most widely used cultivars of each species.
[2] Mean rates of water use based on research by Aronson et al. 1987; Aronson, Gold, and Hull 1987; Beard 1985; Biran et al. 1981; Carrow 1991; Gibeault et al. 1985; Johns, Beard, and van Bavel 1983; Kim and Beard 1988; Kim, Beard, and Sifers 1988; Kneebone and Pepper 1982; Kneebone and Pepper 1984; Kopec et al. 1988; Krans and Johnson 1974; Meyer, Gibeault, and Youngner 1985; O'Neil and Carrow 1983; Pruitt 1964; Shearman and Beard 1973; Tovey, Spencer, and Muckell 1969; van Bavel 1966; Youngner et al. 1981.

Table 5.3. Turfgrass morphological, anatomical, and physical characteristics contributing to drought resistance

Term	Definition
Drought resistance	Various mechanisms that a turfgrass plant may have to withstand periods of drought. Two major types are drought avoidance and tolerance.
1. Drought avoidance	Ability of a plant to avoid tissue damage in a drought period by postponement of dehydration. This avoidance may be either through limiting ET or factors that influence soil water uptake (i.e., deep rooting, root viability, resistance to edaphic stresses). The plant is then able to maintain adequate tissue water content and thus avoid or postpone the stress.
2. Drought tolerance	Ability of a turfgrass to tolerate a drought period. Two potential mechanisms are by escape and hardiness.
a) Escape	The plant has a life cycle such that it lives through the drought in a dormant state or as seed.
b) Hardiness	The plant develops a greater hardiness to low tissue water deficits. This process normally involves a greater drought tolerance of protoplasm and protoplasmic membranes from alterations in their properties, and binding of water to protoplasmic constituents. Osmotic adjustments to maintain adequate tissue water content may also be involved during long-term or short-duration moisture stress periods.

Additional stress mechanism studies on these 11 turfgrasses revealed that specific types of plant morphology affect the resistance to ET. The major factors discovered were low leaf area and high canopy resistance. These characteristics, in addition to leaf firing and shoot recovery, can be used as guidelines when selecting cultivars possessing low water use rates and drought resistance (Beard, Green, and Sifers 1992).

Turfgrass breeders have used these characteristics to make field selections that produce turfgrasses that use less water and survive extended periods of drought. Other factors that contribute to drought avoidance are rooting depth and resistance to low soil oxygen, high temperature, soil strength, salinity, adverse pH, and toxic elements (Carrow 1995). Simple visual evaluation is not enough because these factors require objective measurement and are influenced by spatial variability. In addition, factor interaction usually complicates single trait selection.

Water Quality and Turfgrass Culture

One of the most difficult challenges in turfgrass culture deals with soils and water that are high in salts. Arid and semiarid regions are the primary areas where salinity and water quality problems are encountered. Because of degradation of some ground and surface water supplies, however, salinity problems appear in humid regions as well. Competition for potable, nonpotable, and effluent water resources has changed turfgrass management and irrigation practices. Salinity sources, turfgrass resistance mechanisms, growth responses, and basic management principles were reviewed by Harivandi, Butler, and Wu (1992). There also are several useful references on irrigation management with effluent water (Feigin, Ravina, and Shalhevet 1991; Pettygrove and Asano 1985).

Compared with turfgrass soils irrigated with potable water, those irrigated with effluent water in the southwestern United States increased in electrical conductivity (EC), nitrate-nitrogen (NO_3-N), phosphorous (P), potassium, sodium (Na), and exchangeable sodium percentage (Hayes, Mancino, and Pepper 1990; Hayes et al. 1990; Mancino and Pepper 1992). Soil pH was not significantly different. Iron, magnesium, copper, and zinc concentrations were within ranges considered normal for agricultural soils. Aerobic bacteria populations in the soil were similar under both of the irrigation waters used, indicating no promotion or inhibition by effluent use. Water collected from a 0.61-meter (m) depth on effluent-irrigated turfgrass had higher EC and Na content than potable leachate collected from the same depth. But the increased soluble salt levels did not exceed the current recommended potable water quality limits. They concluded that effluent water did not radically alter soil quality, but was a good source of available nitrogen. The increasing Na accumulation could be managed through annual applications of calcium sulfate or sulfur.

Establishment and quality of common bermudagrass and perennial ryegrass were evaluated under effluent and potable irrigation waters. There was significantly lower seed emergence, but improved seed establishment under effluent irrigation. The improved seed establishment was attributed to effluent water providing more accessible nutrients to initially shallow-rooted seedlings. Under a leaching fraction (water moved through the soil profile) of approximately 20%, established effluent-irrigated turfgrasses did not exhibit signs of salt stress. No single nitrogen fertilization rate or irrigation water consistently produced a superior quality turfgrass. But plots receiving additional nitrogen fertilizer and effluent water on overseeded perennial ryegrass showed signs of excess growth, heat stress, and chlorosis. These studies indicated that turf establishment, seeding rates, and supplemental nutrient applications need to be adjusted depending on effluent water quality, irrigation amount, and turf quality desired.

The salt tolerance of turfgrass species has become more important as the poor quality nonpotable and effluent water use on recreational turfgrass has increased. Horst and colleagues ranked several of the major turfgrass species in order of salt resistance (Table 5.4). Mass screening methods for turfgrass salt resistance were developed to evaluate turfgrasses such as buffalograss, zoysiagrass, bentgrass, seashore paspalum, and inland saltgrass (Horst et al. 1996; Marcum et al. 2003; Pessarakli, Marcum, and Kopec 2005).

Table 5.4. Relative salt resistance of several turfgrass species used in the United States

Cool-season	Warm-season	Ranking
Alkaligrass	Seashore paspalum	Excellent
	Zoysiagrass	Good
	St. Augustinegrass	
	Bermudagrass hybrids	
Creeping bentgrass	Bermudagrass	
Tall fescue	Bahiagrass	Fair
Perennial ryegrass	Centipedegrass	Poor
Fine fescues	Carpetgrass	
Kentucky bluegrass	Buffalograss	

*Based on the most used cultivars of each species.

Turfgrass Pest Management and Fertilization Effects on Water Quality

Pesticides and fertilizers are applied to turfgrass and, depending on an array of processes, these chemicals break down into biologically inactive by-products. Two major concerns are whether pesticides and nutrients leach or runoff from turfgrass areas. The downward movement of pesticides or nutrients through the turfgrass–soil system by water is called leaching. Runoff is the portion of precipitation (rainfall) that leaves the area over the turfgrass–soil surface. There are several interacting processes that influence the fate of pesticides and fertilizers applied to turf. For the purposes of this section, the following seven categories that influence the fate of pesticides and nutrients will be discussed: volatilization, water solubility, sorption, plant uptake, degradation, runoff, and leaching. The roles these processes play in the likelihood that the pesticides will reach ground or surface water will be addressed by runoff.

Volatilization

Volatilization is the process through which chemicals transform from a solid or liquid into a gas. The vapor pressure of a chemical is the best indicator of its potential to volatilize. Pesticide volatilization increases as the vapor pressure increases. As temperature increases, so do vapor pressures and the chance for volatilization loss. Volatilization loss generally is less after a late afternoon or an early evening pesticide application rather than in the late morning or early afternoon when temperatures are increasing. Volatilization also will increase with air movement and can be greater from unprotected areas than from areas with windbreaks. Immediate irrigation is usually recommended for highly volatile pesticides to decrease loss.

Only a few studies have evaluated nitrogen volatilization from turfgrass. Nitrogen volatilization depends on the degree of irrigation after the application of fertilizer (Bowman, Devitt, and Miller 1995; Joo, Christians, and Blackmer 1991; Joo, Christians, and Bremner 1987). When no irrigation was used, as much as 36% of the nitrogen volatilized. A water application of 1 centimeter (cm) decreased volatilization to 8%. Nitrogen volatilization was higher for the split irrigation applications (four 6-mm per week) compared with a single irrigation (one 25-mm per week), but this difference was not significant (Starrett, Christians, and Austin 1995). Miltner and colleagues (1996) could not account for 36% of the nitrogen applied one year after a spring application. It was suggested that volatilization and denitrification could be responsible.

Pesticide volatilization can take place for several days after application. Reported volatile losses over a one- to four-week period, expressed as a percentage of the total applied, ranged from less than 1 to 16% (Murphy, Cooper, and Clark 1996a, b; Snyder and Cisar 1996; Yates 1995; Yates, Green, and Gan 1996). Results of volatilization studies showed that maximum loss occurred when surface temperature and solar radiation were highest, and that volatile losses were directly related to the vapor pressure characteristics of the pesticide. Thus, examining the physical and chemical properties of the pesticide is a good way to determine if volatilization losses are likely to occur under particular weather and application conditions.

Post-application irrigation had an effect on the volatilization of some insecticides (Murphy, Cooper, and Clark 1996b). The insecticide was applied once, followed by 13 mm of irrigation, and applied again separately with no post-application irrigation. Without post-application irrigation, the insecticide (trichlorfon) volatile loss totaled 13% compared with 9% when irrigated. Also, withholding post-application irrigation resulted in less conversion of trichlorfon to its more toxic breakdown product, DDVP (or dichlorvos). It seems that light post-application irrigation may have a small, positive effect on preventing volatile loss of pesticides.

Water Solubility

The extent to which a chemical will dissolve in a liquid is referred to as solubility. Although water solubility is usually a good indicator of mobility (Smith and Bridges 1997), it is not necessarily the only criterion. In addition to pesticide solubility, the affinity of a pesticide to adhere to soils must be considered (Carroll and Hill 1997).

Sorption

The tendency of a pesticide to leach or runoff is strongly dependent on the interaction of the pesticide with solids in the soil. Sorption includes the process of adsorption and absorption. Adsorption refers to the binding of a pesticide to the soil particle surface. Absorption implies that the pesticide penetrates into a soil particle or is taken up by plant leaves or roots.

This difference is important because pesticides may become increasingly absorbed with time (months to years), and desorption (or release) of the absorbed pesticide may be decreased with time. The unavailable or undetachable pesticide is often referred to as bound residue and is generally unavailable for microbial degradation or pest control.

Factors that contribute to sorption of pesticides on soil materials include (1) chemical and physical characteristics of the pesticide, (2) soil composition, and (3) nature of the soil solution. In general, sandy soils offer little in the way of sorptive surfaces. Soils containing higher amounts of silt, clay, and organic matter provide a rich sorptive environment for pesticides. Research indicates that turfgrass leaves and thatch adsorb a significant amount of pesticide (Carroll and Hill 1997; Lickfeldt and Branham 1995; Turco 1997).

Adsorption of pesticides is affected by the partition coefficient, which is reported as K_d or more accurately as K_{oc}. A K_{oc} less than 300 to 500 is considered low. The strength of adsorption is inversely related to the pesticide's solubility in water and directly related to its partition coefficient. For example, chlorinated hydrocarbons are strongly adsorbed, whereas phenoxy herbicides like 2,4-D are much more weakly adsorbed.

Plant Uptake

Plants can directly absorb pesticides or influence pesticide fate by altering the flow of water in the root zone. Turfgrasses with higher rates of transpiration can decrease the leaching of water-soluble pesticides. In situations where the turf is not actively growing or root systems are not well developed, pesticides are more likely to migrate into the soil profile with percolating water.

Degradation

Degradation occurs because of the presence of soil microorganisms and chemical processes in the turfgrass–soil system. Pesticides are broken down in a series of steps that eventually lead to the production of carbon dioxide, water, and some inorganic products (e.g., nitrogen, P, sulfur, etc.). Microbial degradation may be either direct or indirect. Some pesticides are directly used as a food source by microorganisms. In most instances, though, indirect microbial degradation of pesticides occurs though passive consumption along with other food sources in the soil.

Chemical degradation is similar to microbial degradation except that pesticide breakdown is not achieved by microbial activity. The major chemical reactions such as hydrolysis, oxidation, and reduction occur in both chemical and microbial degradation. Photochemical degradation is an entirely different breakdown process driven by solar radiation. It is the combined pesticide degradation that results from chemical, microbial, and photochemical processes under field conditions that is of the most interest.

Degradation rates are also influenced by factors like pesticide concentration, temperature, soil water content, pH, oxygen status, prior pesticide use, soil fertility, and microbial population. These factors change dramatically with soil depth and greatly decrease microbial degradation as pesticides migrate below the soil surface. With some pesticides, microbial degradation is enhanced because of microbial buildup after the first application, thereby decreasing the amount of material available for leaching after the second application (Cisar and Snyder 1993, 1996; Snyder and Cisar 1993).

In the instance of degradation rates, the average days to 90% degradation in turf soils generally is significantly less than established values based on agricultural systems. Thus, leaching potential for most pesticides is less in turfgrass systems because turfgrass thatch plays an important role in adsorbing and degrading applied pesticides (Horst et al. 1996).

Persistence of a pesticide, expressed as half-life (DT_{50}), is the time required for 50% of the original pesticide to degrade. DT_{50} measurements are commonly made in the laboratory under uniform conditions. In turfgrass, soil temperature, organic carbon, and moisture content change constantly. These factors dramatically influence the rate of degradation. Consequently, DT_{50} values should be considered as guidelines rather than absolute values.

Leaching

The downward movement of nutrients and pesticides through the turfgrass–soil system by water is called leaching. Compared with some agricultural crops, the research demonstrates that leaching is decreased in turfgrass systems. This decrease occurs because of the increase in adsorption on leaves, thatch, and soil organic matter; a high level of microbial and chemical degradation; and decreased percolation because of an extensive root system, greater plant uptake, and high transpiration rates. Separate discussions on nitrogen and pesticide leaching follow.

University research indicates that very little nitrogen leaching occurs when nitrogen is applied properly, i.e., according to the needs of the turf and in consideration of soil types, irrigation regimes, and anticipated rainfall. In putting green construction, mixing peat moss with sand at a level of 3% significantly decreased nitrogen leaching compared with pure sand root zones during the year of establishment (Brauen and Stanke 1995). Light applications of slow-release nitrogen sources on a frequent interval provided excellent protection from nitrate leaching.

Properly maintained turf grown in a loam soil allowed less than 1% of the nitrogen applied to leach to a depth of 1.2 m (Miltner et al. 1996; Yates 1995). Most of the nitrogen was recovered in clippings, thatch, and soil. The researchers suggested that the remaining amount volatilized or was lost through denitrification. Nitrogen leaching losses can be greatly decreased by irrigating lightly and frequently, rather than heavily and less frequently (Starrett, Christians and Austin 1995). Nitrogen was applied at moderate rates and irrigated with one 25-mm versus four 6-mm applications. Up to 30 times more nitrogen (Table 5.5) was leached after a single 25-mm irrigation application, perhaps in part because of macropore flow caused by earthworm activity.

Table 5.5. Percentage of nitrogen recovered from undisturbed soil columns with turfgrass cover

Category	Heavy irrigation[1]		Light irrigation[2]		Probability[5]
	Mean[3]	Std. dev.	Mean[4]	Std. dev.	
	Percentage of nitrogen applied				
Volatilization	0.9	0.6	2.3	1.5	0.053
Clippings, verdure	14.3	5.5	37.3	29.9	0.095
Thatch mat	11.3	4.2	16.7	11.6	0.315
0–4 in.	13.4	7.6	12.6	5.8	0.834
4–8 in.	7.7	5.2	6.4	4.4	0.666
8–12 in.	7.2	7.3	5.6	4.9	0.674
12–16 in.	7.8	8.0	6.7	7.8	0.826
16–20 in.	7.6	8.1	2.2	1.6	0.180
Leachate	12.3	4.7	0.4	0.7	<0.001

[1]Heavy irrigation, four one-inch applications of water
[2]Light irrigation, sixteen 0.25-inch applications of water
[3]Heavy irrigation, six replications
[4]Light irrigation, five replications
[5]t-test, probability that a difference exists (lower the value, more likely there is a true difference)

Irrigating bermudagrass and tall fescue with adequate amounts (no drought stress) of moderately saline water did not increase the concentration or amount of nitrate leached (Bowman, Devitt, and Miller 1995). Higher amounts of salinity in the root zone, drought, or the combination of these two stresses caused high concentrations and larger amounts of nitrate to leach from both a tall fescue and bermudagrass turf. This evidence suggests that drought, high salinity, or both impair the capacity of the root system of the turf and that management modification may be needed to prevent nitrate leaching.

The physical and chemical properties of the pesticides proved to be good indicators of the potential for leaching, runoff, and volatilization (Kenna 1994). Products that exhibit high water solubility, low soil adsorption potential, and greater persistence are more likely to leach. For example, fenamiphos, a commonly used nematicide, has high water solubility, a low adsorption potential, and a toxic breakdown metabolite that tends to persist in the soil. As expected, 18% of fenamiphos and its metabolite leached from a sand-based green in Florida. When all studies are considered, however, the average loss was 5% (Cisar and Snyder 1993, 1996; Snyder and Cisar 1993).

Soil type and precipitation/irrigation amount also are important factors in leaching losses (Table 5.6). Mecoprop (MCPP) and triadimefon are two pesticides whose chemical and physical properties indicate a relatively high potential for leaching. Results show significant leaching from coarse sand profiles, especially under high precipitation, and much less leaching from sandy loam and silt loam soils (Petrovic 1995).

Table 5.6. Percentage of applied pesticide found in drainage water from experimental fairways

| Soil | Precipitation | Pesticide | | | |
		Isazofos	MCPP	Trichlorfon	Triadimefon
		Percentage of pesticide leached			
Sand	Normal	10.40	51.00	1.18	1.00
	Above	5.60	62.12	3.44	2.44
Sandy loam	Normal	0.04	0.79	1.13	0.06
	Above	0.09	0.46	4.41	0.01
Silt loam	Normal	0.68	0.44	0.63	0.24
	Above	0.30	1.25	3.33	0.28

Several simulation models currently are used to predict the downward movement of pesticides through soil. A good review on pesticide transport models was conducted by Cohen and colleagues (1995). In time, many of these models will need adjustments that take into account the role of a dense turf canopy and thatch layer (Smith 1995; Smith and Tilloston 1993).

In summary, the dense turf cover decreased the potential for leaching losses of pesticides; conversely, more leaching occurred from newly planted turf stands. Generally, sandy soils are more prone to leaching losses than soils with silt and clay. The physical and chemical properties of the pesticides were good indicators of leaching potential. Finally, current pesticide models tend to over-predict the leaching loss of most pesticides applied to turf if valid adjustments are not made to account for the roles of turf canopy, thatch, and root system.

Runoff

Pesticide and nutrient runoff pose a greater threat to water quality than leaching. Runoff refers to the portion of precipitation (rainfall) that is discharged from the area through stream channels. The water lost without entering the soil is called surface runoff and that which enters the soil before reaching the stream is called groundwater runoff or seepage flow from groundwater. Pesticides and nutrients applied to turfgrass, under some circumstances, can be transported offsite in surface runoff.

In Pennsylvania, runoff experiments were conducted on plots characterized by slopes of 9 to 13%, good quality loam soil, and turf cover consisting of either creeping bentgrass or perennial ryegrass cut at a 12.5 mm (1/2-inch) height (Linde 1993; Linde et al. 1995). Nitrate concentrations in the runoff or leaching samples did not differ significantly from the nitrate concentration in the irrigation water. The study was conducted on excellent quality turf and on soil with a high infiltration rate.

Nitrogen runoff also was measured in Georgia using a simulated storm event (25 mm applied at a rate of 50 mm hour (hr)$^{-1}$ 24 hrs after nitrogen was applied (Smith and Bridges 1997). As much as 40 to 70% of the rainfall water left the plots as runoff. A total of 16% (12.5 milligrams (mg) liter (L)$^{-1}$) of the NO_3-N applied at 24 kilograms (kg) hectare (ha)$^{-1}$ to actively growing bermudagrass was found in surface runoff water (Table 5.7). But, 64% (24.8 mg L^{-1}) of the NO_3-N applied at 24 kg ha^{-1} to dormant bermudagrass was found in surface runoff water.

In Oklahoma, the effects of buffer strips and cultivation practices on pesticide and nitrogen runoff were investigated. It was concluded that soil moisture was the major factor influencing runoff. During the first simulated rainfall event in July, soil moisture conditions were low to moderate. After a 50 mm (2 inches) rainfall event, less than 1% of the applied nitrogen was collected in the runoff (Cole et al. 1997). In August, when the simulated rainfall occurred after 150 mm (6 inches) of actual rainfall the previous week (i.e., high soil moisture), the amount of nitrogen collected after the simulated rainfall averaged more than 8%. When soil moisture was moderate to low in the Oklahoma study, the presence of a 2.4 to 4.9 m (8 to 16 feet) untreated buffer strip significantly decreased nitrogen runoff, whereas when soil moisture was high, the buffer strips made no difference. In both instances, less runoff occurred when sulfur-coated urea was applied compared with straight urea.

In Georgia, studies were conducted on plots with a 5% slope and a sandy clay soil typical of that region (Smith and Bridges 1997). Pesticides were applied and 25-mm (1-inch) simulated rainfall events occurred 24 and 48 hrs afterward. At a rainfall rate of 50 mm (2 inches) per hr, as much as 40 to 70% of the rainfall left the plots as runoff during simulated storm events. The collected surface water contained moderately high concentrations of treatment pesticides having high water solubility (Table 5.7). For example, under these conditions only very small amounts (<1%) of chlorthalonil and chlorpyrifos could be detected in the runoff. But, between

10 and 13% of the 2,4-D, MCPP, and dicamba was transported off the plots over an 11-day period. About 80% of this transported total moved off the plots with the first rainfall event 24 hrs after pesticide application. Finally, the runoff loss of several herbicides was greater when applied to dormant turf as compared with an actively growing turf.

Table 5.7. The percentage of applied pesticide and concentration of pesticide transported from runoff plots during a storm event that occurred 24 hours after application

Pesticide or fertilizer treatment	Application rate	Percentage transported	Conc. at 24 hrs after application
	kg ha^{-1}	%	Tg L^{-1}
Nitrate-N	24.40	16.4	12,500
Nitrate-N (dormant bermuda)	24.40	64.2	24,812
Dicamba	0.56	14.6	360
Dicamba (dormant bermuda)	0.56	37.3	360
Mecoprop	1.68	14.4	810
Mecoprop (dormant bermuda)	1.68	23.5	1,369
2,4-D DMA	2.24	9.6	800
2,4-D DMA (dormant bermuda)	2.24	26.0	1,959
2,4-D DMA (pressure injected)	2.24	1.3	158
2,4-D DMA (2 m buffer strip)	2.24	7.6	495
2,4-D LVE	2.24	9.1	812
Trichlorfon[1]	9.15	32.5	13,960
Trichlorfon[1] (pressure injected)	9.15	6.2	2,660
Chlorothalonil[2]	9.50	0.8	290
Chlorpyrifos[3]	1.12	0.1	19
Dithiopyr	0.56	2.3	39
Dithiopyr (granule)	0.56	1.0	26
Benefin	1.70	0.01	3
Benefin (granule)	1.70	0.01	6
Pendimethalin	1.70	0.01	9
Pendimethalin (granule)	1.70	0.01	2

[1] Trichlorfon + dichlorvos metabolite.
[2] Total for chlorothalonil and OH-chlorothalonil.
[3] Total for chlorpyrifos and OH-chlorpyrifos.

In Oklahoma, soil moisture was a significant factor in determining how much pesticide ran off the plot areas (Cole et al. 1997). Where soil moisture was low to moderate, buffer zones were effective in decreasing pesticide runoff. When soil moisture was high, they were not effective except for the insecticide chlorpyrifos. In both Oklahoma and Georgia, best management practices were investigated on how cutting heights and buffers of different lengths could minimize fertilizer and pesticide runoff. The effect of soil cultivation (core aerification) on runoff potential also was studied. In Oklahoma, a 4.9-m buffer cut at 5 cm (3 inches) significantly decreased the amount of 2,4-D found in runoff water from a 4.9-m treated bermudagrass fairway. But the results in Georgia that used smaller buffer strips indicated no decrease in the amount of pesticide transported in the surface-water solution (Smith and Bridges 1997).

Among the conclusions, or trends, observed from the pesticide runoff studies were the following: (1) dense turf cover decreases the potential for runoff losses of pesticides; (2) the physical and chemical properties of pesticides are good indicators of potential runoff losses; (3) heavy textured, compacted soils are much more prone to runoff losses than sandy soils; (4) moist soils are more prone to runoff losses than drier soils; (5) buffer strips at higher cutting heights tend to decrease runoff of pesticides when soil moisture is low to moderate before rainfall events; and (6) the application of soluble herbicides on dormant turf can produce high levels of runoff losses.

Biological Control

Biological control is intended to decrease the amount of pesticide needed to maintain turfgrasses. But the development of effective biological methods of pest control has been difficult. The microorganisms that inhabit the turfgrass root zone are just starting to be characterized in a way that will lead to positive developments in biological control. It is important that the mechanisms of biological control be understood thoroughly before products are commercialized. We are just starting to understand why some of these organisms fail when used in field situations.

Biological control of turfgrass pests is generally accomplished with a living organism that either lowers the population density of the pest problem or decreases its ability to cause injury to the turf. Once laboratory, or greenhouse, evidence demonstrates that a specific predator or microbial antagonist controls a turfgrass pest problem, additional field research must be performed to determine whether it is a functional biocontrol agent acceptable for commercial use (Couch 1995). Specifically, sound scientific research must

1. Determine if the biocontrol agent readily establishes in the turf or surrounding areas.

2. Evaluate the effects of pesticides on the growth and development of the antagonist.

3. Estimate the likelihood of resistance to the effects of the biological agent.

4. Decrease the need for conventional pesticides by providing an adequate level of control.

5. Provide evidence that the biological agent is safe for people, wildlife, and the environment.

6. Develop methods for producing commercial quantities with an acceptable shelf life and cost.

Wildlife Programs

Two wildlife programs have emerged within the turfgrass industry over the last 15 years to help in understanding a large turfgrass area's impact on wildlife. The first is the Audubon Cooperative Sanctuary Program (ACSP) administered by Audubon International. The second is the Wildlife Links research program, which is coordinated by the National Fish and Wildlife Foundation (NFWF). Both programs have had a positive effect on how water is used for turfgrass irrigation, as well as how water is managed in streams, ponds, and wetlands for native flora and fauna.

A cooperative effort between the United States Golf Association and the Audubon International, the ACSP promotes ecologically sound land management and the conservation of natural resources. Its positive impact extends beyond the boundaries of the golf course and helps benefit the local community.

There are six different categories in which golf courses apply for certificates of recognition: environmental planning, wildlife and habitat management, member/public involvement, integrated pest management, water conservation, and water quality management. Audubon International provides each golf course with one-on-one assistance in devising an appropriate environmental plan.

The Wildlife Links Program investigates how golf courses and urban areas are used by wildlife. It was established in early 1995 to fund research, management, and education projects needed to provide the game of golf with state-of-the-art information on wildlife management issues. The program is administered by the NFWF. Congress established the NFWF in 1984 as a nonprofit organization dedicated to the conservation of natural resources—fish, wildlife, and plants. Among its goals are species habitat protection, environmental education, public policy development, natural resource management, habitat and ecosystem rehabilitation and restoration, and leadership training for conservation professionals. It meets these goals by forging partnerships between the public and private sectors.

The overall goal of the Wildlife Links program is to protect and enhance the wildlife, fish, and plant resources found on golf courses. This aim includes providing golf course designers and superintendents with the information they need to promote the wildlife on their golf facilities, while still providing quality playing conditions for the game of golf. This information includes how golf courses can be maintained as biologically productive sites for wildlife, solid recommendations regarding wildlife issues that can be incorporated into long-term management strategies, and education for golfers and the general public about these issues.

Literature Cited

Aronson, L. J., A. J. Gold, and R. J. Hull. 1987. Cool-season turfgrass response to drought stress. *Crop Sci* 27:1261–1266.

Aronson, L. J., A. J. Gold, R. J. Hull, and J. L. Cisar. 1987. Evapotranspiration of cool-season turfgrasses in the humid northeast. *Agron J* 79:901–904.

Beard, J. B. 1985. An assessment of water use by turfgrasses. Pp. 45–60. In V. A. Gibeault and S. T. Cockerham (eds.). *Turfgrass Water Conservation*. Publ. 21405. University of California, Riverside.

Beard, J. B. 2002. *Turf Management for Golf Courses*. 2nd ed. Ann Arbor Press, Chelsea, Michigan. 793 pp.

Beard, J. B. and W. H. Daniel. 1965. Effect of temperature and cutting on the growth of creeping bentgrass (*Agrostis palustris* Huds.) roots. *Agron J* 57(3):249–250.

Beard, J. B., R. L. Green, and S. I. Sifers. 1992. Evapotranspiration and leaf extension rates of 24 well-watered, turf-type Cynodon genotypes. *HortSci* 27(9):986–988.

Biran, I., B. Brando, I. Bushkin-Harav, and E. Rawitz. 1981. Water consumption as growth rate of 11 turfgrasses as affected by mowing height, irrigation frequency, and sod moisture. *Agron J* 73:85–90.

Bowman, D. C., D. A. Devitt, and W. W. Miller. 1995. The effect of salinity on nitrate leaching from turfgrass. *USGA Green Section Record* 33(1):45–49.

Brauen, S. E. and G. Stahnke. 1995. Leaching of nitrate from sand putting greens. *USGA Green Section Record* 33(1):29–32.

Carroll, M. J. and R. L. Hill. 1997. Modeling pesticide transport in turfgrass thatch and foliage. Pp. 70–71. In M. P. Kenna and J. T. Snow (eds.). *USGA 1997 Turfgrass and Environmental Research Summary*. United States Golf Association, Far Hills, New Jersey.

Carrow, R. N. 1991. Turfgrass water use, drought resistance, and rooting patterns in the Southeast. ERC-91. Project completion report, USDI/USGS, Project 02. University of Georgia, Griffin.

Carrow, R. N. 1995. Drought resistance aspects of turfgrasses in the southeast: Evapotranspiration and crop coefficients. *Crop Sci* 35(6):1685–1690.

Christensen, D. and Y. Qian. 2004. Development of Stress-tolerant, Turf-type Saltgrass Varieties. *2004 USGA Turfgrass Environmental Research Summary*, pg. 19. United States Golf Association, Far Hills, New Jersey.

Cisar, J. L. and G. H. Snyder. 1993. Mobility and persistence of pesticides in a USGA-type green. I. Putting green facility for monitoring pesticides. *Int Turfgrass Soc Res J* 7:971–977.

Cisar, J. L. and G. H. Snyder. 1996. Mobility and persistence of pesticides applied to a USGA Green. III. Organophosphate recovery in clippings, thatch, soil, and percolate. *Crop Sci* 36:1433–1438.

Cohen, S. Z., R. D. Wauchope, A. W. Klein, C. V. Eadsforth, and R. Graney. 1995. Offsite transport of pesticides in water: Mathematical models of pesticide leaching and runoff. *Pure Applied Chem* 67:2109–2148.

Cole, J. H., J. H. Baird, N. T. Basta, R. L. Hunke, D. E. Storm, G. V. Johnson, M. E. Payton, M. D. Smolen, D. L. Martin, and J. C. Cole. 1997. Influence of buffers on pesticide and nutrient runoff from bermudagrass turf. *J Environ Qual* 26:1589–1598.

Couch, H. B. 1995. *Diseases of Turfgrasses*. Krieger Publishing Company, Malabar, Florida. 421 pp.

Cuany, R. L. 1992. Development of dryland western turfgrass cultivars. *USGA Turfgrass Environ Res Summ*, pp. 31–32. United States Golf Association, Far Hills, New Jersey.

Duncan, R. R. 2000. *Seashore Paspalum: The Environmental Turfgrass*. Ann Arbor Press, Chelsea, Michigan. 281 pp.

Engelke, M. C. and S. Anderson. 2003. Zoysiagrasses (*Zoysia* spp.). Pp. 271–285. In M. D. Casler and R. R. Duncan (eds.). *Turfgrass Biology, Genetics, and Breeding*. John Wiley & Sons, Hoboken, New Jersey.

Feigin, A., I. Ravina, and J. Shalhevet. 1991. *Irrigation with Treated Sewage Effluent*. Springer-Verlag, New York. 224 pp.

Gibeault, V. A., J. L. Meyer, V. B. Youngner, and S. T. Cockerham. 1985. Irrigation of turfgrass below replacement of evapotranspiration as a means of water conservation: Performance of commonly used turfgrasses. Pp. 347–356. In F. Lemaire (ed.). *Proc 5th Int Turfgrass Res Conf, Avignon, France. 1–5 July*. Inst Natl de la Recherche Agron, Paris.

Harivandi, M. A., J. D. Butler, and L. Wu. 1992. Salinity and turfgrass culture. Pp. 207–229. In D. V. Waddington, R. N. Carrow, and R. C. Shearman (eds.). *Turfgrass. Agronomy Monograph No. 32*. ASA-CSSA-SSSA, Madison, Wisconsin.

Hayes, A. R., C. F. Mancino, and I. L. Pepper. 1990. Irrigation of turfgrass with secondary sewage effluent: I. Soil and leachate water quality. *Agron J* 82:939–943.

Hayes, A. R., C. F. Mancino, W. Y. Forden, D. M. Kopec, and I. L. Pepper. 1990. Irrigation of turfgrass with secondary sewage effluent: II. Turf quality. *Agron J* 82:943–946.

Horst, G. L., P. J. Shea, N. E. Christians, D. R. Miller, C. Stuefer-Powell, and S. K. Starrett. 1996. Pesticide dissipation under golf course fairway conditions. *Crop Sci* 36:362–370.

Huang, B., X. Liu, and J. D. Fry. 1998a. Shoot physiological responses of two bentgrass cultivars to high temperature and poor soil aeration. *Crop Sci* 38(5):1219–1224.

Huang, B., X. Liu, and J. D. Fry. 1998b. Effects of high temperature and poor soil aeration on root growth and viability of creeping bentgrass. *Crop Sci* 38(6):1618–1622.

Johns, D., J. B. Beard, and C. H. M. van Bavel. 1983. Resistances to evapotranspiration from a St. Augustinegrass turf canopy. *Agron J* 75:419–422.

Joo, Y. K., N. E. Christians, and J. M. Bremner. 1987. Effect of N-(n-Butyl) thiophosphoric triamide (NBPT) on growth response and ammonia volatilization following fertilization of Kentucky bluegrass (*Poa pratensis* L.) with urea. *J Fert Issues* 4(3):98–102.

Joo, Y. K., N. E. Christians, and A. M. Blackmer. 1991. Recovery by Kentucky bluegrass turf of urea-derived N-15 amended with N-(n-Butyl) thiophosphoric triamide (NBPT). *Soil Sci Soc Am J* 55:528–530.

Kenna, M. P. 1994. Beyond appearance and playability: Golf and the environment. *USGA Green Section Record* 32(4):12–15.

Kim, K. S. and J. B. Beard. 1988. Comparative turfgrass evapotranspiration rates and associated plant morphological characteristics. *Crop Sci* 28:328–331.

Kim, K. S., J. B. Beard, and S. I. Sifers. 1988. Drought resistance comparisons among major warm-season turfgrasses. *USGA Green Section Record* 26(5):12–15.

Kneebone, W. R. and I. L. Pepper. 1982. Consumptive water use by sub-irrigated turfgrasses under desert conditions. *Agron J* 74:419–423.

Kneebone, W. R. and I. L. Pepper. 1984. Luxury water use by bermudagrass. *Agron J* 76:999–1002.

Kopec, D. M., P. W. Brown, D. C. Slack, C. F. Mancino, and L. F. Salo. 1988. Desert turfgrass crop coefficients. Pp. 153. In *Agronomy Abstracts*. ASA, Madison, Wisconsin.

Krans, J. V. and G. V. Johnson. 1974. Subirrigation and fertilization of bentgrass during prolonged heat stress. Pp. 527–533. In E. C. Roberts (ed.). *Proc 2nd Int Turfgrass Res Conf, Blacksburg, Virginia. 19–21 June 1973*. ASA and CSSA, Madison, Wisconsin.

Lehman, V. G. and M. C. Engelke. 1991. Heritability estimates of creeping bentgrass root systems grown in flexible tubes. *Crop Sci* 31(6):1680–1684.

Lehman, V. G. and M. C. Engelke. 1993. Heritability of creeping bentgrass shoot water content under soil dehydration and elevated temperatures. *Crop Sci* 33(5):1061–1066.

Lickfeldt, D. W. and B. E. Branham. 1995. Sorption of nonionic organic compounds by Kentucky bluegrass leaves and thatch. *J Environ Qual* 24:980–985.

Linde, D. T. 1993. *Surface Runoff and Nutrient Transport Assessment on Creeping Bentgrass and Perennial Ryegrass Turf.* M.S. Thesis, Pennsylvania State University, The Graduate School College of Agricultural Sciences. 84 pp.

Linde, D. T., T. L. Watschke, A. R. Jarrett, and J. A. Borger. 1995. Surface runoff assessment from creeping bentgrass and perennial ryegrass turf. *Agron J* 87:176–182.

Mancino, C. F. and I. L. Pepper. 1992. Irrigation of turfgrass with secondary sewage effluent: Soil quality. *Agron J* 84:650–654.

Marcum, K. B., G. Wess, D. T. Ray, and M. C. Engelke. 2003. Zoysiagrass, salt glands, and salt tolerance: Observing the density of salt glands may make selecting for salt-tolerant grasses a lot easier. *USGA Green Section Record* 41(6):20–21.

Meyer, J. L., V. A. Gibeault, and V. B. Youngner. 1985. Irrigation of turfgrass below replacement of evapotranspiration as a means of water conservation: Determining crop coefficient of turfgrasses. Pp. 357–364. In F. Lemaire (ed.). *Proc 5th Int Turfgrass Res Conf, Avignon, France. 1–5 July.* Inst Natl de la Recherche Agron, Paris.

Miltner, E. D., B. E. Branham, A. E. Paul, and P. E. Rieke. 1996. Leaching and mass balance of 15N-labeled urea applied to a Kentucky bluegrass turf. *Crop Sci* 36:1427–1433.

Murphy, K. C., R. J. Cooper, and J. M. Clark. 1996a. Volatile and dislodgeable residues following trichlorofon and isazofos application to turfgrass and implications for human exposure. *Crop Sci* 36:1446–1454.

Murphy, K. C., R. J. Cooper, and J. M. Clark. 1996b. Volatile and dislodgeable residues following triadimefon and MCPP application to turfgrass and implications for human exposure. *Crop Sci* 36:1455–1461.

O'Neil, K. J. and R. N. Carrow. 1983. Perennial ryegrass growth, water use, and soil aeration status under soil compaction. *Agron J* 75:177–180.

Pessarakli, M., K. B. Marcum, and D. M. Kopec. 2005. Growth responses and nitrogen-15 absorption of desert saltgrass under salt stress. *J Plant Nutr* 28(8):1441–1452.

Petrovic, A. M. 1995. The impact of soil type and precipitation and nutrient leaching from fairway turf. *USGA Green Section Record* 33(1):38-41.

Pettygrove, G. S. and T. Asano (eds.). 1985. *Irrigation with Reclaimed Municipal Wastewater—A Guidance Manual.* Lewis Publishing, Chelsea, Michigan.

Pruitt, W. O. 1964. Evapotranspiration—A guide to irrigation. *Calif Turfgrass Culture* 14(4):27-32.

Riordan, T. P. and S. J. Browning. 2003. Buffalograss, Buchloe dactyloides (Nutt.) Engelm. Pp. 257-270. In M. D. Casler and R. R. Duncan (eds.). *Turfgrass Biology, Genetics, and Breeding.* John Wiley & Sons, Hoboken, New Jersey.

Shearman, R. C. and J. B. Beard. 1973. Environmental and cultural pre-conditioning effects on the water use rate of *Agrostis palustris* Huds., cultivar 'Penncross.' *Crop Sci* 13:424-427.

Smith, A. E. 1995. Potential movement of pesticides following application to golf courses. *USGA Green Section Record* 33(1):13-14.

Smith, A. E. and D. C. Bridges. 1997. Pp. 74-75. In M. P. Kenna and J. T. Snow (eds.). *USGA 1997 Turfgrass and Environmental Research Summary.* United States Golf Association, Far Hills, New Jersey.

Smith, A. E. and T. R. Tilloston. 1993. Pp. 168-181. In K. D. Racke and A. R. Leslie (eds.). *Pesticides In Urban Environments.* ACS Symposium Series 522, American Chemical Society, Washington, D.C.

Snyder, G. H. and J. L. Cisar. 1993. Mobility and persistence of pesticides in USGA-type green. II. Fenamiphos and fonofos. *Int Turfgrass Soc Res J* 7:978-983.

Snyder, G. H. and J. L. Cisar. 1996. Mobility and persistence of turfgrass pesticides in a USGA green. Pp. 109-140. In M. P. Kenna (ed.). *USGA Environmental Research Program: Pesticide and Nutrient Fate 1996 Annual Project Reports.* United States Golf Association, Green Section Research. Stillwater, Oklahoma.

Starrett, S. K., N. E. Christians, and T. A. Austin. 1995. Fate of nitrogen applied to turfgrass-covered soil columns. *J Irrg Drain Eng* 121:390-395.

Taliaferro, C. M. 2003. Bermudagrass (*Cynodon* (L.) Rich). Pp. 235-256. In M. D. Casler and R. R. Duncan (eds.). *Turfgrass Biology, Genetics, and Breeding.* John Wiley & Sons, Hoboken, New Jersey.

Tovey, R., J. S. Spencer, and D. C. Muckell. 1969. Turfgrass evapotranspiration. *Agron J* 61:863-867.

Turco, R. 1997. Degradation of fungicides in turfgrass systems. Pp. 79-82. In M. P. Kenna and J. T. Snow (eds.). *USGA 1997 Turfgrass and Environmental Research Summary.* United States Golf Association, Far Hills, New Jersey.

Turgeon, A. J. 1996. *Turfgrass Management.* 4th ed. Prentice Hall, Upper Saddle River, New Jersey.

van Bavel, C. H. M. 1966. Potential evaporation: The combination concept and its experimental verification. *Water Resource Res* 2:455–467.

Yates, M.V. 1995. The fate of pesticides and fertilizers in a turfgrass environment. *USGA Green Section Record* 33(1):10–12.

Yates, M. V., R. L. Green, and J. Gan. 1996. Measurement and model prediction of pesticide partitioning in a field-scale turfgrass plot. Pp. 80–94. In M. P. Kenna and J. T. Snow (eds.). *USGA Environmental Research Program: Pesticide and Nutrient Fate 1996 Annual Project Reports.* USGA Green Section Research, Stillwater, Oklahoma.

Youngner, V. B., A. W. Marsh, R. A. Strohman, V. A. Gibeault, and S. Saulding. 1981. Water use and quality of warm-season and cool-season turfgrass. Pp. 251–257. In R. W. Sheard (ed.). *Proceedings of the 4th International Turfgrass Research Conference.* Ontario Agricultural College, University of Guelph, Canada. 19–23 July.

6

Soil Water in Managed Turfgrass Landscapes

Ed McCoy

Introduction

Water flow through soil is controlled by local weather conditions: rainfall places water at the soil surface and its intensity and duration dictates which portion will infiltrate or run off. Solar radiation, relative humidity, and wind control the rate of water evapotranspiration. Water flow through soil is controlled by the characteristics and current growth stage of the plant. The atmosphere's evaporative demand is tempered by the plant that draws water for transpiration from the soil. Consequently, intra- and inter-species differences in canopy resistance and variations in turfgrass cultural practices affect soil water uptake (see Sections 11 and 12). Water flow through soil is controlled by the retention and transmission capabilities of the soil pore space. These capabilities often are inversely related across the range of soil textures where coarser-textured soils show greater transmission capabilities and finer-textured soils show greater retention capabilities. Finally, antecedent soil water content affects the rate of water infiltration and flow through soil.

Soil Components and Their Associated Pores

With regard to its physical behavior, soil is frequently viewed as a disperse, 3-phase system consisting of solid, liquid, and gaseous phases. The solid phase is composed of discrete mineral and organic particles together with mineral and organic coatings often associated with the particles. The mineral particles are designated by size into sand (0.05 to 2.0 mm), silt (0.002 to 0.05 mm), and clay (< 0.002 mm) fractions. The sand and silt fractions are roughly spherical particles that have limited chemical reactivity, whereas the clay fraction often contains plate-like particles that, due to their layered phylosilicate mineralogy, possess substantial chemical reactivity. Various mineral coatings (iron and aluminum oxides and carbonates, for example) also exist in the soil, adhering to particle surfaces. These mineral coatings impart color to the soil or may bind particles together, the degree of either being dependent on the coating concentration.

No widely held classification for the organic solids in soil currently exists. Yet for the purpose of this paper, the organic solids consist of dead plant, animal, and microbial remains and excreta. Ever changing in their chemical composition due to progressive biochemical decomposition, these organic solids can range in size from visible particles to colloids. It is the highly decomposed, darkly colored colloidal fraction often referred to as humus that wields the greater impact on the physical and chemical behavior of a native soil. This humus can exist in soil as either coatings on mineral particles or discrete entities within the collection.

Rather capriciously, the weight fractions of sand, silt, and clay particles in soil total 1.0 kg kg^{-1}, whereas the weight fraction of organic matter is expressed relative to the total weight of dry soil—mineral and organic combined. Because of variations in the weight fraction of sand, silt, and clay across the wide range of native soils, a classification scheme has been formed whereby a soil is referred to by its texture. Specifically, soil texture is the weight fraction of sand, silt, and clay within a given soil such that it is assigned a name based on the weight dominance among these 3 fractions. Thus a silt loam soil contains mostly silt with smaller and roughly equal contributions of sand and clay, whereas a sandy loam soil contains mostly sand with smaller and roughly equal contributions of silt and clay. Interestingly, however, a soil containing equal parts of sand, silt, and clay is referred to as a clay loam soil—seemingly an exception to the former two examples—because clay is sufficiently reactive as to impart its influence on soil behavior at proportionally lower concentrations. More general than the formal textural classification, soils also are spoken of with regard to texture as being fine, medium, and coarse. A fine-textured soil is dominated by the smaller-sized silt and clay particles and lacks any grittiness between the fingers. Conversely, a coarse-textured soil is dominated by sand and, not surprisingly, feels like sandpaper.

In a particulate media such as soil, pores exist between individual particles. The volume of all such pores expressed relative to the total soil volume is referred to as the porosity of the soil. Soil porosity typically ranges from 0.6 to 0.4 m^3 m^{-3}. Larger porosity values are found in finer-textured soils simply because the overall soil volume is more highly divided when composed of smaller-sized particles. In addition, a collection of smaller-sized particles generates characteristically smaller pores, whereas a collection of larger particles generates larger pores. For example, from tight packing of identical spheres it can be calculated that the largest diameter pore is approximately 0.4 times the diameter of the spheres. This is true regardless of whether the particles are sand sized or silt sized; the ratio remains the same. Thus, a finer-textured soil, while having a larger total porosity, is dominated by smaller pore sizes whereas a coarser-textured soil containing larger-diameter pores has a lesser total porosity. Texture, however, is not the only component or aspect of a soil that influences pore sizes; organic matter also plays a role.

Organic matter contents in most native soils range from nil to no more than approximately 0.08 kg kg^{-1}, yet generally being more chemically reactive than clay, retaining water many times in excess of its own weight, and serving as the food source for virtually all life in the soil. Consequently, only incremental changes in organic matter content are needed to substantially change a soil's overall behavior. The organic fraction in soil essentially impacts pores in soil by increasing the soil porosity and by extending the pore size distribution, for a given soil texture, adding both smaller and larger pore sizes. (This last statement calls for a bit more explanation:

Native soils never exist as a collection of uniformly sized mineral particles. The textural components of sand, silt, and clay are themselves composed of a range of particle sizes and even a soil classified as having a sand texture also may contain some silt and clay. Because soils contain a range of particle sizes, they also contain a range or distribution of pore sizes. And, as stated before, this distribution is shifted to overall smaller pore sizes for fine-textured soils and in the opposite direction for coarse-textured soils. Thus, organic matter acts on this overall pattern by broadening the distribution whether it is centered by texture on smaller or larger pores.) The smaller pores are added because organic matter is itself porous with these internal pores being rather tiny. The larger pores are added because of how humus coatings contribute to gluing mineral particles together.

A view of a soil's physical nature that was limited to texture and organic matter content alone would be rather short sighted. With cycles of wetting and drying and freezing and thawing, as well as through bioturbation, soil particles assemble themselves into larger units called aggregates. And, when a large proportion of individual soil particles is thus self-assembled, the soil is said to have structure. Factors promoting the assembly of particles into aggregates include fineness of texture, adequate organic matter content, and a sufficient time (years to decades) exposed to natural processes without man's disturbance. A coarsely textured, sandy soil containing little organic matter would not naturally form aggregates; or, if they were to form, they would be weak and ephemeral.

Like texture, soil aggregation has important consequences for the physical behavior of soil. By transforming a uniform collection of particles into larger, discrete units, fissures are created within the soil body. These fissures, although many fewer in number than the pores between individual particles, often are much larger. Thus, through aggregation, a fine-textured soil may become interwoven by number of sizable fissures. Conversely, a poorly structured soil is one where individual soil particles do not exhibit any association, one with another—a soil lacking a significant degree of aggregation and few inter-aggregate fissures. It is, therefore, the coarseness or fineness of texture, the concentration of organic matter, and the degree of soil aggregation that mostly define the distribution of pores within a soil and, consequently, much of soil physical behavior.

Soil Pores and Water

At any instant in time, a given soil pore may either be water-filled or air-filled. When, within a soil body, all soil pores are water-filled, the soil is said to be saturated. In this instance, the soil water content, defined as the volume of water divided by the total soil volume, equals the total porosity value. The saturated water content reflects the maximum volume of water a soil can hold. More commonly, a soil in the natural environment is unsaturated, having some pores water-filled and some pores air-filled. Its water content value will be less than the total porosity value. Thus water content values range in soil from around 0.05 to 0.5 m^3 m^{-3}, and essentially quantify its dryness (small values) or wetness (large values). Even though water content values are useful to describe the wetness state of a soil, they do not describe how different soils interact with water and the hydrologic environment.

Water generally interacts with soil pores akin to how water behaves in glass capillary tubes. In a glass capillary tube, when one end is dipped into water, water will rise into the tube to a final height that is inversely proportional to the tube's internal diameter. In other words, the height of capillary rise is greater in smaller-diameter glass capillaries and lesser in larger-diameter capillaries. Because of this relationship between internal diameter and height of rise, a soil containing mostly smaller-sized pores will have a greater height of capillary rise than a soil with larger-sized pores. Consequently, a fine-textured soil, owing to its smaller-diameter pores, will show greater heights of capillary rise and a coarse-textured soil containing larger pores will show smaller heights of rise. It also is said that the finer-textured soil retains more water against the downward force of gravity than a coarse-textured soil. This is demonstrated in Figure 6.1 (see page I-9), which shows soil water content as a function of height above a water table in three different soils. The coarse-textured soil among these three is the sandy loam, the medium-textured is the loam, and the fine-textured is the clay loam. Thus, the coarse-textured soil does not retain water as well as the other two soils and, consequently, is rather dry throughout much of its height. Alternatively, the finer-textured soil retains rather large water contents even to the top.

The ability of a soil to retain water against the force of gravity is central to its ability to provide water to plants between irrigation or rain events. A coarse-textured soil, having mostly larger sized pores, does not retain much water and is consequently viewed as droughty. Turfgrass transpiration and soil evaporation may, within only 1 day, exhaust this reserved water and the plant would experience drought stress. Alternatively, a fine-textured soil, having mostly smaller-sized pores, retains a greater volume of water that the plant can use. In this instance, onset of drought stress could be delayed by 6 or more days due to the larger moisture reserve. Yet some pores are sufficiently tiny, mostly in fine- and medium-textured soils, that the water held therein is not available to the plant. Essentially a plant is incapable of performing the work required to extract water from these very tiny pores. Consequently, there is no direct relation between retained water content and plant survival. Rather, the term "available water"—taking into account retained water that is not available to the plant—denotes that volume of water retained within a given soil that also the plant can use. Tables of available water for various soil textures are available in a variety of references (Water Management Committee 2005).

In addition to being retained by soil pores, water also can be transmitted through the interconnected collection of pores that exist in a soil. Generally speaking, a soil having mostly larger pore sizes transmits water more readily than soils with smaller pores. Consequently, coarse-textured soils transmit water more readily than fine-textured soils. So under identical water supply conditions, water moves at a higher velocity and penetrates the earth more deeply in coarse-textured than in fine-textured soils. This is demonstrated in Figure 6.2 (see page I-10), which shows the velocity of water flow and depth of water penetration in the same three soils as before. The view is at 12 hours after establishing a shallow but inexhaustible ponding of water at the soil surface. Thus, the larger pore sizes of the sandy loam soil are able to transmit water more rapidly and to a deeper depth than a clay loam soil. The precise term used to express the capability of soil to transmit water is the hydraulic conductivity given, for example, in units of mm h^{-1}. Saturated hydraulic conductivity values can exceed 200 mm h^{-1} for sands and range to less than 0.05 mm h^{-1} for clay soils (Rawls et al. 1993).

Available water and saturated hydraulic conductivity are commonly used indexes to address the suitability of the physical soil environment in supporting plants. Another aspect of the soil physical environment that is equally important but much less frequently discussed is the ability of atmospheric gases to diffuse into the soil, providing needed oxygen to the roots. Gaseous diffusion through soil depends principally on the proportion of air-filled pores in the soil, the so-called air-filled porosity. Thus, it has been observed that a soil should contain at minimum 0.1 m^3 m^{-3} air-filled porosity to allow sufficient gaseous diffusion for long-term plant survival. Interestingly, pore sizes only have an indirect effect on gaseous diffusion. Because gravity removes water more readily from larger-diameter pores, it is these larger pores that more readily become air-filled and serve the needs of soil aeration. Thus, poor soil aeration driven by inadequate gas diffusion in soils can occur whether the texture is fine or coarse, whether the soil contains much or little organic matter, and whether the soil is well or poorly aggregated. It is only when an adequate volume of pores are air-filled rather than water-filled that gaseous diffusion occurs at a sufficient rate for long-term plant survival.

The Fates of Water in Soil

Water falling to the earth can be intercepted by surface vegetation, infiltrate into the soil, or run off. Once in the soil, water can wet the surface soil, redistribute itself between soil layers, and eventually be removed from the system by evapotranspiration or drainage. Principally, these fates are influenced by

1. the quantity and rate of water delivery,
2. the path of water flow relative to the pull of gravity,
3. the water transmission and retention properties of the soil, and
4. the antecedent conditions—whether the soil is initially wet or dry.

The principal paths of water flow in soil are illustrated in Figure 6.3.

Rainfall occurs over a wide range of intensities and storm durations. Depending on these properties of the rainstorm, varying proportions of the water delivered can be intercepted or infiltrated or can run off. Essentially, low intensity and short duration storms may result in much of the water being intercepted by the plant canopy, delivering little to the soil, and contributing no runoff. On the other hand, high intensity and long duration storms (a somewhat rare occurrence, fortunately) result in much of the water volume directed to runoff. It is the intermediate intensity and duration storms that deliver the greatest proportion to the soil via infiltration at the surface.

Infiltration is best thought of as a rate. So if rainfall occurs at a rate less than the soil infiltration rate, all water arriving at the surface will be taken in. On the other hand, if the rain rate exceeds the soil infiltration rate, water arriving at the surface in excess of its capacity of being taken in will accumulate and run off. Thus, the soil infiltration rate essentially determines what portion of a given rain storm will enter the soil and what portion will run off. The soil infiltration rate, however, is not a constant value for a given soil. It is controlled both by the soil hydraulic conductivity and antecedent conditions at the surface. As expected, therefore, coarse-textured soils having larger hydraulic conductivity values also will tend to have larger infiltration rates. But if any soil, coarse or fine, is initially dry at the surface, then water can be absorbed rapidly by the soil, a process that yields a large infiltration rate. As the surface wets during a longer duration storm, however, this sorption process diminishes and the infiltration rate declines. Consequently, the infiltration path that controls water entry into the soil is both soil and event dependent.

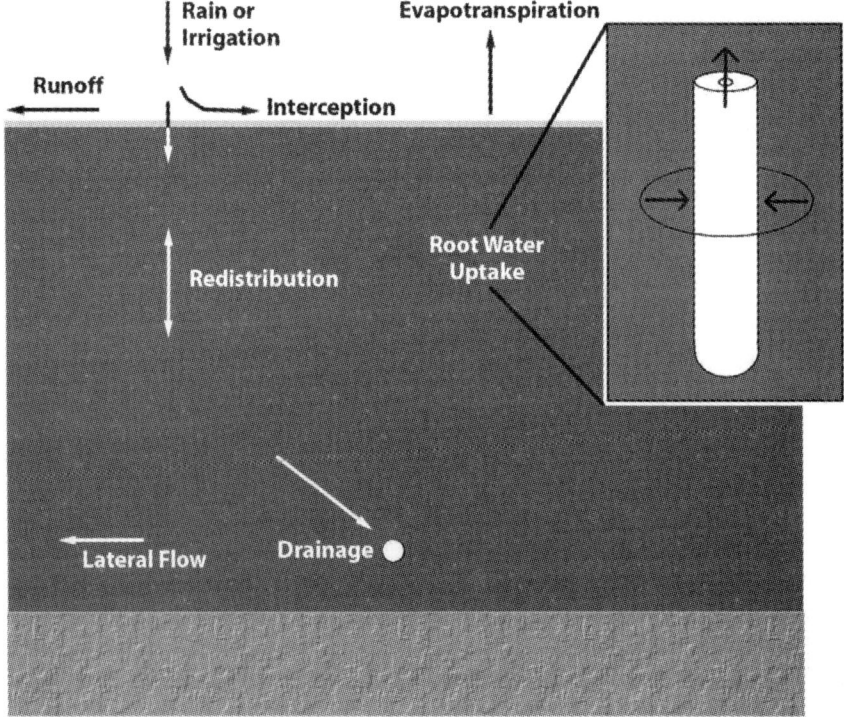

Figure 6.3. The principal paths of water flow in a soil containing a uniformly vegetated surface and with a slowly permeable layer at depth.

After water enters the soil, the main tendency is for it to continue flowing downward under the influence of gravity. Its rate of flow is largely influenced by the soil hydraulic conductivity. Yet when this flowing water encounters unfilled retention capacity (air-filled pores that can retain water against the force of gravity), a portion of the flow is retained, lessening the quantity redistributing downward. Eventually, the downward flowing water is entirely consumed by the surface soil layers, or if some portion remains, it continues on to below the rooting zone as drainage. Thus, wetting of a coarse-textured soil in excess of its retention capacity will yield a sizable volume of water rapidly draining below the root zone. The same volume of water added to a fine-textured soil, however, may yield little drainage water that travels quite slowly. Soil hydraulic conductivity and water retention capacity, balanced by antecedent conditions, will determine the drainage from a soil for a given event.

Soil properties can vary substantially with increasing depth. This aspect principally impacts drainage where a slowly permeable layer at depth in the soil can limit downward flow and inhibit natural drainage. Consequently water will flow either laterally downslope or to man-installed drainage elements. In either instance, however, the flow path is at some angle to vertical, not making full use of the pull of gravity, and the flow rates are decreased proportionally.

Water retained in the rooting zone can then be taken up by roots and removed from the soil via transpiration. Because most plants store little water, it is necessary that the rate of root water uptake sufficiently matches the rate of canopy transpiration in response to an evaporative demand. If water uptake falls behind, the plant will experience water stress. This begins to happen as water uptake exhausts the available water in the vicinity of individual roots. The overall process of water uptake by roots is, however, a bit more complex than simply exhausting the available supply. Because water is taken up first from the soil directly adjacent to the root, as uptake continues, water must flow to the root from progressively further distances (ranging from a few millimeters to a few centimeters). But, the soil through which this water must flow has had some water removed already and it becomes ever more difficult to convey water through this drier soil. The reason is that as soil dries, its hydraulic conductivity declines as well, sometimes by several orders of magnitude. So due to the progressively smaller hydraulic conductivity and the diminishing supply, water uptake may—particularly in the afternoon—fall behind transpiration and the plant will wilt. It often will recover overnight, however, as continued water flow at least partially replenishes the dry soil adjacent to the root. Only after this process repeats for several days will virtually all available water be exhausted and the plant fail to recover.

Finally, it is important to remember that water uptake by roots is a biologically mediated process that must be accompanied by oxygen diffusion from the atmosphere. This is why a proper balance between air-filled and water-filled pores in the soil is required for the plant-life sustaining process of transpiration.

Sound recommendations for turfgrass and landscape soil water management are available (McIntyre and Jakobsen 2000; Water Management Committee 2005). There are, however, certain features of a managed turfgrass landscape that strongly influence water flow yet are less well documented and incompletely incorporated in management recommendations. These features are the inevitable soil layering that occurs in the urban landscape, the use of root zones to avoid soil compaction, and the effect of sloping terrain on subsurface lateral water flow.

Soil Layering in the Urban Landscape

Native soils commonly exhibit physical property variation with depth. This variation results from natural soil-forming processes that create soil horizons in the vertical plane. Soil properties within these naturally formed layers commonly transition gradually from one horizon to the next, and the distinction between adjacent horizons may be rather subtle. In many urban soil instances, however, soil layers can exhibit very dramatic differences in their properties with sharp interfaces between adjacent layers. This characteristic of urban soils commonly results from the re-creation of a soil profile using soil materials that were not individually formed at the location of interest.

Site preparation for residential and commercial construction commonly consists of excavation, segregation of soil layers, soil material stockpiling, and respreading. These steps are taken to clear the site for subsequent construction and to create a surface topography that is both functional and aesthetically pleasing. Excavation commonly is performed according to the existing soil horizons whereby the equipment operator distinguished between topsoil and subsoil resources. These individual soil resources often are stockpiled separately. Also, contractors building a housing development may choose to strip the topsoil and sell it to a soil-blending company, which then markets this topsoil to another customer presently needing the soil. Consequently, site preparation transforms the native soil to an urbanized form that can exhibit distinct layers with contrasting physical properties and sharp interfaces in between.

Soil layering in site preparation for home lots often is less extreme, consisting simply of a relatively shallow layer of stockpiled or imported topsoil placed on the existing grade. These topsoil layers are observed to range from 60 to 350 mm. Repeated construction and demolition as would occur in a city center, however, can result in multiple, distinct layers ranging to a substantial soil depth (Craul 1992). Even though the topsoil layer may be perfectly suited to support the growth of intended plantings, there is an influence of the layering itself on water flow within the soil profile. Essentially this influence is an interruption of downward water flow at the layer interface following rainfall or irrigation. More specifically, when the saturated hydraulic conductivity of subsoil is 5-fold or less than the surface-soil layer, then the subsoil behaves, temporarily, as an impermeable barrier to downward water flow (Fausey and Brehm 1976). Thus, soil layering can create problems in water drainage from the profile and water can accumulate quickly within the surface-soil layer during frequent rainfall periods. The result is periodic waterlogging of plant roots and an increased opportunity to generate runoff.

In many instances, however, these consequences of soil layering are of minor importance. Periodic waterlogging alone rarely kills turfgrass, and the increased runoff is still substantially less than that occurring on adjacent paved or roofed surfaces. Thus, layering that occurs in many residential sites is not a serious problem, mainly because site use and turfgrass quality expectations generally are low. Soil layering has a much greater impact within the managed turfgrass realm when the site is exposed to frequent foot traffic and/or when turfgrass quality expectations are great. This is because an excessively wet surface soil exposed to foot traffic often can suffer from compaction—a situation that can damage the turfgrass system seriously.

Soil Compaction

Frequent human contact is an important feature of managed turfgrass. Foot and vehicle traffic from both frequent maintenance and play apply an external stress to the soil surface. Further, these activities commonly occur over a much wider range of soil moisture conditions than found with agricultural traffic. Thus, managed turfgrass landscapes can be prone to soil compaction. Defined simply as an increase in soil bulk density, compaction is an interaction between the level of applied stress and the strength of the soil in resisting deformation. If the applied stress exceeds the soil compressive strength, compaction will occur.

The principal factors that influence soil strength are texture and water content. In engineering applications, soil textures are classified broadly as either cohesive or granular. Cohesive soils contain an appreciable quantity of silt and clay. Granular soils consist mostly of clean sands and gravels. Cohesive soils, when moist, have a lower compressive strength and are subsequently prone to greater compaction. That is, soils with appreciable quantities of silt and clay are particularly prone to compaction when in a moist state. Granular soils typically have higher compressive strength and can resist compressive stresses. This is the main reason that higher sand content soil materials are preferred as root zones for soils exposed to high levels of foot traffic.

By definition, soil compaction results in a loss of total porosity. More significant is that not all soil porosity is influenced to the same degree. Soil compaction results in the collapse of the larger-sized pores with a corresponding increase in the volume percentage of smaller pores. A compacted soil, therefore, will exhibit decreased infiltration rates, decreased drainage, poor soil aeration, and a platy or massive soil structure. These marginalized soil physical conditions result in a less favorable environment for turfgrass roots, and for many beneficial soil microbes, earthworms, and arthropods. For this reason, compaction is probably the most serious damage that can occur in soils of managed turfgrass landscapes. Thus, soil layering combined with cohesive soil textures leads to the downfall of trafficked turfgrass via compaction.

Root Zones for High-Traffic Areas

Historically, the principal approach to protecting a soil against compaction is to replace the existing soil that typically exhibits cohesive behavior with a root zone having properties of a granular media. This goal is achieved by establishing sufficiently high sand contents in the root zone. Sand contents exceeding 0.75 kg kg^{-1} generally assure both inter-particle contact of the sand grains to resist compaction and a sufficient proportion of pores between grains that are not occluded by finer-textured materials. For example, sand contents exceeding 0.75 kg kg^{-1} resulted in more than 15% of the total pore volume being composed of larger-sized (> 0.05-mm diameter) pores (McCoy 1998). These pores that are greater than 0.05 mm are more readily drained and responsible for rapid water flow through the soil. Thus, a soil should contain at least 10 to 20% of these larger-sized pores to allow adequate water infiltration, drainage, and gas exchange with the atmosphere. Correspondingly, a soil containing this minimum sand content was shown to yield saturated hydraulic conductivity values of approximately 50 mm h^{-1}, a commonly recommended minimum value for high-traffic turfgrass root zones.

As sand content increases above this minimum, the porosity from larger pores and saturated hydraulic conductivity values increase as well. Yet there is a practical or agronomic upper limit to saturated hydraulic conductivity values estimated to be between 1300 to 1500 mm h^{-1}. This is particularly true for cool-season turfgrass where, above this limit, a viable turfgrass can be established and maintained only with great difficulty. Finally, sand content of a root zone does not alone dictate its saturated hydraulic conductivity. Sand texture, the uniformity of particle sizes, and any added amendments also influence root zone hydraulic conductivity to varying degrees (Baker 1990; McCoy 1992; Zhang and Baker 1999). Yet given this range of root zone materials, hydraulic conductivity values generally are confined to the range of 50 to 1500 mm h^{-1}.

Again, depending on the performance expectations, degree of use, and foot traffic levels, modified by conventional practices, it is suggested that the sand content for high-traffic turfgrass areas exceeds 0.75 kg kg^{-1}. In terms of physical behavior, this range from sandy loam to sand textures allows rapid water infiltration, drainage, and gas exchange with the atmosphere. Further, these properties will not decline as a result of compressive stresses from foot traffic and maintenance equipment.

Soil Profiles Found in High-Traffic Areas

The principal motivation for the use of high-sand-content (> 0.75 kg kg^{-1} sand) root zones is to resist soil compaction from frequent foot traffic. To achieve this goal, however, depth of the sand-enriched root zone need only be relatively shallow because foot traffic stresses do not penetrate deeply into the soil. Minimum depths of this modified surface layer are in the range of 75 to 100 mm. Yet a high sand content surface layer, with its associated large hydraulic conductivity, also allows a greater proportion of rainfall infiltration compared with runoff. The result, for a shallow root zone overlying a slowly permeable subsoil, is rapid water accumulation and saturation of the root zone layer during frequent rainfall periods. Subsequently, the air-filled pore space needed for oxygen diffusion is eliminated.

Thus, some provisions need to be made to maintain a suitable air-filled and water-filled porosity within the turfgrass rooting zone. Otherwise, if inadequate air-filled pore space exists through some reasonable depth in the root zone, the turfgrass will undergo aeration stress (i.e., become waterlogged), the turfgrass will become shallow rooted, and invasion by shallow-rooted species will occur. Of course there is a climatic influence where periods of high rainfall serve to exacerbate this problem. During mid-summer conditions, turfgrass aeration stress typically will be less severe (given that the turfgrass is not over irrigated).

There are three possible steps for maintaining a favorable air and water balance within the root zone of a layered soil containing a sandy surface media. All three steps involve controlling the height of the capillary fringe in the soil profile. A capillary fringe is a region of saturated or nearly saturated soil that exists above a water table. It forms due to capillary rise in soil where virtually all soil pores are sufficiently small to remain water-filled throughout its thickness. Within a high-sand-content root zone, the thickness of the capillary fringe can range from just a few to 200 mm or more, depending mostly on the sand texture. Consequently, within the capillary fringe, there are few air-filled pores. An illustration of capillary fringe formation in layered soils is shown in Figure 6.4 (see page I-11).

The steps to control the height of the capillary fringe are to (1) limit occurrence of the capillary fringe in the first place, (2) allow the presence of the capillary fringe to be only ephemeral, or (3) displace the capillary fringe deeply into the soil. All three approaches or their combinations are present in successful systems.

Limiting the occurrence of a capillary fringe is achievable in climates having infrequent rainstorms delivering small rainfall amounts. Because these climates will be dry with respect to water needs of the turfgrass, this scenario must be coupled with judicious water management so as to avoid excessive irrigation. By limiting the quantity of water, the formation of a capillary fringe can be avoided. Although exceptional in its low cost and simplicity, this scenario is reliant on somewhat rare climate conditions. In most regions of the United States, seasonal rainstorms are either frequent or deliver larger amounts, or both.

During frequent or high intensity rainstorms and when shallow, sand-enriched root zones are placed over a slowly permeable subsoil, the capillary fringe can form quickly and extend throughout the turfgrass rooting zone. The need is then to remove this excess water in a timely fashion to minimize turfgrass impacts. Timely removal of this capillary fringe suggests the implementation of a subsurface drainage methodology. Yet, implementation of conventional subsurface drainage (a spaced array of drainpipes placed from 450 to 600 mm deep) often is of little aid in these profiles due to the slowly permeable subsoil layer between the water-accumulating root zone and the drainage elements. On the other hand, recent adoption of capillary style drainage has shown some benefit in the timely drawdown of this capillary fringe. Capillary drainage is characterized by shallow (225 to 300 mm deep), closely spaced (0.9 to 1.2 m) elements associated with a narrow, permeable channel extending to the base of the sand-enriched root zone. By intercepting the root zone layer and creating a hanging-water-column suction, water within the capillary fringe can be removed. Also, deep cultivation frequently is applied to these areas to loosen or otherwise modify the subsoil layers and improve their drainage potential. The effect, however, is shown to be temporary (Lodge and Baker 1993).

Root zone air and water balance issues associated with the hydraulic behavior of layered soils are also addressed by displacing the capillary fringe deeper into the soil profile. Simply stated, this constitutes increasing the thickness of the sandy root zone to depths of 250 and 350 mm, well beyond that needed to resist foot traffic compaction. Yet this also results in a suitable air-filled porosity in the surface layers where the majority of turfgrass roots reside. Further, as root zone depths increase, there is a general tendency for increasing the sand contents within the root zone layer (Figure 6.5, see page I-12). The goal here is to balance the interplay between the proportion of rainfall infiltrating the soil, the height of the resultant capillary fringe, and the expense of deeper excavations and increased quantity of root zone sand or mix. Finally, and perhaps most important, a deeper root zone depth would allow a more direct hydraulic connection with subsurface drainage elements, assisting greatly in the elimination of free water within the profile.

With deeper root zone depths, the turfgrass area is constructed as an excavation into the soil native to the site. The high-sand-content root zone and, in some instances, coarse sand or fine gravel sublayers subsequently are placed within this excavation. Subsurface drainage from this essentially closed basin obviously is necessary and typically is provided by drainage pipe spaced from 3 to 6 m and placed in shallow trenches in the subsoil. The resultant soil profile ranges from two to multiple layers.

The simplest of these soil profiles is the University of California method of putting green construction (Davis, Paul, and Bowman 1990; Harivandi 1998) where a minimal 300 mm depth of specified sand is placed within the native soil excavation. The large hydraulic conductivity of the sand yields nearly complete infiltration of rain or irrigation that generates a free water surface within the root zone above the root zone/soil interface. The thickness of the capillary fringe is controlled via lateral water movement to the gravel backfill of the drainpipe trench. This soil profile also is used for other intense traffic turfgrass situations with sand selection based on the intended use of the area.

One of the most complicated of these multiple-layer soil profiles is that suggested for use in putting greens by the U.S. Golf Association (USGA 1993). Depending on the availability of suitable root zone and gravel materials, this profile consists minimally of three layers including the native subsoil as the lowermost layer. In this configuration, a 300-mm-thick, high-sand-content root zone mix is positioned above a minimally 100-mm-thick, specified but predominately fine-gravel zone. The gravel subsequently rests on the subsoil except when adjacent to drain line trenches where the same gravel also fills the trench. The particle size distribution of the gravel must conform to engineering specifications for a drainage filter. This requirement helps to ensure maintenance of layer integrity and suitable hydraulic performance of the gravel.

Hydraulically, the gravel blanket of a USGA putting green promotes rapid drainage of the root zone. This occurs because excess water exiting the root zone follows a nearly vertical path, employing the maximum extent of the gravitational gradient. Further, the maximal distance drainage water must travel to exit the root zone is virtually the root zone depth, or 300 mm. Lateral flow to the drainage elements occurs predominantly within the very high conductivity gravel layer. The gravel drainage blanket beneath the finer-textured root zone also creates a large difference in the pore size distribution across this interface. This large separation of predominant pore sizes within these adjacent media yields a capillary break in the vertical direction. Consequently, the lower portions of the root zone will develop a capillary fringe and remain saturated (or nearly so) after drainage has virtually ceased. The root zone texture and depth suggested for a USGA green helps ensure that the capillary fringe will extend only into in the lower portion of the root zone and that an adequate air-filled porosity is maintained near the soil surface. Overall, this results in a root zone that retains more moisture than the same media having an infinite depth. Much of this retained water subsequently is taken up by evapotranspiration.

Parenthetically, use of a sandy loam planting mix to depths of 150 to 200 mm is a component of a unique set of recommendations for woody species planting in the urban environment (Urban 1989). Layered soils—by design—are not limited simply to putting greens and athletic fields.

Slope Effects on Water within Layered Soils

Surface slopes found on turfgrass landscapes also occur on interfaces between soil layers within the profile. This is because profile layers typically are built to a uniform thickness across the area. When the interface between layers is well defined and there is a wide disparity between soil textures of adjacent layers, the inevitable accumulation of water is subject to lateral flow. This down-slope movement of subsurface water is particularly evident in profiles with highly permeable root zone media and deeper root zone depths (Prettyman and McCoy 2003a). Naturally, the reasons here are that only a highly permeable root zone would allow sufficient rates of flow for the modest slopes of these systems. Also, a deeper profile depth would provide a greater reservoir of soil water available for such flow. Consequently, the capillary fringe of such soil profiles would in a reasonable time (1 to 2 days) migrate downslope resulting in lower soil water contents at higher elevation locations and higher water contents at lower elevation locations (Figure 6.6, see page I-13). This lateral flow process is responsible, within some turfgrass landscapes, for excessively wet root zones at low elevation locations and droughty conditions at high elevation locations (Prettyman and McCoy 2003b).

Impact of Layered Soils on Available Water

The available water concept as published in textbooks and used in irrigation scheduling is expressed as the volume of water retained in the soil from field capacity to the permanent wilting point. Further, this available water is subdivided into readily available water, whose uptake would not place the plant at risk for water stress, and less available water, where the plant is expected to show some drought symptoms in its use. The convention (Water Management Committee 2005) is that 50% of the total available water is readily available, although some have suggested that for fine turfgrass this value should be less. Because available water is influenced strongly by soil texture, values of available water for various soil texture classes have been constructed. Thus, using the published value of available water for sand, it is possible to determine that there is 18 mm of available water in a 300-mm root zone with, at most, half of this usable without placing the turfgrass at risk of drought stress symptoms.

Yet the available water concept (among other weaknesses) presumes that the soil in question extends to an essentially infinite depth below the soil surface or the maximum extent of turfgrass rooting. For layered soils this clearly is not the case. As mentioned previously, layered soils with a sandy surface layer result in an interruption of downward water flow such that when drainage becomes negligible (a definition of field capacity), a sizable volume of water is retained in the root zone layer. Thus, the available water concept and published available water values are open to question for layered turfgrass soils.

An alternative view of available water was offered in McCoy and McCoy (2005). The basis for this redefinition results from a 2-year field study wherein a complete water balance was performed on experimental greens supporting a bentgrass turfgrass. The experimental greens consisted of a 300-mm-deep root zone placed above a 100-mm-thick gravel drainage blanket, all contained within a non-weighing lysimeter. The study employed six root zones: two containing pure sand, two containing sand +10% (vol.) sphagnum peat, and two containing sand + 10% peat + 10% (vol.) topsoil. Two different sands were used with one being slightly finer and one being slightly coarser, and with both containing approximately 74% medium and coarse particles.

This field research recorded all rainfall and irrigation inputs and all drainage losses; from daily soil moisture measurements, the researchers calculated daily turfgrass evapotranspiration. For one instance each during the years 2000 and 2001, irrigation was withheld to impose drought stress on the turfgrass to the point at which first wilt or "footprinting" became visually apparent. These dry-down periods were initiated by a heavy irrigation or rainfall. Thus, from tracking soil moisture changes and drainage losses during the dry-down period, a field-based estimation was available of water actually used by the turfgrass from a well-watered condition to first wilt. Following the procedure described above, available water for a pure sand root zone, a sand + 10% peat root zone, and a sand + 10% peat + 10% soil root zone was 23, 31, and 39 mm, respectively. These values represent the depth of readily available water within a 300-mm root zone characteristic of a modern green and are substantially larger than the 9 mm estimate using conventional available water concepts.

Other observations from this study also are of interest. First, the rate of water uptake during the drydown period was consistent at all root zone depths even though end-of-season measurements expectedly showed much denser turfgrass rooting at shallow soil depths. This indicates a disconnect between turfgrass rooting and water uptake patterns in this particular system. Also, the appearance of footprinting was associated with root zone water contents approaching 0.1 m^3 m^{-3}, regardless of the soil water suction value of this water content from our water retention measurements. Both of these additional observations bring to question the available water concept in layered soils.

Conclusion

To resist compaction within a high-traffic turfgrass soil, the surface 75- to 100-mm layer should consist of a media containing at minimum 0.75 kg kg^{-1} sand. Textural discontinuity within such a soil profile commonly results in more than a five-fold difference in saturated hydraulic conductivity between the surface and subsoil layers. When layer permeability differences exceed this limit, there is an interruption of downward water flow, and the water accumulates and is retained within the surface layer. Coupling of shallow layering with the increased infiltration of the sandy root zone results in a waterlogging hazard for the turfgrass. As dictated by the thickness of the capillary fringe for a sandy root zone, deeper root zone depths and higher sand content root zones aid in avoiding this waterlogging hazard. Even modest slopes within such a system, however, can lead to subsurface lateral flow, confounding this simple solution.

These layered soil systems also befuddle present attempts at irrigation water conservation. This is due to the inability of conventional available water concepts to deliver meaningful estimates of the soil water reservoir capacity. Yet once these issues—and the lateral flow problem—are addressed adequately, purposeful creation of a layered soil can serve as a long-term water conservation measure. This is because such a system would (1) contain a sufficiently permeable surface layer to decrease runoff and perhaps even accept run-on from adjacent impermeable surfaces and (2) retain this water in the vicinity of plant roots for subsequent evapotranspiration. At least this may occur in areas where there is some rainfall.

Literature Cited

Baker, S. W. 1990. *Sands for Sports Turfgrass Construction and Maintenance*. The Sports Turfgrass Research Institute, Bingley, England.

Craul, P. J. 1992. *Urban Soil in Landscape Design*. John Wiley & Sons, New York. 416 pp.

Davis, W. B., J. L. Paul, and D. Bowman. 1990. The sand putting green: Construction and management. Publication no. 21448. University of California Division of Agriculture and Natural Resources, Davis.

Fausey, N. R. and R. D. Brehm. 1976. Shallow subsurface drainage-field performance. *Trans ASAE* 19:1082–1084, 1088.

Harivandi, M. A. 1998. Golf green construction: A review of the University of California method. *California Turfgrass Culture* 48(3&4):17–19.

Lodge, T. A. and S. W. Baker. 1993. Porosity, moisture release characteristics and infiltration rates of three golf green rootzones. *J Sports Turfgrass Res Inst* 69:49–58.

McCoy, E. L. 1992. Quantitative physical assessment of organic materials used in sports turfgrass rootzone mixes. *Agron J* 84:375–381.

McCoy, E. L. 1998. Sand and organic amendment influences on soil physical properties related to turfgrass establishment. *Agron J* 90:411–419.

McCoy, E. L. and K. R. McCoy. 2005. Putting green root zone amendments and irrigation water conservation. *USGA Turfgrass and Environmental Research Online* 4(8):1–9.

McIntyre, K. and B. Jakobsen. 2000. *Practical Drainage for Golf, Sportsturf and Horticulture.* Ann Arbor Press, Chelsea, Michigan. 202 pp.

Prettyman, G. W. and E. L. McCoy. 2003a. Profile layering, root zone permeability and slope affect soil water contents during putting-green drainage. *Crop Sci* 43:985–994.

Prettyman, G. W. and E. L. McCoy. 2003b. Localized drought on sloped putting greens with sand-based rootzones. *USGA Turfgrass and Environmental Research Online* 2(4):1–8.

Rawls, W. J., L. R. Ahuja, D. L. Brakensiek, and A. Shirmohammadi. 1993. Infiltration and soil water movement. Chapter 5. In D. R. Maidment (ed.). *Handbook of Hydrology.* McGraw-Hill, New York.

Urban, J. 1989. New techniques in urban tree plantings. *J Arboric* 15:281–284.

U.S. Golf Association, Green Section Staff (USGA). 1993. The 1993 revision, USGA recommendations for a method of putting green construction. *USGA Green Section Record* 32(2):1–3.

Water Management Committee of the Irrigation Association. 2005. Landscape irrigation scheduling and water management, http://www.irrigation.org/gov/default.aspx?r=18pg=bmps.htm (21 February 2007)

Zhang, J. and S. W. Baker. 1999. Sand characteristics and their influence on the physical properties of rootzone mixes used for sports turfgrass. *J Turfgrass Sci* 75:66–73.

7

Leaching of Pesticides and Nitrate in Turfgrasses

Bruce Branham

Introduction

Turfgrasses for amenity and sports use have been grown for more than five centuries, but during the last 50 years Americans have pushed the field of turfgrass management to new heights. Or perhaps it would be more accurate to say that new lows have been achieved, as cutting heights on highly maintained turf have been lowered to below 0.1 inch (2.5 millimeters [mm]). The more intensive management developed during the last 50 years has increased the frequency and severity of pest outbreaks, particularly diseases, resulting in widespread use of fungicides and other pest management products on golf courses and, to a lesser extent, athletic fields. The frequent use of fungicides on golf courses makes turfgrass one of the largest users of fungicides in the United States with an estimated cost of $130 million in 2004 (Fungicides 2005). In 2001, the U.S. Environmental Protection Agency (USEPA 2004) estimated total fungicide use at $835 million. Home lawns under professional management may receive one to two herbicide applications per year and potentially one insecticide application as well. Home lawns rarely receive fungicide applications.

The frequent use of pesticides on highly maintained turf, along with their use on amenity plants, has focused much public scrutiny on pesticide movement off of treated plant surfaces. This section will focus on the data of pesticide and nitrate leaching from turf.

Pesticide Leaching

Pesticide leaching is a complex process that has been studied intensively (Cheng 1990; Helling, Kearney, and Alexander 1971; Wauchope et al. 2002). Investigators have found that the major factors that control pesticide leaching are pesticide sorption to organic carbon, described by the organic carbon sorption coefficient, K_{oc}; pesticide soil half-life, $t_{1/2}$; application rate; amount and intensity of rainfall and irrigation; and the presence and distribution of macropores (Gustafson 1989; Malone et al. 2004). Of those factors, K_{oc} and half-life are the most dominant factors in determining the propensity of a pesticide to leach (Gustafson 1989; Wauchope et al. 2002).

Factors that Attenuate Leaching in Turf

Although the majority of research on pesticide leaching has focused on row crop agriculture, turfgrasses present some unique features that dramatically influence pesticide fate processes. In most other agriculture cropping systems, the majority of the pesticide application is deposited on the soil surface. A well-maintained turf, however, presents a continuous plant cover and a layer of living and dead organic matter—thatch or mat—which provides the principal site for interaction with the applied pesticide. Several researchers have documented that thatch has a higher sorptive capacity than soil (Dell et al. 1994; Lickfeldt and Branham 1995). Raturi, Carroll, and Hill (2003) measured the leaching of two pesticides through soil columns with thatch present or removed. Under extreme conditions (26 or 80 centimeters [cm] of water applied 24 hours after pesticide application), leaching removed 84 and 76% of triclopyr and carbaryl applied to the soil-only column and 77 and 61% of triclopyr and carbaryl applied to a column containing turf. The researchers concluded that the presence of thatch lessened the leaching potential of the two pesticides.

A well-maintained turf also develops considerable organic matter in the turfgrass rootzone. Researchers at the University of Nebraska documented the development of organic matter in sand rootzones (Kerek et al. 2002). Organic matter contents were determined in sand rootzones varying in age from 2 to 28 years (yr). They found a linear increase in organic matter content with putting green age, reaching 4 to 6% after 20+ yr of growth. Organic matter content increased by 0.15% per yr in the sand rootzones.

Porter and colleagues (1980) measured nitrogen (N) contents of turf soils ranging from recently established to more than 100 yr old. They found that N levels, and associated organic matter, increased during the first 10 yr after turf establishment and then tended to plateau. Because pesticide absorption is correlated strongly to organic carbon in soil, turf soils should possess a significant capacity to sorb pesticides, resulting in less pesticide mobility than would be expected in the same soil under continual crop production.

The persistence of a pesticide is a critical determinant of pesticide mobility. Pesticide degradation in soils and organic matter is driven mainly by microbial degradation. Because of the high levels of surface organic matter present in turf and the concomitant high levels of microbial activity, pesticide half-lives in turf may be substantially different than when applied to bare soil. Horst and colleagues (1996) measured the half-life of four pesticides applied to Kentucky bluegrass turf. They determined that metalaxyl, pendimethalin, chlorpyrifos, and isazophos had half-lives of 16, 12, 10, and 7 days (d), respectively, when applied to turf. The generally accepted half-lives of these products applied to soils, as listed in the Soil Conservation Service/Agricultural Research Service/Cooperative Extension Service pesticide properties database (Wauchope et al. 1992), are 70, 90, 30, and 34 d for metalaxyl, pendimethalin, chlorpyrifos, and isazophos, respectively.

Gardner and Branham (2001a, b) and Gardner, Branham, and Lickfeldt (2000) went one step further and directly compared pesticide persistence in soil with turf to the same soil with the turf physically removed. Decreases in pesticide $t_{1/2}$ ranged from minor for mefanoxam to quite substantial for ethofumesate (Table 7.1). Decreased soil $t_{1/2}$ values in turf will result in less pesticide available for leaching.

Table 7.1. Comparison of half-lives of pesticides applied to turf or to soil with turf removed

Pesticide	$t_{1/2}$ in turf (days)	$t_{1/2}$ in soil (days)	SCS/ARS/CES
Ethofumesate	3[†]	51	30
Cyproconazole	8 – 12	128	90
Propiconazole	12 – 15	29	110
Halofenozide	>64	>64	--
Mefanoxam	5 – 6	7 – 8	70

[†] (Gardner and Branham 2001a)

Roy and colleagues (2001) studied the degradation of dicamba in soil or thatch and reported that $t_{1/2}$ was 5.9 to 8.4 times shorter in thatch than in soil. Thatch $t_{1/2}$ ranged from 23 d at 4°C and low moisture content to 5.5 d at 20°C and higher moisture content, whereas soil values for $t_{1/2}$ ranged from 136 d to 36.2 d under similar temperature and moisture conditions.

The high levels of surface organic matter in turf lead to increased pesticide sorption and decreased pesticide half-lives, both factors that tend to mitigate pesticide leaching. Perhaps because of the high levels of organic matter, earthworm populations are high in turf soils (Potter 1993). Earthworm channels can increase macroporosity as well as root development. In addition, turfgrasses often are cultivated using machines that create holes in or remove the turf. The average depth of penetration is 7 to 8 cm, but some machines can create holes to a depth of 25 cm.

Macropore flow should be expected to have an impact on pesticide leaching in turfgrasses, but research on this topic is limited. Roy, Parkin, and Wagner-Riddle (2000) examined water movement in soil covered with turfgrass. They observed that the saturated hydraulic conductivity (K_{sat}) of the 5-cm turf layer was 126 times that of the layer immediately below the turf. They attributed this large difference to the lower bulk density of the thatch portion of the turf and the presence of macropores.

Jarvis, Bergstrom, and Dik (1991) examined macropore flow in monolith lysimeters of the same soil under continual barley production or rotation in its fourth year of grass cover. The K_{sat} of the two cropping systems was similar in the top 15 cm, which was attributed to macropores in the grass system and a recent cultivation in the barley system. In the next 15 cm, however, K_{sat} in the grass system was 40 times K_{sat} in the barley cropping system. Although this example is not a turf, it does illustrate that a grass crop that is not tilled will develop macropores that result in rapid preferential flow. Unlike pastures, however, most highly maintained turfgrasses are used for sport, with significant foot and vehicular traffic that may result in compaction and a subsequent decrease in macropores. Further, many athletic turfs are artificial rootzones constructed from sand. Measurements of K_{sat} in turf as affected by cultivation, "topdressing," and other common cultural practices would be of interest.

Turf culture produces high levels of surface organic matter and associated organic carbon that is the primary sorbent of pesticides. The high levels of surface organic matter also increase microbial activity, the primary degrader of pesticides. The benefits of turf in decreasing pesticide leaching due to high levels of surface organic matter could be partly offset by the effects of macropore flow, especially when it occurs within a short time of a solute application.

Almost all recently constructed golf course putting greens, and occasionally tees, as well as many sports fields, use a rootzone constructed from sand. These rootzones are, by design, porous to provide rapid drainage and resistance to compaction. A considerable amount of the research on pesticide leaching in turf has been conducted on sand rootzones, which may represent a worst-case scenario for pesticide movement.

Pesticide Leaching in Turf

Several pesticides have been studied by more than one group of researchers, providing additional insight into leaching under these conditions. Petrovic and colleagues (1996) studied the leaching of several pesticides through turfgrass. They examined the effect of removing 33 and 66% of the turf by cultivation, as well as the effect of adding peat to a pure sand rootzone. They found very high levels of metalaxyl leaching with 16.5% of the applied metalaxyl collected in the leachate when a 100% turf cover was present. Removing the turf cover resulted in successively higher levels of metalaxyl leaching with 14, 26.7, and 36.4% applied metalaxyl leaching when 33, 66, and 100% of the turf cover was removed.

These levels of leaching are quite high compared with data collected by Wu and colleagues (2002b) in California. They applied metalaxyl to sand rootzones managed under putting green conditions and monitored the leachate for 150 d after application. The maximum concentration of metalaxyl in the leachate was 7.2 micrograms per liter ($\mu g\ L)^{-1}$ and the total recovered during the course of the experiment was 0.71% of the applied metalaxyl. How can these results be so different? The southern California experiment was managed as a putting green with irrigation applied as needed to prevent drought stress. This approach, which is a standard turf management practice, applies smaller quantities of irrigation, allowing the pesticide to stay in the turfgrass rootzone where microbial activity is high and degradation may be relatively rapid. The Cornell study (Petrovic et al. 1996) purposely over-watered to increase the likelihood of pesticide leaching. They applied irrigation daily at a rate of 1.95 cm d^{-1}, which exceeds typical daily evapotranspiration rates by a factor of three. This irrigation was applied regardless of natural rainfall, which would further increase leaching.

A commonly used turfgrass pesticide with a propensity to leach is dicamba, which has been studied by several different turfgrass research groups. Roy and colleagues (2001) studied dicamba leaching in a Kentucky bluegrass turf instrumented with suction lysimeters that permit sampling of soil water at various depths. They applied dicamba at 0.6 kilograms (kg) active ingredient (ai) per hectare (ha)$^{-1}$ three times during the growing season: May 5, September 28, and November 27. They detected relatively large dicamba concentrations, mostly in late fall and early winter samples, exceeding 1,000 µg L^{-1} at depths of 17, 29, and 43 cm. These results are much higher than those reported by other researchers. Smith and Bridges (1996) reported a maximum concentration of dicamba of 2.6 µg L^{-1}, whereas Gold and colleagues (1988) reported a maximum concentration of 38 µg L^{-1} after an application of 0.11 kg ai ha^{-1}. Snyder and Cisar (1997) reported concentrations of 1.68 and 2.53 µg L^{-1} of dicamba after two applications to a sand putting green, whereas Harrison and colleagues (1993) reported a maximum concentration of dicamba of 118 µg L^{-1}.

The drinking water standard for dicamba in the United States is 200 µg L^{-1}. Total losses of dicamba ranged from 10.8% of the applied in the Florida study (Snyder and Cisar 1997) to 0.2 to 0.5% in the work of Smith and Bridges (1996). Dicamba is an organic acid that is not appreciably sorbed in soils and has a low K_{oc} of 2. Leaching can be expected if irrigation occurs immediately after application, or if biological activity is low, and thus degradation is decreased. Two factors contributed to the high concentration detected in the Roy study. First, one of the applications was made on November 27, when soils in Canada would be cold and biological degradation would be at a minimum. A rain event within several days could be expected to cause rapid downward movement of dicamba via preferential flow because soils are usually moist to wet at this time of year. The application date of the Canadian study was late but not outside the realm of practice, although most turf agronomists would discourage applications that late in November in Canada. Second, the rate was four to five times higher than common usage rates in the United States. These two factors, a high application rate and a very late application, combined to produce the high concentrations detected.

The leaching of two insecticides, trichlorfon and chlorpyrifos, have been studied by several research groups. Petrovic and colleagues (1993) studied the leaching of trichlorfon in three soil types using two irrigation regimes. They reported a maximum concentration of 467 µg L^{-1} and cumulative losses of 3.33 to 4.41 % of the total applied for the high irrigation regime and 0.63 to 1.18 % of the total applied for the low irrigation regime. Of the three soil types, the two sandy soils yielded higher leaching levels than the silt loam soil, but the differences were not as large as might be expected. Interestingly, no leachate was generated in the low irrigation treatment until 4.4 inches (11.1 cm) of irrigation were applied during a 2-d period, which produced almost all the trichlorfon that was recovered in the leachate.

Wu and colleagues (2002a) studied the leaching and persistence of trichlorfon and chlorpyrifos in sand greens in southern California. The highest concentration of trichlorfon in the leachate in their study was 71.3 µg L^{-1}. In the Wu study, trichlorfon applications were made in two consecutive years, and the cumulative recovery of trichlorfon in the leachate was 0.05 and 0.0001% of the total applied in 1996 and 1997, respectively. Chlorpyrifos was less mobile than trichlorfon with the highest concentration in the leachate of 1.9 µg L^{-1}. Chlorpyrifos recovery in the leachate was 0.01 and 0.0006% of the total applied in 1996 and 1997, respectively.

These studies show that turf is a system that generally results in low levels of pesticide leaching, even when applying potentially mobile pesticides. But in the worst-case scenario—large leaching events soon after application—significant leaching can occur.

Nitrogen

Nitrogen is applied frequently to turfgrass to maintain color, density, and growth. Typical yearly application rates for turfgrasses range from 1 to 4 pounds (lbs) N/M/yr (50 to 200 kg N/ha/yr) for cool-season turfgrasses and from 2 to 6 lbs N/M/yr (50 to 300 kg N/ha/yr) for warm-season turfgrasses. Unlike N application to other cropping systems, N applications to turf are made periodically throughout the growing season. General guidelines are to apply no more than 1 lb N/M/application/growing month. Applications are made as often as every 3 to 4 weeks (wk). The trend on high maintenance turf such as putting greens, however, is to fertilize lightly, 0.1 to 0.2 lbs N/M/application, but frequently, every 1 to 2 wk.

The main environmental concern with N fertilization is the potential leaching of nitrate (NO_3^-). Nitrate is essentially non-sorbed by soils and is free to move with downward flowing water. Nitrate leaching is a natural occurrence, and some level of nitrate leaching will occur on all soils. The goal in agricultural systems is to minimize the amount of nitrate leaching attributable to N fertilization. A major concern with nitrate leaching is drinking water contamination. The EPA has set a limit of 10 parts per million (PPM) NO_3-N as the drinking water standard.

The 10 PPM NO_3-N standard is valid to protect human health, but this value does not give any direction in determining environmental impacts. A common assumption is that leaching losses of N go directly to groundwater. Most golf courses and many agricultural lands are tile drained, so leachate intercepted by these drains is deposited in surface waters. Fresh water lakes and rivers are generally nutritionally limited by phosphorus, so nitrate levels have not been the main focus of concern. But, in estuaries and coastal systems, N often is the limiting nutrient to growth. Because many of these coastal system watersheds consist of highly permeable sandy soils, groundwater delivers the majority of land-derived N to these estuaries (Valiela et al. 1992). In these situations, the EPA drinking water standard provides no guidance on appropriate levels of nitrate in groundwater.

Many of the same factors that influence pesticide leaching are important in nitrate leaching. High levels of surface organic matter in a turf create a zone of intense microbial activity that will compete for available N, lowering the potential for downward transport of N. Turfgrasses have extremely high root densities in the surface 15 cm, which results in rapid uptake of available N. Preferential, or macropore, flow may move nitrate quickly from the turf surface to below the rootzone.

Fate of Nitrogen in Turf

What happens to a N fertilizer after its application to turf? How much of the applied N leaches to groundwater? These questions have received considerable scientific study, and answers to these questions are possible, but not always complete.

The age of the turf and fertilization history have much to do with the fate of applied N. A number of comprehensive studies of fertilizer N fate have shown that leaching losses are negligible (Miltner et al. 1996; Morton, Gold, and Sullivan 1988; Starr and DeRoo 1981). One of the earliest studies of N fate in turf was conducted by Starr and DeRoo (1981) in Connecticut. Their study focused on the distribution of N, applied as ^{15}N labeled $(NH_4)_2SO_4$, in the clippings, soil, thatch, and leachate, and focused on the effects of returning clippings on N dynamics. Starr and DeRoo used suction lysimeters installed 180 to 240 cm below the turf surface to sample for nitrate leaching. They found that nitrate concentrations ranged from 0.3 to 10 PPM NO_3-N during the 3-yr study. Average NO_3-N concentrations were 2 and 1.9 PPM where clippings were returned or removed, respectively. Samples from wells upstream of the plot area averaged 0.9 and 2.7 PPM NO_3-N, leading the authors to conclude that there was little leaching of NO_3-N from the turf plots. During the course of the study, only one leachate sample contained any ^{15}N, indicating that a very small portion of the applied fertilizer N was lost by leaching. The ^{15}N labeling allowed the authors to complete a mass balance of fertilizer N. They recovered between 64 and 76% of the applied N and attributed the losses to either ammonia volatilization or denitrification. Interestingly, more of the applied N was recovered from thatch than from soil, indicating that microbial activity and plant uptake were responsible for the immobilization of N in thatch. The mass balance for the labeled fertilizer reported leaching losses as a trace, i.e., too low to contribute to the mass balance.

Another comprehensive study of fertilizer fate was conducted by Miltner and colleagues (1996). They also observed that thatch immobilized much of the applied N, and leaching of labeled and nonlabeled N was very low, averaging less than 1 PPM NO_3-N for the 2-yr study. Miltner and colleagues also studied the effect of application timing on nitrate leaching, examining the popular practice of applying N as turf enters winter dormancy, so-called late fall fertilization. Labeled urea was applied on November 8 in Michigan, with rapid immobilization in the thatch and plant material. Of the 39 kg N/ha of quick-release urea applied on November 8, only 4.8 kg N/ha was recovered from the soil 18 d after application. The balance of the applied fertilizer, almost 88% of the applied N, was recovered in the thatch and plant tissue. In the Miltner study, ^{15}N recoveries gradually dropped with time, and at 750 d after treatment, the authors recovered 81 and 64% of the initial ^{15}N application. Like Starr and DeRoo, Miltner attributed the recovery losses to denitrification and/or ammonia volatilization.

Clipping Removal

Clipping removal was, and is, practiced by many homeowners, primarily to remove the debris after mowing. With the development of improved mulching mowers, however, there are not many reasons to continue this practice. Clipping disposal is costly, no longer accepted in landfills, and a waste of valuable nutrients that should be recycled. There are still a few situations in which clipping removal is justified—for example, golf course putting greens—but in general, this practice should fade away as more consumers understand that there are no benefits to this practice. Many fertilization recommendations are made, however, as if clippings are being removed; that is, N fertilizer recommendations are not sufficiently detailed to provide guidance for professionals or homeowners. The research of Starr and DeRoo (1981) showed that returning clippings provided considerable boost to the N economy of the turf. Returning clippings for 3 yr provided as much N to the turf as was derived from the soil fraction. By using ^{15}N labeled fertilizer, which subsequently labeled the clippings, they were able to determine whether N uptake came from soil organic matter or from clippings. They estimated that returning clippings for 3 yr was providing 0.23 kg N/ha/d for the turf. Based on a typical cool-season growth pattern, this would equate to 48 kg N/ha/yr (1 lb N/M/yr) from returning clippings.

The studies of Starr and DeRoo (1981), Miltner and colleagues (1996), and others (Gold et al. 1990; Morton, Gold, and Sullivan 1988) indicated that nitrate leaching from fertilized turf, when managed appropriately, was a minor occurrence. When N is applied to turf, its fate is limited to leaching, volatile losses, storage as soil organic matter, and use for plant growth. When clippings are returned, plant growth should require minimal N, leaving leaching, volatile losses, and storage as the primary fates of fertilizer N. Storage is an important mechanism for N utilization in turf and can explain much of the research data that shows minimal nitrate leaching from turf.

Porter and colleagues (1980) showed that turfs build soil organic matter, and thus organic N levels, for up to 30 yr after turf establishment. In particular, the first 10 yr of regular turf maintenance will see a substantial increase in soil organic matter, and thus soil N. Fertilization rates will need to be higher during the years after turf establishment to account for the increasing soil organic matter, which can be considered N storage. Porter estimated that up to 3 lbs N/M/yr could be stored in organic matter in the first years after establishment. After years of consistent maintenance, the storage function would slow and eventually stop. Porter reported that after 10 yr, soil N storage slowed to less than 1 lb N/M/yr reaching near zero, or equilibrium, after 30 yr. Therefore, many studies that report extremely low levels of nitrate leaching from relatively high N application rates do so because the storage function is immobilizing much of the applied N.

How much nitrate leaching should be expected if a turf is maintained at a uniform level of fertilization for a long period of time? Frank and colleagues (2006) attempted to answer that question. Beginning in 1998, N was applied at a rate of 2 or 5 lbs N/M/yr (98 or 245 kg N/ha/yr) to Kentucky bluegrass turf that had been established in 1989 and fertilized consistently since establishment. Starting in 2000, the leachate was monitored continually for nitrate levels. Nitrate levels in the leachate of the turf receiving 5 lbs N/M/yr were consistently above 10 PPM NO_3-N, and in the fall of 2001, nitrate concentrations went above 20 PPM and were not below that value for any subsequent sampling, yielding a flow weighted average leachate nitrate concentration of 21 PPM NO_3-N during the study. The turf receiving 2 lbs N/M/yr never had nitrate concentrations that exceeded the 10 PPM drinking water standard, but had a flow weighted average of 4 PPM NO_3-N. Although this is a low value, it is much higher than previous studies, even though the N rate was relatively low. Studies cited earlier would suggest that 5 lbs N/M/yr should not cause excessive nitrate leaching, but in the study by Frank and colleagues (2006), it is the cumulative effect of fertilizing at a high rate of N for several years that eventually leads to high levels of nitrate leaching. As was hypothesized by Porter, after a period of building organic matter, equilibrium is reached and further net immobilization of N becomes negligible. When this point is reached, nitrate leaching becomes the likely avenue for removal of N beyond the needs of the plant community.

How much of a factor are volatile losses of N? This is a difficult question to answer, and one where more research is needed. As mentioned earlier, every mass balance study of N has failed to achieve a quantitative recovery of applied N. The researchers have generally attributed the lack of complete recovery to volatile N losses. Horgan, Branham, and Mulvaney (2000), however, measured denitrification in the field and concluded that denitrification losses could not account for the incomplete recovery of N. Another loss mechanism, perhaps gradual losses of ammonia from plants or decaying residues, must be responsible for the decreased recovery of N with time in field experiments.

What about the form of the fertilizer? Should slow-release fertilizers be used exclusively? Does the use of slow-release fertilizers substantially decrease nitrate-leaching losses? Common sense would suggest that nitrate-leaching losses will be lower when slow-release fertilizer sources are used, but the issue is more complicated. Several studies have shown increased leaching losses when quick-release fertilizers are used compared with slow-release fertilizers (Engelsjord and Singh 1997; Guillard and Kopp 2004; Petrovic, Hummel, and Carroll 1986). For example, Guillard and Kopp (2004) compared ammonium nitrate, a polymer-coated fertilizer, and an organic fertilizer (Sustane, a composted poultry manure product) all applied at a rate of 3 lbs N/M/yr. The ammonium nitrate source produced a flow weighted leachate nitrate concentration of 4.6 mg L^{-1}, whereas the polymer-coated fertilizer and the organic N sources produced only 0.57 and 0.31 mg L^{-1}, respectively. The annual percentage loss of each fertilizer was 16.8, 1.7, and 0.6% for ammonium nitrate, polymer-coated, and organic N sources, respectively. These data clearly indicate the value of using slow-release fertilizers. Yet, the leaching losses observed from ammonium nitrate must be tempered by the findings of other researchers. Miltner and colleagues (1996) observed very low levels of nitrate leaching from fertilization with urea, a quick-release source, at a rate of 4 lbs N/M/yr. Bowman and colleagues

(1989) showed that turfgrasses rapidly immobilized inorganic N sources, stating that inorganic N is depleted within 2 to 4 d after application. Bowman's data indicate that fertilizer source should affect nitrate leaching if a significant rain event occurs within 2 d of a fertilizer application. After this period, the inorganic N has been immobilized by the turfgrass, and associated soil microorganisms should be no more susceptible to leaching than slow-release sources of N.

The data of Guillard and Kopp (2004) and Mangiafico and Guillard (2006) call into question the practice of late fall fertilization when practiced with water-soluble N sources. They saw a steady increase in nitrate leaching when fertilizer was applied after September 15. During 2 yr of monitoring, between 30 and 41% of the N applied between October 15 and December 15 was recovered in the leachate.

The data of Guillard and Kopp in Connecticut are in sharp contrast with the data of Frank and colleagues (2006) and Miltner and colleagues (1996), who saw low and very low levels, respectively, of nitrate leaching from late season fertilizer applications in Michigan. When the Guillard and Frank data are viewed together, however, it should be clear that although the source of N can influence episodic nitrate leaching events, from a longer-term perspective, the annual rate of N fertilization will have the biggest impact on nitrate leaching. In other words, turfs fertilized at 4 lbs of N/M/yr with all slow-release N will produce more nitrate leaching than turfs fertilized at 2 lbs N/M/yr with a quick-release N source when viewed for more than just a 1- or 2-yr study. The Guillard data make a convincing argument that quick-release fertilizers should not be used in late fall fertilization, but recall that where clippings are returned, added N has three possible fates—leaching, storage, or volatilization. With time, the storage function will approach zero, leaving only leaching and volatilization as the possible fates of applied N. Using slow-release forms of N will decrease episodic nitrate leaching events, but in fragile environments, the best way to decrease nitrate leaching is to apply less N.

Summary

Pesticide and nutrient leaching can cause considerable harm to groundwater and, to some extent, surface water supplies. Turfgrass managers must adopt management practices that lessen the potential for leaching. A healthy turf provides considerable protection against leaching because of the high levels of organic matter and associated microbial activity that serve to immobilize and degrade applied pesticides and nitrates. Excessive irrigation or large rain events, which lead to preferential or macropore flow, can mitigate these advantages and push solutes below this zone of microbial activity. It is dangerous to generalize when discussing pesticides because each pesticide has different characteristics that affect its distribution and fate, but most pesticides currently used in turf present fairly low risks of producing significant groundwater contamination. Exceptions will occur, but pesticide applications in turf are on a small scale—two home lawns in a neighborhood or 3 acres of putting green on a 150-acre property—and therefore, the total quantity of pesticide reaching groundwater, even in a worst-case scenario, generally is not significant.

A healthy turf has a great capacity to use applied nutrients. Nitrate leaching may, however, present problems in some segments of the turf industry where N fertilization rates have not been lowered to account for turf age and clipping return.

Literature Cited

Bowman, D. C., J. L. Paul, W .B. Davis, and S. H. Nelson. 1989. Rapid depletion of nitrogen applied to Kentucky bluegrass turf. *J Amer Soc HortSci* 114:229-233.

Cheng, H. H. (ed.). 1990. *Pesticide in the Soil Environment: Processes, Impacts, and Modeling.* SSSA Book Ser, No. 2, Soil Science Society of America (SSSA), Madison, Wisconsin.

Dell, C. J., C. S. Throssell, M. Bischoff, and R. F. Turco. 1994. Estimation of sorption coefficients for fungicides in soil and turfgrass thatch. *J Environ Qual* 23:92-96.

Engelsjord, M. E. and B. R. Singh. 1997. Effects of slow-release fertilizers on growth and on uptake and leaching of nutrients in Kentucky bluegrass turfs established on sand-based root zones. *Can J Plant Sci* 77:433-444.

Frank, K. W., K. M. O'Reilly, J. R. Crum, and R. N. Calhoun. 2006. The fate of nitrogen applied to a mature Kentucky bluegrass turf. *Crop Sci* 46:209-215.

Fungicides hit growth spurt in T&O markets. 2005. *Grounds Maintenance* 40(8):6.

Gardner, D. S. and B. E. Branham. 2001a. Mobility and dissipation of ethofumesate and halofenozide in turfgrass and bare soil. *J Agric Food Chem* 49:2894-2898.

Gardner, D. S. and B. E. Branham. 2001b. Effect of turfgrass cover and irrigation on soil mobility and dissipation of mefanoxam and propiconazole. *J Environ Qual* 30:1612-1618.

Gardner, D. S., B. E. Branham, and D. W. Lickfeldt. 2000. Effect of turfgrass on soil mobility and dissipation of cyproconazole. *Crop Sci* 40:1333-1339.

Gold, A. J., T. G. Morton, W. G. Sullivan, and J. McClory. 1988. Leaching of 2,4-D and dicamba from home lawns. *Water Air Soil Pollut* 37:121-129.

Gold, A. J., W. R. DeRagon, W. M. Sullivan, and J. L. Lemunyon. 1990. Nitrate-nitrogen losses to groundwater from rural and suburban land uses. *J Soil Water Cons* 45:305-310.

Guillard, K. and K. L. Kopp. 2004. Nitrogen fertilizer form and associated nitrate leaching from cool-season lawn turf. *J Environ Qual* 33:1822-1827.

Gustafson, D. I. 1989. Groundwater Ubiquity Score — A simple method for assessing pesticide leachability. *Env Tox Chem* 8:339–357.

Harrison, S. A., T. L. Watschke, R. O. Mumma, A. R. Jarrett, and G. W. J. Hamilton. 1993. Nutrient and pesticide concentrations in water from chemically treated turfgrass. Pp. 191–207. In K. D. Racke and A. R. Leslie (eds.). *Pesticides in Urban Environments: Fate and Significance*. American Chemistry Society, Washington, D.C.

Helling, C. S., P. C. Kearney, and M. Alexander. 1971. Behavior of pesticide in soils. *Adv Agron* 23:147–239.

Horgan, B. P., B. E. Branham, and R. L. Mulvaney. 2002. Mass balance of ^{15}N applied to Kentucky bluegrass including direct measurement of denitrification. *Crop Sci* 42:1595–1601.

Horst, G. L., P. J. Shea, N. Christians, D. R. Miller, C. Stuefer-Powell, and S. K. Starrett. 1996. *Crop Sci* 36:362–370.

Jarvis, N. J., L. Bergstrom, and E. Dik. 1991. Modeling water and solute transport in macroporous soil. II. Chloride breakthrough under non-steady flow. *J Soil Sci* 42:71–81.

Kerek, M., R. A. Drijber, W. L. Powers, R. C. Shearman, R. E. Gaussoin, and A. M. Streich. 2002. Accumulation of microbial biomass within particulate organic matter of aging golf greens. *Agron J* 94:455–461.

Lickfeldt, D. W. and B. E. Branham. 1995. Sorption of nonionic organic compounds by Kentucky bluegrass leaves and thatch. *J Environ Qual* 24:980–985.

Malone, R. W., M. J. Shipitalo, R. D. Wauchope, and H. Sumner. 2004. Residual and contact herbicide transport through field lysimeters via preferential flow. *J Environ Qual* 33:2141–2148.

Mangiafico, S. S. and K. Guillard. 2006. Fall fertilization timing effects on nitrate leaching and turfgrass color and growth. *J Environ Qual* 35:163–171.

Miltner, E. D., B. E. Branham, E. A. Paul, and P. E. Rieke. 1996. Leaching and mass balance of 15N-labeled urea applied to a Kentucky bluegrass turf. *Crop Sci* 36:1427–1433.

Morton, T. G., A. J. Gold, and W. M. Sullivan. 1988. Influence of overwatering and fertilization on nitrogen losses from home lawns. *J Environ Qual* 17:124–130.

Petrovic, A. M., N. W. Hummel, and M. J. Carroll. 1986. Nitrogen source effects on nitrate leaching from late fall nitrogen applied to turfgrass. Pg. 137. In *Agronomy Abstracts*. ASA, Madison, Wisconsin.

Petrovic, A. M., R. G Young, C. A Sanchirico, and D. J. Lisk. 1993. Downward migration of trichlorfon insecticide in turfgrass soils. *Chemosphere* 27:1273–1277.

Petrovic, A. M., W. C. Barrett, I. Larsson-Dovach, C. M. Reid, and D. J. Lisk. 1996. The influence of a peat amendment and turf density on downward migration of metalaxyl fungicide in creeping bentgrass sand lysimeters. *Chemosphere* 33:2335–2340.

Porter, K. S, D. R. Bouldin, S. Pacenka, R. S. Kossack, C. A. Shoemaker, and A. A. Pucci, Jr. 1980. *Studies to Assess the Fate of Nitrogen Applied to Turf: Part I. Research Project Technical Complete Report.* OWRT Project A-086-NY, Cornell University Center for Environmental Research, Ithaca, New York.

Potter, D. A. 1993. Pesticide and fertilizer effects on beneficial invertebrates and consequences for thatch degradation and pest outbreaks in turfgrass. ACS Symposium Series 522:331–343.

Raturi, S., M. J. Carroll, and R. L. Hill. 2003. Turfgrass thatch effects on pesticide leaching: A laboratory and modeling study. *J Environ Qual* 32:215–223.

Roy, J. W., G. W. Parkin, and C. Wagner-Riddle. 2000. Water flow in unsaturated soil below turfgrass: Observations and LEACHM (within EXPRESS) predictions. *Soil Sci Am J* 64:86–93.

Roy, J. W., J. C. Hall, G. W. Parkin, C. Wagner-Riddle, and B. S. Clegg. 2001. Seasonal leaching and biodegradation of dicamba in turfgrass. *J Environ Qual* 30:1360–1370.

Smith, A. E. and D. C. Bridges. 1996. Movement of certain herbicides following application to simulated golf course greens and fairways. *Crop Sci* 36:1439–1445.

Snyder, G. H. and J. L. Cisar. 1997. Mobility and persistence of turfgrass pesticides in a USGA-type green: IV. Dicamba and 2,4-D. *Int Turfgrass Soc Res J* 8:205–211.

Starr, J. L. and H. C. DeRoo. 1981. The fate of nitrogen fertilizer applied to turfgrass. *Crop Sci* 21:531–536.

U.S. Environmental Protection Agency (USEPA). 2004. Pesticide Industry Sales and Usage. http://www.epa.gov/oppbead1/pestsales/01pestsales/market_estimates2001.pdf (30 November 2006)

Valiela, I., K. Foreman, M. LaMontagne, D. Hersh, J. Costa, P. Peckol, B. DeMeo-Anderson, C. D'Avanzo, M. Babione, C. H. Sham, J. Brawley, and K. Lajtha. 1992. Couplings of watersheds and coastal waters: Sources and consequences of nutrient enrichment in Waquiot Bay, Massachusetts. *Estuaries* 15:443–457.

Wauchope, R. D., T. M. Butler, A. G. Hornsby, P. W. M. Augustijn-Beckers, and J. P. Burt. 1992. The SCS/ARS/CES pesticide properties database for environmental decision-making. *Rev Environ Contam Tox* 123:1–155.

Wauchope, R. D., S. Yeh, J. B. Linders, R. Kloskowski, K. Tanaka, B. Rubin, A. Katayama, W. Kördel, Z. Gerstl, M. Lane, and J. B. Unsworth. 2002. Pesticide soil sorption parameters: Theory, measurement, uses, limitations and reliability. *Pest Manag Sci* 58:419–455.

Wu, L., R. L. Green, G. Liu, M. V. Yates, P. Pacheco, J. Gan, and S. R. Yates. 2002a. Partitioning and persistence of trichlorfon and chlorpyrifos in a creeping bentgrass putting green. *J Environ Qual* 31:889–895.

Wu, L., G. Liu, M. V. Yates, R. L. Green, P. Pacheco, J. Gan, and S. R. Yates. 2002b. Environmental fate of metalaxyl and chlorothalonil applied to a bentgrass putting green under southern California conditions. *Pest Manag Sci* 58:335–342.

8

Nutrient and Pesticide Transport in Surface Runoff from Perennial Grasses in the Urban Landscape

K. W. King and J. C. Balogh

Introduction

Turfgrass provides benefits to society and the environment that are functional (e.g., erosion and air pollution control, wildlife habitat, dust prevention, noise abatement, and heat dissipation), recreational (e.g., a safe sport/entertainment surface), and aesthetic (e.g., increased property values and quality of life) (Beard and Green 1994). But both the perception (Kohler et al. 2004; Peacock, Smart, and Warren-Hicks 1996; CAST 1985; Shuman 2002; Smith and Bridges 1996) and the potential (Balogh and Walker 1992) for turfgrass systems to degrade the natural resource base do exist. Water quality, resource allocation, and environmental issues specifically related to turfgrass management on existing and proposed developments include

- Excessive use of potable water for irrigation, especially in arid and semiarid climates

- Potential movement of nutrients and pesticides to surface and groundwater

- Direct exposure of beneficial soil organisms, wildlife, and aquatic systems to pesticides and fertilizers

- Loss of soil and sediment during renovation and construction periods (especially true of urbanizing developments)

- Disturbance of the water balance during development or renovation by converting vegetation, changing topography, and increasing the use of water for irrigation

- Disturbance and loss of wetlands, and

- Disturbance and change of existing land use patterns.

Hydrology and Runoff

To better grasp the surface water issues related to turfgrass, an understanding of the factors that govern water flow and transport is required. The composite of these factors is known as hydrology, defined as the study of the physical and chemical properties of water (primarily on or near the earth's surface) and its circulation through the hydrologic cycle (Maidment 1993). The hydrologic cycle (Figure 8.1) includes precipitation, evaporation, infiltration, groundwater flow, runoff, streamflow, and the transport of dissolved or suspended pollutants in the flowing water. The focus from this point forward will be on the surface runoff component of the hydrologic cycle as it relates to nutrient and pesticide transport from turfgrass.

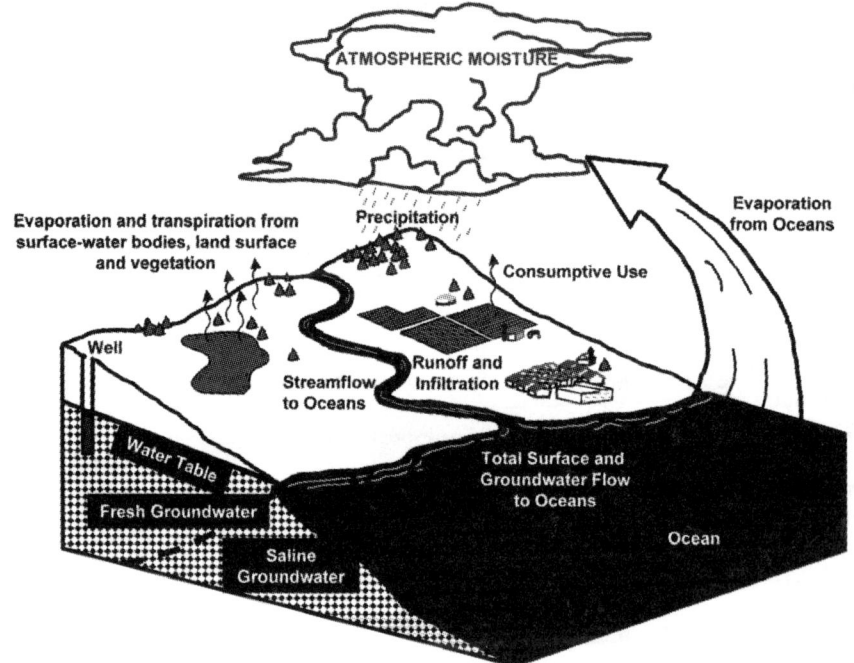

Figure 8.1. The hydrologic cycle. (USGS 1984)

The factors impacting surface runoff and pollutant transport include climate, site and soil conditions, and management. The most significant climate factors are precipitation (volume, intensity, and duration), evapotranspiration, and temperature. Site and soil conditions also are critical to the potential for offsite movement of sediment, nutrients, and pesticides. The most significant site and soil conditions are soil texture and organic matter content, bulk density, hydraulic conductivity, thatch layer presence, landscape slope, and proximity to water resources. The most critical factor is management. Management includes irrigation, drainage, fertilizer and pesticide application, and cultural practices.

Climate

Runoff occurs when the rate of precipitation is greater than the infiltration rate of the soil. The duration and total volume of rain is directly related to runoff volume. Increasing intensity of rain raises the flow of runoff water and, consequently, the energy available for sediment, nutrient, and pesticide extraction and transport. The more intense the rainfall, the less time required to initiate storm runoff. Rainfall intensity also may affect depth of surface interaction with nutrients and pesticides.

The greatest nutrient and pesticide concentration in runoff generally occurs during the first major runoff event following application (Smith and Bridges 1996). The greatest surface loss of nutrients and pesticides occurs when storm runoff occurs shortly after application (Schueler 1995). The time between application and runoff is critical in determining total nutrient and pesticide losses in surface runoff. As time between application and runoff increases, a greater proportion of the nutrient and/or pesticide will be unavailable for transport.

Evapotranspiration (ET) is the amount of applied water transpired by plants and evaporated from the soil surface. ET is regulated by meteorological conditions, including solar radiation, wind speed, air temperature, and moisture gradients. Soil and plant resistances also play a major role in regulating ET under dry soil conditions. ET impacts runoff by influencing the quantity of water in the soil. Higher antecedent soil water content may (1) increase the potential for runoff, (2) decrease the time to onset of runoff, and (3) decrease subsurface leaching of soluble nutrients and pesticides prior to runoff starting (Baird et al. 2000; Balogh and Anderson 1992; Cole et al. 1997; Walker and Branham 1992).

Temperature affects transport of nutrients and pesticides by impacting the microbial activity in the soil. Higher temperatures generally increase microbial-mediated processes. Depending on soil moisture, aeration, and plant growth conditions, increased temperatures may or may not increase exposure to losses in runoff water.

Soil and Site Conditions

Soil texture and organic matter content affect runoff potential (Becker et al. 1989; Walker et al. 1990). Texture affects erodibility, the potential for particle transport, and chemical enrichment factors. Texture and organic matter content affect adsorption and mobility of nutrients and pesticides. Runoff is usually greater from finer-textured soils. Initiation of runoff takes longer and volumes of runoff are generally less on sandy soils. This decreases initial surface losses of soluble and sediment-bound nutrients and pesticides. In the case of golf courses, fairways and roughs with finer-textured soils have greater potential for runoff than sandy greens and tees. Drainage from greens and tees, however, may contribute to runoff on fairways and roughs (King et al. 2006). High rates of infiltration, rapid percolation, and high hydraulic conductivity decrease the volume and velocity of surface runoff. A decrease in runoff and the elimination of sediment movement from turfgrass systems decreases potential chemical losses to surface water.

Soil structure and stability of soil aggregates affect infiltration rates, crusting, effective depth for chemical entrainment in runoff, potential for chemical movement on sediment, and chemical enrichment in sediment. Surface crusting and compaction of soil by traffic decreases infiltration rates and time to runoff, and increases runoff volume and initial concentrations of soluble and sediment-bound nutrients and pesticides in runoff water. Decreasing porosity or increasing bulk density heightens runoff losses. This process occurs by decreasing infiltration of water and chemicals into the soil and increasing the potential for sediment-bound losses from exposed soil surfaces.

Thatch is a combination of living, dead, and partially decomposed plant residues located above the soil surface. Thatch has been shown to influence infiltration rates (Taylor and Blake 1982) and water storage capacity (Zimmerman 1973). Thatch also has been reported to play a significant role in the sorption of pesticides (Branham and Wehner 1985; Hurto, Turgeon, and Cole 1979), nutrient sequestration (Engelsjord, Branham, and Horgan 2004), and soil fertility relationships (Nelson, Turgeon, and Street 1980). The thatch organic layer intercepts applied chemicals and water before they enter the mineral soil. The rate and extent of movement through the thatch layer is decreased in many instances by the increased immobilization of (1) soil microorganisms, (2) uptake by turfgrass roots, and (3) adsorption. Thatch decreases chemical movement in runoff water.

Both proximity to surface water and land slope have a direct relationship with runoff potential. Increasing slope may increase (1) runoff velocity and volume, (2) soil detachment and erosion, and (3) the effective surface depth for extraction of nutrients and pesticides from the root zone. Distance affects whether chemicals sorbed to sediment or dissolved in runoff water reach receiving surface water. The size and type of the surface water body influences the impact of the incoming contaminant. Surface waters at greatest risk of contamination with nutrients and pesticides are (1) small, shallow ponds and streams immediately adjacent to managed turfgrass areas, (2) surface waters unprotected by vegetated buffer zones or structural barriers that divert or slow the flow of runoff discharging into the receiving water, and (3) surface waters directly in contact with drainage channels and drain tiles.

Turfgrass Management

Management is an integral part of developing and maintaining high quality, efficient turfgrass systems. Judicious irrigation, fertilization, mowing, aeration, and pest control all contribute to maintenance of healthy turfgrass (Balogh and Anderson 1992; Balogh and Watson 1992; Turgeon 1996; Walker and Branham 1992; Witteveen and Bavier 1999). Turfgrass is the most intensively managed system in the urban landscape (Shuman, Smith, and Bridges 2000; Smith and Bridges 1996; Walker et al. 1990). Management is essential for maintaining the desired plant growth. The intensity of turfgrass management, however, is dependent on the type of turfgrass and its intended use (Rieke and Lyman 2002). Types of turfgrass include warm-season and cool-season species. The properties of cool- and warm-season grasses are provided by Beard (1973) and DiPaola and Beard (1992). Selecting the appropriate species is important for minimizing the environmental impacts from management (Bowman, Cherney, and Rufty 2002). The primary turfgrass management factors that impact runoff are water management, nutrient and pesticide application, and cultural practices such as mowing and aeration.

Water management is one of the dominant factors controlling turfgrass management decisions and turfgrass quality. Irrigation and drainage practices have a direct relationship with cultural, nutrient, pest, and structural management practices. The annual cycle of water and sediment movement on turfgrass reflects the integrated soil, water, plant, and climate continuum. Controlling the quantity and timing of irrigation is a major factor in conserving the quality and quantity of water resources (Balogh and Watson 1992; Watson 1994). The technical aspects of irrigation and water conservation practices have been reviewed in detail by Murphy (2002); Balogh and Watson (1992); Carrow (1994); Carrow, Shearman, and Watson (1990); Kneebone, Kopec, and Mancino (1992); and Pira (1997). Integrated water-use plans for turfgrass must consider water stress, water use, and water conservation. Water use affects both water quality and quantity when water resources are limited or drought conditions occur.

Periodic applications of nutrients and pesticides are essential to maintaining high-quality turfgrass (Branham, Kandil, and Meuller 2005). Runoff losses of pesticides and fertilizers are directly related to their timing and rate of application, formulation, chemical properties, and placement. Several guidelines for basic turfgrass management are available through local extension services. Application of nutrients should be made based on the ability of the plant to use the nutrient. Application of nutrients and pesticides should be avoided when precipitation is imminent. Nutrient and pesticide formulation (liquid versus granular) and properties (adsorption, water solubility, etc.) also can impact the potential for offsite transport. Application placement also can have a marked effect on transport potential.

Cultural practices, such as mowing and aeration, also affect the potential for runoff. Turfgrass canopies and thatch layers can decrease soluble pollutant runoff losses by (1) increasing infiltration, (2) increasing the time to runoff, (3) decreasing runoff volumes, (4) decreasing erosion and sediment transport, and (5) increasing adsorption. Pollutant losses in runoff may be increased when soluble forms are leached from decomposing clippings or dead turfgrass surfaces directly into runoff water. Clippings should be evenly spread in areas where nitrogen released during decomposition will be taken up by growing vegetation. Cultivation and aeration of turfgrass soils will maintain high rates of infiltration and soil drainage. High rates of infiltration and percolation decrease losses in surface flow. Movement of soluble pollutants into subsurface drainage systems, however, may reemerge in drainage effluent (King et al. 2006).

Runoff and Pollutant Discharges

Discharge Volume

Often, runoff is presented in terms of a runoff coefficient. The runoff coefficient is defined as the fraction of precipitation measured as runoff. A few studies present runoff coefficients or substantial data to calculate the runoff coefficient (Table 8.1). The objectives of these studies varied, accounting for the range in scale (plots to watersheds) and land use/vegetation. The studies represent a range of locations in the United States and abroad. Studies from wooded areas and grazed pasture/rangeland have been included for comparison, but do not represent a comprehensive review of the data available for those land uses. Studies conducted with rainfall simulators as well as natural events have been reported.

The range of runoff coefficients from turfgrass and residential classification varies from 0.0028 to 0.77. Larger coefficients are consistently reported from studies in which antecedent soil moisture is high. Based on the studies reviewed, the impact of scale on runoff coefficient is not conclusive. The average runoff coefficient for all turfgrass/residential studies (excluding forested, agricultural, commercial, and industrial areas) conducted on areas less than 500 m^2 is 0.16, whereas the runoff coefficient is 0.20 for study areas greater than 500 m^2. The 4% difference might be a result of impervious areas found in the larger study areas.

Nutrients

Nutrient concentrations from turfgrass managed as a golf course or home lawn are limited (Table 8.2). These studies clearly indicate that mean nitrogen concentrations in the form of ammonium, nitrate, or total nitrogen are well below any level of major concern. Phosphorus concentrations reported in surface runoff from lawns and golf courses, however, generally are greater than concentrations associated with eutrophication. The average reported concentrations from these studies on home lawns and golf courses is 0.4 mg L^{-1} NH$_4$-N, 0.4 mg L^{-1} NO$_3$-N, 3.4 mg L^{-1} Total Nitrogen, 0.2 mg L^{-1} Dissolved Reactive Phosphorus, and 1.0 mg L^{-1} Total Phosphorus. These concentrations are comparable to those reported for native grasses and woodlands. Cohen and colleagues (1999) reported that a survey of runoff on 17 golf courses in the United States did not contain any instances of NO$_3$-N exceeding the MCL standard of 10 mg L^{-1}. The median NO$_3$-N value recorded in that survey was 0.38 mg L^{-1}, a value comparable to the mean value reported here.

Nutrient loads in surface water have been reported by several researchers (Table 8.3). Quantifying loads from turfgrass systems is critical from both an environmental/regulatory perspective, especially in light of total maximum daily loads (TMDL), and an economic standpoint. Loads from turfgrass and residential areas reported in these studies generally are less than those from agriculture, but greater than those reported for forested catchments or native prairies. The loads reported here generally increased as the scale of study increased.

Pesticides

Approximately 2 to 4 kilograms of pesticides per hectare are applied to managed turfgrass each year (Schueler 1995). Pesticides applied on golf courses account for approximately 1% of the pesticide use in the United States (Cohen 1995). Only recently have advances in science and technology permitted the detection of pesticide residues in the part per billion range.

When pesticide application is necessary, it is important to consider the properties of the chemical. Pesticide properties of the most significance with respect to runoff/transport potential are water solubility, soil adsorption, and persistence in the soil (Mugaas et al. 1991). As summarized in Table 8.4, the most mobile and persistent pesticides used in the urban landscape are being measured only in small quantities and rarely exceed any health standard. Similarly, Cohen (1998) compiled surface runoff data from 36 golf courses. In the Cohen study, only 31 of 90 organic compounds tested for were actually detected, and of those only 0.29%, or 8 of the 2731 samples, exceeded the health standard. Based on the available data, application of pesticides on healthy turfgrass poses minimal risk to surface water bodies.

Turfgrass Management System Plans

Turfgrass management system (TMS) plans are designed to maintain high-quality turfgrass and protect water and soil resources. TMS plans are multiple, integrated, best management practices involving irrigation, fertilization, pest and disease control, soil and water conservation practices, and other agronomic practices related to turfgrass management. A best management practice (BMP) may be defined as any practice, method, criteria, or structure that controls, prevents, or decreases the offsite transport of pollutants such as sediment, nutrients, and/or pesticides. Several BMPs specific to the turfgrass industry have been studied (Table 8.5).

Turfgrass managers often are faced with multiple options for managing turfgrass. They are asked to balance turfgrass quality and growth with climate, soil, vegetative conditions, and management practices. Their choice of practice is critical for controlling and/or decreasing surface runoff and pollutant transport. The most significant management practices used to control or decrease runoff losses from established turfgrass are (1) maintenance of healthy turfgrass (Watschke 1990), (2) control of irrigation scheduling and volume based on plant requirement (Balogh and Watson 1992; Bastug and Buyuktas 2003; Murphy 2002), (3) establishment and maintenance of buffer zones (Bell and Moss 2005; Cole et al. 1997), and (4) protection of trees and wetlands (Kohler et al. 2004; Reicher et al. 2005).

Directions for the Future

Water quality research and data collection programs often form the basis from which related legislation is derived. In the absence of real data, computer models often are used to populate water quality databases. Mathematical computer models offer a convenient and economical means to evaluate alternative land management practices without, in theory, physically collecting data from the site of interest. Models allow for environmental impact analysis before construction, as well as evaluating impacts of long-term management decisions before implementation. National Pollutant Discharge Elimination Systems and TMDL legislation depend on accurate model prediction for assessment.

Several modeling studies of turfgrass systems using existing agricultural models and simulation tools developed specifically for turfgrass have been documented (Durborow et al. 2000; Haith

2001, 2002; Haith and Rossi 2003; King and Balogh 1997, 1999; Ma et al. 1999a, b; Mankin 2000; Morioka and Cho 1992; Roy et al. 2001; Schwartz and Shuman 2005; Smith et al. 1993; Starrett and Starrett 2000). The results of these studies show promise, but gaps in basic science of turfgrass systems and model deficiencies limit the robustness of the predictions. Long-term reliable data from a range of geographic locations are necessary to populate the database for model development, enhancement, and reliability.

Understanding the role of turfgrass systems in the larger watershed continues to be important. Previous studies have addressed runoff volume and nutrient and pesticide loss, but these studies generally focused on small areas from plots up to individual greens or fairways (Cohen et al. 1999). Studies on this scale are valuable, but they may or may not represent the diversity and interconnectedness associated with a complete turfgrass system and how that system functions in the watershed. As presented here, a handful of studies containing data from golf course and/or urban watersheds do exist. Yet more often than not, these studies are based on limited grab samples or conducted for short durations. Long-term (more than five years) data from diverse geographic locations are required before any more thorough understanding of turfgrass management impacts on the environment can be expected.

Conclusions

The body of available knowledge on surface runoff quantity and chemistry from urban landscapes has improved over the last two decades, but more information is required before any overarching, widespread conclusions can be made. Of the studies referenced here, a reasonable case could be made that runoff volume is generally small, and losses of nutrients and pesticides are less than those from agriculture. Yet, more geographically diverse, long-term data sets on cool- and warm-season grasses in well-defined catchments under natural conditions would strengthen the argument.

Table 8.1. Rainfall-runoff coefficients from turfgrass, residential, and forest land use classifications

Reference	Land use	Area	Slope	C	Duration	Study location
Gaudreau et al. 2002	Common bermudagrass	6 m²	8.5%	0.65	8 events	College Station, TX
Chichester 1977	Poverty grass Bluegrass Bluegrass Alfalfa/orchardgrass mix	8 m²	23%	0.0046 0.0065 0.0028 0.0028	4 years	Coshocton, OH
Cole et al. 1997‡	Bermudagrass	8.8 m²	6%	0.04–0.16 dry soil 0.49–0.80 wet soil	2 events	Stillwater, OK
Schwartz and Shuman 2005	Tifway bermudagrass	25.2 m²	5%	0.073	4 years	Griffin, GA
Shuman 2002‡	Tifway bermudagrass	25.2 m²	5%	0.25	4 events	Griffin, GA
Smith and Bridges 1996‡	Tifway bermudagrass	27.4 m²	5%	0.48	4 events	Griffin, GA
Morton, Gold, and Sullivan 1988	90% Kentucky bluegrass; 10% red fescue	32 m²	2–3%	0.0098	2 years	Kingston, RI
Easton and Petrovic 2004	80% Kentucky bluegrass; 20% perennial ryegrass	37.2 m²	7–9%	0.083	33 events	Ithaca, NY

Continued on next page

Table 8.1. (continued) Rainfall-runoff coefficients from turfgrass, residential, and forest land use classifications

Reference	Land use	Area	Slope	C	Duration	Study location
Erickson et al. 2001	St. Augustine grass	47.5 m^2	10%	0.004	1 year	Ft. Lauderdale, FL
Timmons and Holt 1977	Native prairie	89.6 m^2	5%	0.051	5 years	Big Stone County, MN
Graczyk et al. 2003	Lawn	90.5 m^2	14%	0.02	24 events	Northern WI
	Lawn	40.9 m^2	17%	0.006	14 events	
	Lawn	7.9 m^2	14%	0.004	11 events	
	Lawn	100.3 m^2	5%	0.10	25 events	
	Woods	66.9 m^2	15%	0.0007	6 events	
	Woods	39.0 m^2	16%	0.0006	14 events	
	Woods	12.1 m^2	16%	0.006	19 events	
	Woods	37.2 m^2	24%	0.0006	11 events	
	Woods	35.8 m^2	23%	0.0003	5 events	
	Woods	33.9 m^2	22%	0.0008	12 events	
Linde and Watschke 1997‡	Bentgrass	123.5 m^2	9-11%	0.0192	10 events	State College, PA
	Perennial ryegrass			0.0468		
Westerstrom and Singh 2000*	Grass	200 m^2	2%	0.615	5 years	Lulea, Sweden
Birdwell 1995	Golf green (bermudagrass)	0.025 ha	10%	0.11	3 months	College Station, TX
	Golf fairway (bermudagrass)	1.57 ha	11%	0.14		

Reference	Land use	Area	%		Events	Location
Owens, Van Keuren, and Edwards 1998	Summer grazed pasture (orchardgrass/tall fescue mix)	0.45 ha	12%	0.082	8 years	Coschocton, OH
	Winter grazed pasture (orchardgrass/tall fescue mix)	1.6 ha	15%	0.063		
Smith et al. 1992	Native tall grasses	1.6 ha	4%	0.091	57 events	El Reno, OK
	Old world bluestems (Ischaemum complex)	2.7 ha	8%	0.151	40 events	Woodward, OK
	Native short/mid grasses	2.9 ha	8%	0.084	27 events	Woodward, OK
Dennis 1986	Residential (43.4% lawns; 41% forest)	3.5 ha	0–20%	0.77	8 events	Augusta, ME
	Forest (97.3% forest; 2.2% lawns)	2.4 ha	0–20%	0.61		
Olness et al. 1975	Grazed rangeland	9.6 ha	2.7%	0.18	15 events	Chichasha, OK
	Grazed rangeland	11.0 ha	2.7%	0.17	23 events	
	Grazed rangeland	7.8 ha	3.2%	0.31	23 events	
	Grazed rangeland	11.1 ha	3.5%	0.27	23 events	
King et al. 2001	Golf course subarea	29 ha	5–8%	0.136	13 months	Austin, TX
Oltmann and Shulters 1989	Industrial	112 ha	8.0%	0.18	47 events	Fresno, CA
	Single dwelling	38 ha	7.9%	0.21	42 events	
	Multiple dwelling	1.7 ha	7.0%	0.39	88 events	
	Commercial	25 ha	13.8%	0.93	49 events	
Selbig et al. 2004	57% ag; 22% woodland; 4% residential; 17% other	19.9 km^2	--	0.052	42 events	Dane County, WI

Continued on next page

Table 8.1. (continued) Rainfall-runoff coefficients from turfgrass, residential, and forest land use classifications

Reference	Land use	Area	Slope	C	Duration	Study location
Pekarova and Pekar 1996	Lesser Carpathians					Slovakia
	10% ag; 90% forest	7.25 km²	265†	0.32	2 years	
	10% ag; 90% forest	37.8 km²	426†	0.52	2 years	
	Strazov highlands					
	85% ag; 4.2% forest; 10.8% other	0.12 km²	74†	0.33	4 years	
	100% forest	0.0864 km²	40†	0.22	6 years	
	100% forest	0.22 km²	70†	0.27	4 years	
	Tatra mountains					
	80% forest; 20% other	23.37 km²	1343†	0.65	4 years	
	49.6% forest; 50.4% other	23.4 km²	1352†	0.54	4 years	
	Ondava highlands					
	50% ag; 50% forest	574.1 km²	569†	0.45	7 years	
	60% ag; 40% forest	167.5 km²	527†	0.46	7 years	
	40% ag; 60% forest	185.8 km²	527†	0.52	7 years	
	50% ag; 50% forest	156.7 km²	499†	0.34	2 years	
	91% ag; 9% forest	0.345 km²	228†	0.53	6 years	
	5% ag; 95% forest	0.195 km²	190†	0.44	6 years	

‡ simulated events
* runoff generated from snowmelt
† represents difference in maximum and minimum elevation above sea level (m)

Table 8.2. Selected studies identifying nutrient and sediment concentrations (mg L^{-1}) in surface waters from grassed and wooded catchments

Reference	Land use	Area	TSS	NH$_4$	NO$_3$+NO$_2$	TN	DRP	TP	Duration	Study location
Gaudreau et al. 2002	Common bermudagrass	6 m²	---	---	1.3	---	4.2	---	8 events	College Station, TX
Chichester 1977	Poverty grass	8 m²	---	---	12.4	---	---	---	4 years	Coshocton, OH
	Bluegrass				20.5					
	Bluegrass				25.7					
	Alfalfa/orchard-grass mix				40.6					
Cole et al. 1997‡	Common bermudagrass turfgrass	17.6 m²	---	6.2	4.51	---	8.04	---	2 events	Stillwater, OK
Morton, Gold, and Sullivan 1988	90% Kentucky bluegrass; 10% red fescue	32 m²	---	---	0.87	---	---	---	2 years	Kingston, RI
Easton and Petrovic 2004	80% Kentucky bluegrass; 20% perennial ryegrass	37.2 m²	---	1.44	10.1	---	0.50	---	18 months	Ithaca, NY
Graczyk et al. 2003	Woods	37.4 m²	---	0.98	0.21	7.4	0.33	1.12	11 events	WI
	Lawns	59.9 m²	---	0.42	0.77	2.8	0.17	0.32	18 events	

Continued on next page

Table 8.2. (continued) Selected studies identifying nutrient and sediment concentrations (mg L^{-1}) in surface waters from grassed and wooded catchments

Reference	Land use	Area	TSS	NH$_4$	NO$_3$+NO$_2$	TN	DRP	TP	Duration	Study location
Linde and Watschke 1997	Bentgrass and ryegrass plots	123.5 m^2	---	---	1.47	---	4.06	---	2 years	State College, PA
Waschbusch, Selbig, and Bannerman 1999	Lawns	---	90.5	---	---	---	0.71	1.27	2 years	Madison, WI
	Streets	---	68.5	---	---	---	0.11	0.27		
Garn 2002	Fertilized lawn	---	---	1.11	0.09	5.9	0.7	2.57	2 years	Lauderdale Lakes, WI
	Nonphosphorus fertilized Lawn	---	---	1.0	0.14	6.5	0.34	1.89		
	Unfertilized lawn	---	---	0.76	0.12	4.1	0.4	1.73		
	Unfertilized woods	---	---	2.95	0.16	12.7	1.04	3.52		
Doran, Schepers, and Swanson 1981	Bromegrass	0.11 ha	182	0.63	3.31	9.2	4.15	4.72	3 years	Clay Center, NE
Smith et al. 1992	Native tall grasses	1.6 ha	---	0.65	1.32	3.4	0.25	0.32	57 events	El Reno, OK
	Old world bluestems (Ischaemum complex)	2.7 ha	---	0.28	1.61	4.2	0.45	1.44	40 events	Woodward, OK
	Native short/mid grasses	2.9 ha	---	0.24	1.67	2.4	0.42	0.98	27 events	Woodward, OK

Study	Land use	Area							Duration	Location
Olness et al. 1975	Grazed rangeland	9.9 ha	---	0.14	0.37	1.85	0.01	0.99	1 year	Chickasha, OK
Winter and Dillon 2004	13 sites on 5 golf courses	---	---	---	0.3	0.94	---	0.03	2 years	Ontario, CA
Starrett and Bhandari 2004	Native prairie	---	477	---	---	1.18	---	0.39	3 months	Manhattan, KS
	Golf course construction	---	2955	---	---	3.94	---	0.93	20 months	
	Golf course	---	604	---	---	1.91	---	0.51	4 years	
King et al. 2001	Golf course storm events	29 ha	---	---	0.3	---	0.00	---	22 events	Austin, TX
	Golf course baseflow	---	---	---	0.86	---	0.01	---	13 months	
Kunimatsu, Sudo, and Kawachi 1999	Golf course	53	9.9	0.3	0.29	1.3	0.05	0.1	2 years	Japan
	Forest	23	2.1	0.02	0.29	0.4	0.00	0.01		
Mallin and Wheeler 2000	Golf course	54 ha	---	0.04	0.32	---	0.019	---	1 year	New Hanover County, NC
	Golf course	53.7 ha	---	0.03	0.32	---	0.008	---	1 year	New Hanover County, NC
	Golf course	---	---	0.23	1.46	---	0.005	---	9 months	Brunswick County, NC
	Golf course	46.4 ha	---	0.03	0.06	---	0.056	---	1 year	New Hanover County, NC
	Golf course	111.7 ha	---	---	0.11	---	0.004	---	1 year	New Hanover County, NC

‡ simulated events

Table 8.3. Selected studies identifying nutrient loads (kg/ha/yr) in surface waters from grassed, wooded, and agricultural catchments

Reference	Land use	Area	NH_4	NO_3+NO_2	TN	DRP	TP	Duration	Study location
Chichester 1977	Poverty grass	8 m²	---	0.53	---	---	---	4 years	Coshocton, OH
	Bluegrass		---	1.53	---	---	---		
	Bluegrass		---	0.70	---	---	---		
	Alfalfa/orchard-grass mix		---	0.77	---	---	---		
Gross, Angle, and Welterlen 1990	60% tall fescue; 40% Kentucky bluegrass, fertilized	10 m²	0.09	0.06	0.17	0.05	0.02	18 events	Upper Marlboro, MD
	60% tall fescue; 40% Kentucky bluegrass, non-fertilized		0.05	0.04	0.08	0.02	0.01		
Schwartz and Shuman 2005	Tifway bermudagrass	25.2 m²	---	3.05	---	---	---	4 years	Griffin, GA
Easton and Petrovic 2004	80% Kentucky bluegrass; 20% perennial ryegrass	37.2 m²	0.35	0.90	---	0.12	---	18 months	Ithaca, NY
Graczyk et al. 2003	Woods	37.4 m²	0.0004	0.0005	0.02	0.001	0.003	11 events	Northern WI
	Lawns	59.9 m²	0.02	0.11	0.16	0.01	0.03	18 events	

Reference	Cover	Size						Duration	Location
Timmons and Holt 1977	Native prairie	89.6 m²	0.14	0.12	0.84	0.02	0.11	5 years	Big Stone County, MN
Birdwell 1995	Golf green (bermudagrass)	0.025 ha	---	0.52	---	---	---	3 months	College Station, TX
	Golf fairway (bermudagrass)	1.57 ha	---	0.96	---	---	---		
Smith et al. 1992	Native tall grasses	1.6 ha	0.34	0.81	1.76	0.22	0.28	57 events	El Reno, OK
	Old world bluestems (Ischaemum complex)	2.7 ha	0.34	1.50	6.20	0.50	1.91	40 events	Woodward, OK
	Native short/mid grasses	2.9 ha	0.16	0.84	2.03	0.34	0.70	27 events	Woodward, OK
Dennis 1986	Residential (43.4% lawns; 41% forest)	3.5 ha	---	---	---	---	1.4	8 events	Augusta, ME
	Forest (97.3% forest; 2.2% lawns)	2.4 ha	---	---	---	---	0.2		
Olness et al. 1975	Grazed rangeland	9.9 ha	0.37	1.18	4.73	0.03	2.94	1 year	Chickasha, OK
King et al. 2001	Golf course storm events	29 ha	---	2.1	---	0.3	---	22 events	Austin, TX
	Golf course baseflow		---	4.3	---	0.05	---	13 months	

Continued on next page

Table 8.3. *(continued)* Selected studies identifying nutrient loads (kg/ha/yr) in surface waters from grassed, wooded, and agricultural catchments

Reference	Land use	Area	NH_4	NO_3+NO_2	TN	DRP	TP	Duration	Study location
Kunimatsu, Sudo, and Kawachi 1999	Golf course	53 ha	1.7	3.7	13.5	1.6	3.04	2 years	Japan
	Forest	23 ha	0.2	4.1	5.4	0.03	0.13		
Coulter, Kolka, and Thompson 2004	95% agriculture; 5% urban	327 ha	0.34	20.4	---	0.28	1.13	1 year	Fayette County, KY
	43% agriculture; 57% urban	506 ha	0.95	10.8	---	0.12	1.14		
	99% urban; 1% agriculture	226 ha	0.52	5.97	---	0.07	0.66		

Table 8.4. Concentrations (µg L^{-1}) of pesticide residues measured in the surface water in and around the urban landscape

Reference	Site characteristics	Parameter reported	Concentration	Location
Wan, Wong, and Mok 1996	Nine golf courses (ponds and runoff discharge)	max. conc.	Pyridalphenthion: 1 Deverino: 0.1	Singapore
Armbrust and Peeler 2002	27.4 m^2 5% slope Tifway bermudagrass	mean conc. and total percent recovered	imidacloprid: 0.12, 1.7% 2,4-D: 0.67, 2.6%	Georgia
Voss et al. 1999	12 sites in ten urban watersheds	max conc.	2,4-D: 1 acetochlor: 0.01 atrazine: 0.02 dicamba: 0.09 dichlobenil: 1.1 dichlorprop: 0.08 MCPA: 0.4 MCPP: 0.8 metolachlor: 0.15 chlorpyrifos: 0.025 diazinon: 0.45	King County, WA
Bailey et al. 2000	Watershed tributaries to the Sacramento and San Joaquin Rivers	range and median conc.	diazinon: nd-1.5; median 0.21 chlorpyrifos: nd-0.19; median 0.05	California
Hong and Smith 1997	27.4 m^2 5% slope golf course fairway Tifway bermudagrass	total percent recovered	dithiopyr: <2%	Georgia

Continued on next page

Table 8.4. *(continued)* Concentrations (µg L^{-1}) of pesticide residues measured in the surface water in and around the urban landscape

Reference	Site characteristics	Parameter reported	Concentration	Location
Starrett, Klein, and Heier 2004	1.2 ha fairway drains	range	glyphosate: <0.1–5.18 AMPA: <0.1–22 glufosinate: <0.1	Manhattan, KS
Lewis et al. 2002	Fairway pond	mean conc.	diazinon: 0.01 chlorpyrifos: 0.02 fenthion: 0.05	Gulf Breeze, FL
Smith and Bridges 1996	27.4 m^2 5% slope Tifway bermudagrass	max conc. and total percent recovered	2,4-D: 810, 9.6% MCPP: 820, 14.4% chlorothalonil: 294, 0.8% dicamba: 279, 14.4% dithiopyr: 22, 1.9% chlorpyrifos: 19, 0.1% pendimethalin: 9, 0.01% benefin: 3, 0.01%	Georgia
Morioka and Cho 1992	Golf course ponds	range	Flutonil: <0.5–81; isoprothiolane: <0.5–39; chlorpyrifos: <0.5–3; simazine: 0.7–41	Japan
Ryals, Genter, and Leidy 1998	Water features on 3 golf courses	number of detections and mean conc.	atrazine: 70/110, 0.05 chlorothalonil: 104/124, 0.30 chlorpyrifos: 105/133, 3.14 2,4-D: 110/132, 1.31	North Carolina

Reference	Site	Measurement	Results	Location
Williams et al. 2000	Surface water from golf course, residential, and ag watershed	range	2,4-D: nd-4.1 diazinon: nd-1.5 metalaxyl: nd-0.9	Kentucky
Tomimori, Nagaya, and Taniyama 1994	Drainage water from golf courses	number of detects within concentration range	% >1 1.0.1 <0.1 propyzamide: 20 45 56 simazine: 27 54 1 napropamide: 9 --- --- flutoluanil: 69 122 1 isoprothiolane: 27 122 2 captan: 3 5 141 tolclophos- methyl: --- 14 8 diazinon: 3 14 2 fenitrothion: 3 14 2	Japan
Wotzka et al. 1994	57.5 ha watershed (94% residential; 4% commercial) 2-year study with 41 events	max conc.	dicamba: 2.6 MCPA: 43 MCPP: 16 2,4-D: 7.4 alachlor: 1.5 atrazine: 3.8 cyanazine: 1.1 metolachlor: 0.8	Minneapolis, MN
Watschke and Mumma 1989	123.5 m^2 9–14% slope Kentucky bluegrass	range	pendimethalin: nd 2,4-D: 0–312 2,4-DP: 0–210 dicamba: 0–251 chlorpyrifos: nd	Pennsylvania

Table 8.5. Selected list of best management practices (BMPs) tested under different turfgrass management scenarios

BMP	Function	Results	Reference
Irrigation shortly after application	Wash off pesticide from leaf surface to thatch and soil	No significant reductions in pesticide transport for chlorothalonil, paclobutrazol, mefanoxam	Branham, Kandil, and Meuller 2005
Lengthening time between application and runoff	Prolong the time before runoff	Significant reductions in lost amounts of Pendimethalin, propiconazole, mefanoxam when runoff occured 24, 48, and 72 hours compared to 12 hours	Branham, Kandil, and Meuller 2005
Clipping management	Export the adsorbed pesticides with the clippings	Surface runoff losses of chlorothalonil, paclobutrazol, mefanoxam were reducd by 34 to 57% by removing clippings	Branham, Kandil, and Meuller 2005; Taylor, Rosen, and White 1990
Wetlands	Retention and degradation of compounds through physical, chemical, and biological processes	Nine of 17 parameters had mass removal efficiencies greater than 50% for storm events	Kohler et al. 2004; Reicher et al. 2005
Graduated buffer height	Reduce runoff	19% runoff reduction; 17% N reduction; 11% P reduction when compared to single buffer height	Bell and Moss 2005
Buffer length	Intercept flow/slow down runoff, increase infiltration	Significant reductions in losses of pesticides and nutrients	Cole et al. 1997
Mowing height	Increase mowing heights/less stress, less chemical requirements	Minimal impact	Cole et al. 1997
Cultivation	Improve drainage and turfgrass quality	No impact	Cole et al. 1997

Practice	Purpose	Effect	Reference
Turfgrass density and health	Improve drainage/infiltration	15 times less runoff	Easton and Petrovic 2004; Leslie and Knoop 1989
Water-use efficiency (irrigation strategy, scheduling)	Conserve water/less runoff	15% water-use reduction with equivalent turfgrass quality for Kentucky bluegrass, red fescue, and perennial ryegrass	Bastug and Buyuktas 2003
Slow release nitrogen formulation	Reduce availability of soluble forms	Significant reduction in leaching losses and runoff losses	Brown, Thomas, and Duble 1982; Easton and Petrovic 2004; Petrovic 1990; Quiroga-Garza, Picchioni, and Remmenga 2001
Post application irrigation (nutrients)	Place nutrients into the thatch and soil	Significant reduction in percentage in runoff of applied phosphorus: 10.4% and nitrate-N: 0.7% using water-in approach	Shuman 2004
Post application irrigation (pesticides)	Wash off pesticide from leaf surface to thatch and soil	No impact on diazinon concentration, but increased mass lot because of increased antecedent moisture	Evans et al. 1998
Granular vs. liquid formulation pesticides	Reduce availability of soluble forms	Runoff concentration of liquid double that of granular for diazinon	Evans et al. 1998
Granular vs. sprayable pesticides	Reduce availability of soluble forms	Significant reduction in losses dependent on chemical properties and formulation	Hong and Smith 1997; Wilson, Whitwell, and Riley 1995
Granular vs. wettable powder pesticides	Reduce availability of soluble forms	1.4% (WP) and 1.9% (GR) of applied lost after four simulated events	Armbrust and Peeler 2002

Literature Cited

Armbrust, K. L. and H. B. Peeler. 2002. Effects of formulation on the run-off of imidacloprid from turf. *Pest Manag Sci* 58:702–706.

Bailey, H. C., L. Deanovic, E. Reyes, T. Kimball, K. Larson, K. Cortright, V. Connor, and D. E. Hinton. 2000. Diazinon and chlorpyrifos in urban waterways in Northern California, USA. *Environ Toxicol Chem* 19:82–87.

Baird, J. H., N. T. Basta, R. L. Hunke, G. V. Johnson, M. E. Payton, D. E. Storm, C. A. Wilson, M. D. Smolen, D. L. Martin, and J. T. Cole. 2000. Best management practices to reduce pesticide and nutrient runoff from turf. Pp. 268–293. In J. M. Clark and M. P. Kenna (eds.). *Fate and Management of Turfgrass Chemicals*. ACS Symposium Series 743. American Chemical Society, Washington, D. C.

Balogh, J. C. and J. L. Anderson. 1992. Environmental impacts of turfgrass pesticides. Pp. 221–353. In J. C. Balogh and W.J. Walker (eds.). *Golf Course Management and Construction: Environmental Issues*. Lewis Publishers, Chelsea, Michigan.

Balogh, J. C. and W. J. Walker. 1992. *Golf Course Management and Construction: Environmental Issues*. Lewis Publishers, Ann Arbor, Michigan. 951 pp.

Balogh, J. C. and J. R. Watson, Jr. 1992. Role and conservation of water resources. Pp. 39–104. In J. C. Balogh and W. J. Walker (eds.). *Golf Course Management and Construction: Environmental Issues*. Lewis Publishers, Chelsea, Michigan.

Bastug, R. and D. Buyuktas. 2003. The effects of different irrigation levels applied in golf courses on some quality characteristics of turfgrass. *Irrig Sci* 22:87–93.

Beard, J. B. 1973. *Turfgrass: Science and Culture*. Prentice-Hall, Englewood Cliffs, New Jersey.

Beard, J. B. and R. L. Green. 1994. The role of turfgrasses in environmental protection and their benefits to humans. *J Environ Qual* 23:452–460.

Becker, R. L., D. Herzfeld, K. R. Ostlie, and E. J. Stamm-Katovich. 1989. Pesticides: Surface runoff, leaching, and exposure concerns. AG-BU-3911. Minnesota Extension Service, University of Minnesota. 32 pp.

Bell, G. and J. Moss. 2005. Managing golf course roughs to reduce runoff. *USGA Turfgrass and Environ Res Online* 4(10):1–9.

Birdwell, B. 1995. Nitrogen and chlorpyrifos in surface water runoff from a golf course. M.S. Thesis, Texas A&M University, College Station.

Bowman, D. C., C .T. Cherney, and T. W. Rufty, Jr. 2002. Fate and transport of nitrogen applied to six warm-season turfgrasses. *Crop Sci* 42:833–841.

Branham, B. E. and D. J. Wehner. 1985. The fate of diazinon applied to thatched turf. *Agron J* 77:101–104.

Branham, B. E., F. Z. Kandil, and J. Mueller. 2005. Best management practices to reduce pesticide runoff from turf. USGA–Green Section Record. 43:26–30.

Brown, K. W., J. C. Thomas, and R. L. Duble. 1982. Nitrogen source effect on nitrate andammonium leaching and runoff losses from greens. *Agron J* 74:947–950.

Carrow, R. N. 1994. A look at turfgrass water conservation. Pp. 24–43. In United States Golf Association (ed.). *Wastewater Reuse for Golf Course Irrigation*. Lewis Publishers, Inc., Chelsea, Michigan.

Carrow, R. N., R. C. Shearman, and J. R. Watson. 1990. Turfgrass. Pp. 889–919. In B. A. Stewart and D. R. Nielsen (eds.). *Irrigation of Agricultural Crops*. Agronomy Monograph 30. American Society of Agronomy, Crop Science Society of America, Soil Science Society of America, Madison, Wisconsin.

Chichester, F. W. 1977. Effects of increased fertilizer rates on nitrogen content of runoff and percolate from monolith lysimeters. *J Environ Qual* 6:211–217.

Cohen, S., A. Svrjcek, T. Durborow, and N. L. Barnes. 1999. Water quality impacts by golf courses. *J Environ Qual* 28(3):798–809.

Cohen, S. Z. 1995. Agriculture and the golf course industry: An exploration of pesticide use. *Golf Course Management*. Pp. 96, 100, 102, 104.

Cohen, S. Z. 1998. Water quality monitoring at golf courses. *GWMR*:58–59.

Cole, J. T., J. H. Baird, N. T. Basta, R. L. Huhnke, D. E. Storm, G. V. Johnson, M. E. Payton, M. D. Smolen, D. L. Martin, and J. C. Cole. 1997. Influence of buffers on pesticide and nutrient runoff from bermudagrass turf. *J Environ Qual* 26:1589–1598.

Council for Agricultural Science and Technology (CAST). 1985. *Agriculture and Groundwater Quality*. CAST Report No. 103. CAST, Ames, Iowa. 62 pp.

Coulter, C. B., R. K. Kolka, and J. A. Thompson. 2004. Water quality in agricultural, urban, and mixed land use watersheds. *J Amer Water Resources Assoc* 40(6):1593–1601.

Dennis, J. 1986. Phosphorus export from a low-density residential watershed and an adjacent forested watershed. Fifth Annual Conference and International Symposium on Applied Lake and Watershed Management. *Lake and Reservoir Management* 2:401–407.

DiPaola, J. M. and J. B. Beard. 1992. Physiological effects of temperature stress. Pp 231–262. In D. V. Waddington, R. N. Carrow, and R. C. Shearman (eds.). *Turfgrass*. Agronomy Monograph No. 32. American Society of Agronomy, Crop Science Society of America, Soil Science Society of America, Madison, Wisconsin.

Doran, J. W., J. S. Schepers, and N. P. Swanson. 1981. Chemical and bacteriological quality of pasture runoff. *J Soil and Water Conservation* 36:166–171.

Durburow, T. E., N. L. Barnes, S. Z. Cohen, G. L. Horst, and A. E. Smith. 2000. Calibration and validation of runoff and leaching models for turf pesticides, and comparison with monitoring results. Pp. 195–227. In J. M. Clark and M. P. Kenna (eds.). *Fate and Management of Turfgrass Chemicals*. ACS Symposium Series 743. American Chemical Society, Washington, D.C.

Easton, Z. M. and A. M. Petrovic. 2004. Fertilizer source effect on ground and surface water quality in drainage from turfgrass. *J Environ Qual* 33:645–655.

Engelsjord, M. E., B. E. Branham, and B. P. Horgan. 2004. The fate of nitrogen-15 ammonium sulfate applied to Kentucky bluegrass and perennial ryegrass turfs. *Crop Sci* 44:1341–1347.

Erickson, J. E., J. L. Cisar, J. C. Volin, and G. H. Snyder. 2001. Comparing nitrogen runoff and leaching between newly established St. Augustine turf and an alternative residential landscape. *Crop Sci* 41:1889–1895.

Evans, J. R., D. R. Edwards, S. R. Workman, and R. M. Williams. 1998. Response of runoff diazinon concentration to formulation and post-application irrigation. *Trans ASAE* 41:1323–1329.

Garn, H. S. 2002. Effects of lawn fertilizer on nutrient concentration in runoff from lakeshore lawns, Lauderdale Lakes, Wisconsin. USGS Water-Resources Investigations Report 02-4130.

Gaudreau, J. E., D. M. Veitor, R. H. White, T. L. Provin, and C. L. Munster. 2002. Response of turf and quality of water runoff to manure and fertilizer. *J Environ Qual* 31:1316–1322.

Graczyk, D. J., R. J. Hunt, S. R. Greb, C. A. Buchwald, and J. T. Krohelski. 2003. Hydrology, nutrient concentrations, and nutrient yields in nearshore areas of four lakes in northern Wisconsin, 1999–2001. USGS Water-Resources Investigations Report 03-4144.

Gross, C. M., J. S. Angle, and M. S. Welterlen. 1990. Nutrient and sediment losses from turfgrass. *J Environ Qual* 19:663–668.

Haith, D. A. 2001. TurfPQ, a pesticide runoff model for turf. *J Environ Qual* 30:1033–1039.

Haith, D. A. 2002. User's manual for TurfPQ: A PC program for estimating pesticide runoff from turfgrass. Cornell University, Ithaca, New York.

Haith D. A. and F. S. Rossi. 2003. Risk assessment of pesticide runoff from turf. *J Environ Qual* 32:447–455.

Harrison, S. A., T. L. Watschke, R. O. Mumma, A. R. Jarrett, and G. W. Hamilton. 1993. Nutrient and pesticide concentrations in water from chemically treated turfgrass. Pp. 191–207. In K. D. Racke and A. R. Leslie (eds.). *Pesticides in Urban Environments: Fate and Significance*. ACS Symp. Ser. 522. American Chemical Society, Washington, D. C.

Hong, S. and A. E. Smith. 1997. Potential movement of dithiopyr following application to golf courses. *J Environ Qual* 26:379–386.

Hurto, K. A., A. J. Turgeon, and M. A. Cole. 1979. Degradation of benefin and DCPA in thatch and soil from a Kentucky bluegrass *(Poa pratensis)* turf. *Weed Sci* 27:154–157.

Kenna, M. P. 1995. What happens to pesticides applied to golf courses? USGA Green Section Record. 33(1):1–9.

King, K. W. and J. C. Balogh. 1997. Evaluation of an agricultural water quality model for use in golf course management. *Amer Soc Agricultural Engineers Paper* 97–2009. ASAE, St. Joseph, Michigan.

King, K. W. and J. C. Balogh. 1999. Modeling evaluation of alternative management practices and reclaimed water for turfgrass systems. *J Environ Qual* 28:187–193.

King, K. W., R. D. Harmel, H. A. Torbert, and J. C. Balogh. 2001. Impact of a turfgrass system on nutrient loadings to surface water. *J Amer Water Resources Assoc* 37(3):629–640.

King, K. W., K. L. Hughes, J. C. Balogh, N. R. Fausey, and R. D. Harmel. 2006. NO_3-N and dissolved reactive phosphorus in subsurface drainage from managed turfgrass. *J Soil and Water Conservation* 61(1):31–40.

Kneebone, W. R., D. M. Kopec, and C. F. Mancino. 1992. Water requirements and irrigation. Pp. 441–467. In D. V. Waddington, R. N. Carrow, and R. C. Shearman (eds.). *Turfgrass*. American Society of Agronomy, Crop Science Society of America, Soil Science Society of America, Madison, Wisconsin.

Kohler, E. A., V. L. Poole, Z. J. Reicher, and R. F. Turco. 2004. Nutrient, metal, and pesticide removal during storm and nonstorm events by a constructed wetland on an urban golf course. *Ecological Engineering* 23:285–298.

Kunimatsu, T., M. Sudo, and T. Kawachi. 1999. Loading rates of nutrients discharging from a golf course and a neighboring forested basin. *Water Sci Tech* 39:99–107.

Leslie, A. R. and W. Knoop. 1989. Societal benefits of conservation oriented management of turfgrass in home lawns. Pp. 93–96. In A. R. Leslie and R. L. Metcalf (eds.). *Integrated Pest Management for Turfgrass and Ornamentals*. Office of Pesticide Programs 1989-625-030, U. S. Environmental Protection Agency, Washington, D.C.

Lewis, M. A., R. G. Boustany, D. D. Dantin, R. L. Quarles, J. C. Moore, and R. S. Stanley. 2002. Effects of a coastal golf complex on water quality, periphyton, and seagrass. *Ecotoxicology and Environ Safety* 53:154–162.

Linde, D. T. and T. L. Watschke. 1997. Nutrients and sediment in runoff from creeping bentgrass and perennial ryegrass turfs. *J Environ Qual* 26:1248–1254.

Ma, Q. L., A. E. Smith, J. E. Hook, R. E. Smith, and D. C. Bridges. 1999a. Water runoff and pesticide transport from a golf course fairway: Observations vs. Opus model simulations. *J Environ Qual* 28:1463–1473.

Ma, Q. L., A. E. Smith, J. E. Hook, and D. C. Bridges. 1999b. Surface transport of 2,4-D from small turf plots: Observations compared with GLEAMS and PRZM-2 model simulations. *Pestic Sci* 55:423–433.

Maidment, D. R. 1993. *Handbook of Hydrology*. McGraw-Hill, New York.

Mallin, M. A. and T. L. Wheeler. 2000. Nutrient and fecal coliform discharge from coastal North Carolina golf courses. *J Environ Qual* 29:979–986.

Mankin, K. R. 2000. An integrated approach for modeling and managing golf course water quality and ecosystem diversity. *Ecological Modeling* 133:259–267.

Morioka, T. and H. S. Cho. 1992. Rainfall runoff characteristics and risk assessment of agrochemicals used in golf links. *Wat Sci Tech* 25:77–84.

Morton, T. G., A. J. Gold, and W. M. Sullivan. 1988. Influence of overwatering and fertilization on nitrogen losses from home lawns. *J Environ Qual* 17:124–130.

Mugaas, R. J., M. L. Agnew, N. E. Christians, and E. Edwards. 1991. *Turfgrass Management for Protecting Surface Water Quality*. Iowa State University Extension and University of Minnesota Extension PM-1446. Iowa State University, Ames.

Murphy, J. A. 2002. Best management practices for irrigating golf course turf. Rutgers Cooperative Extension, New Jersey Agricultural Experiment Station. E278. 11 pp.

Nelson, K. E., A. J. Turgeon, and J. R. Street. 1980. Thatch influence on mobility and transformation on nitrogen carriers applied to turf. *Agron J* 72:487–492.

Olness, A., S. J. Smith, E. D. Rhoades, and R. G. Menzel. 1975. Nutrient and sediment discharge from agricultural watersheds in Oklahoma. *J Environ Qual* 4:331–336.

Oltmann, R. N. and M. V. Shulters. 1989. Rainfall and runoff quantity and quality characteristics of four urban-use catchments in Fresno, California, October 1981 to April 1983. U.S. Geological Survey Water-Supply Paper 2335. 114 pp.

Owens, L. B., R. W. Van Keuren, and W. M. Edwards. 1998. Budgets of non-nitrogen nutrients in a high fertility pasture system. *Ag Ecosystems Environ* 70:7–18.

Peacock, C. H., M. M. Smart, and W. Warren-Hicks. 1996. Best management practices and integrated pest management strategies for protection of natural resources on golf course watersheds. Pp. 335–338. *Proc EPA Watershed 96 Conference.* July 1996, Baltimore, Maryland.

Pekarova, P. and J. Pekar. 1996. The impact of land use on stream water quality in Slovakia. *J Hydrol* 180:333–350.

Petrovic, A. M. 1990. The fate of nitrogenous fertilizers applied to turfgrass. *J Environ Qual* 19(1):1–14.

Pira, E. S. 1997. *A Guide to Golf Course Irrigation System Design and Drainage.* Ann Arbor Press, Chelsea, Michigan.

Quiroga-Garza, H. M., G. A. Picchioni, and M. D. Remmenga. 2001. Bermudagrass fertilized with slow-release nitrogen sources. I. Nitrogen uptake and potential leaching losses. *J Environ Qual* 30:440–448.

Reicher, Z. J., E. A. Kohler, V. L. Poole, and R. F. Turco. 2005. Constructed wetlands on golf courses help manage runoff from the course and surrounding areas. *USGA Turfgrass and Environmental Research Online* 4(2):1–14.

Rieke, P. E. and G. T. Lyman. 2002. *Fertilizing Home Lawns to Preserve Water Quality.* Michigan State University Extension Bulletin E05TURF. 2 pp.

Roy, J. W., J. C. Hall, G. W. Parkin, C. Eagner-Riddle, and B. S. Clegg. 2001. Seasonal leaching and biodegradation of dicamba in turfgrass. *J Environ Qual* 30:1360–1370.

Ryals, S. C., M. B. Genter, and R. B. Leidy. 1998. Assessment of surface water quality on three Eastern North Carolina golf courses. *Environ Toxicol and Chem* 17:1934–1942.

Schueler, T. 1995. Urban pesticides: From the lawn to the stream. *Watershed Protection Techniques* 2:247–253.

Schwartz, L. and L. M. Shuman. 2005. Predicting runoff and associated nitrogen losses from turfgrass using the root zone water quality model (RZWQM). *J Environ Qual* 34:350–358.

Selbig, W. R., P. L. Jopke, D. W. Marshall, and M. J. Sorge. 2004. *Hydrologic, Ecologic, and Geomorphic Responses of Brewery Creek to Construction of a Residential Subdivision, Dane County, Wisconsin, 1999–2002*. USGS Scientific Investigations Report 2004–5156. 33 pp.

Shuman, L. M. 2002. Phosphorus and nitrate nitrogen in runoff following fertilizer application to turfgrass. *J Environ Qual* 31:1710–1715.

Shuman, L. M. 2004. Runoff of nitrate nitrogen and phosphorus from turfgrass after watering-in. *Communications in Soil Science and Plant Analysis* 35:9–24.

Shuman, L. M., A. E. Smith, and D. C. Bridges. 2000. Potential movement of nutrients and pesticides following application to golf courses. Pp 78-93. In J. M. Clark and M. P. Kenna (eds.). *Fate and Management of Turfgrass Chemicals*. ACS Symposium Series 743. American Chemical Society, Washington, D.C.

Smith, A. E. and D. C. Bridges. 1996. Movement of certain herbicides following application to simulated golf course greens and fairways. *Crop Sci* 36:1439–1445.

Smith, A. E., O. Weldon, W. Slaughter, H. Peeler, and N. Mantripragada. 1993. A greenhouse system for determining pesticide movement from golf course greens. *J Environ Qual* 22:864–867.

Smith, S. J., A. N. Sharpley, W. A. Berg, J. W. Naney, and G. A. Coleman. 1992. Water quality characteristics associated with southern plains grasslands. *J Environ Qual* 21:595–601.

Starrett, S. and A. Bhandari. 2004. Measuring nutrient losses via runoff from an established golf course. *2004 Turfgrass and Environmental Research Summary*. Far Hills, New Jersey. 38 pp.

Starrett, S. K. and S. K. Starrett. 2000. KTURF: Pesticide and nitrogen leaching model. Pp. 180–194. In J. M. Clark and M. P. Kenna (eds.). *Fate and Management of Turfgrass Chemicals*. ACS Symposium Series 743. American Chemical Society, Washington, D.C.

Starrett, S. K., J. Klein, and T. Heier. 2004. Roundup runoff from zoysiagrass fairways. *USGA Turfgrass and Environ Res Online* 3:1–6.

Taylor, D. H. and G. R. Blake. 1982. The effect of turfgrass thatch on water infiltration rates. *Soil Sci Soc Am Proc* 46:616–619.

Taylor, D. H., C. J. Rosen, and D. B. White. 1990. *Fertilizing Lawns*. Minnesota Extension Service. University of Minnesota, AG-FO-3338-B. 4 pp.

Timmons, D. R. and R. F. Holt. 1977. Nutrient losses in surface runoff from a native prairie. *J Environ Qual* 6:369–373.

Tomimori, S., Y. Nagaya, and T. Taniyama. 1994. Water pollution caused by agricultural chemical and fertilizers in the drainage from golf links. *Jpn J Crop Sci* 63:442–451.

Turgeon, A. J. 1996. *Turfgrass Management*. 4th ed. Regents/Prentice Hall, Englewood Cliffs, New Jersey.

U. S. Geological Survey (USGS). 1984. *National Water Summary 1983—Hydrologic Events and Issues*. U. S. Geological Survey Water-Supply Paper 2250. 243 pp.

Voss, F. D., S. S. Embry, J. C. Ebbert, D. A. Davis, A. M Frahm, and G. H Perry. 1999. Pesticides detected in urban streams during rainstorms and relations to retail sales of pesticides in King County, Washington. U. S. Geological Survey Fact Sheet 097-99.

Walker, W. J. and B. Branham. 1992. Environmental impacts of turfgrass fertilization. Pp. 105-219. In J. C. Balogh and W. J. Walker (eds.). *Golf Course Management and Construction: Environmental Issues*. Lewis Publishers, Chelsea, Michigan.

Walker, W. J., J. C. Balogh, R. M. Tietge, and S. R. Murphy. 1990. *Environmental Issues Related to Golf Course Construction and Management*. USGA-Green Section, Far Hills, New Jersey. 378 pp.

Wan, H. B., M. K. Wong, and C. Y. Mok. 1996. Pesticides in golf course waters associated with golf course runoff. *Bull Environ Contam Toxicol* 56:205-209.

Waschbusch, R. J., W. R. Selbig, and R. T. Bannerman. 1999. *Source of Phosphorus in Stormwater and Street Dirt from Two Urban Residential Basins in Madison, Wisconsin, 1994-95*. USGS Water-Resources Investigations Report 99-4021.

Watschke, T. L. 1990. The environmental fate of pesticides. *Golf Course Mgt* 2:18, 22, and 24.

Watschke, T. L. and R. O. Mumma. 1989. *The Effect of Nutrients and Pesticides Applied to Turf on the Quality of Runoff and Percolating Water*. USGS Final Report ER-8904.

Watson, J. R., Jr. 1994. Water —Where will it come from? Pp. 2-23. In United States Golf Association (ed.). *Wastewater Reuse for Golf Course Irrigation*. Lewis Publishers, Inc., Chelsea, Michigan.

Westerstrom, G. and V. P. Singh. 2000. An investigation of snowmelt runoff on experimental plots in Lulea, Sweden. *Hydrological Processes* 14:1869-1885.

Williams, R. M., J. S. Dinger, A. J. Powell, and D. R. Edwards. 2000. *The Effect of Turfgrass Maintenance on Surface Water Quality in a Suburban Watershed, Inner Bluegrass, Kentucky*. Report of Investigations 5, Series XII. Kentucky Geological Survey, University of Kentucky, Lexington.

Wilson, P. C., T. Whitwell, and M. B. Riley. 1995. Effects of groundcover and formulation on herbicides in runoff water miniature nursery sites. *Weed Sci* 43:671-677.

Winter, J. G. and P. J. Dillon. 2004. Effects of golf course construction and operation on water chemistry of headwater streams on the Precambrian Shield. *Environ Pollution* 133:243-253.

Witteveen, G. and M. Bavier. 1999. *Practical Golf Course Maintenance.* Ann Arbor Press, Chelsea, Michigan.

Wotzka, P. J., J. Lee, P. Capel, and M. Lin. 1994. Pesticide concentrations and fluxes in an urban watershed. Pp 135–145. American Water Resources Association, *National Symposium on Water Quality.* AWRA TPS-94-4. Chicago, Illinois.

Zimmerman, T. L. 1973. The effect of amendment, compaction, soil depth, and time on various physical properties of physically modified Hagerstown soil. Ph.D. diss. Pennsylvania State University, University Park (Diss. Abstr. 73-24053).

9

Pesticide and Nutrient Modeling

Stuart Z. Cohen, Qingli Ma, N. LaJan Barnes, and Scott Jackson

Introduction

The main purposes of this section are to summarize briefly the key practices and research regarding the techniques and applications of mathematical models that can predict the off-site transport of turfgrass chemicals to water resources and to offer suggestions for improvement. These models are important tools for risk assessment and risk management of turfgrass chemicals, but their use also has strong potential to produce results that deviate significantly from reality. To put the subject area in context, the first discussion indicates the significant extent to which turfgrass is managed with chemicals in the United States. A definition of mathematical models in the current context provides a basis for the rest of the paper. Examples of the implications of using inappropriate input parameters are given.

The Importance of Turfgrass Pesticides in the United States

It is estimated that there are 20,235,000 hectares (ha) (50,000,000 acres [ac]; 78,125 square miles) of managed turfgrass in the United States (see Section 1). Approximately 6,542,000 of these ha (16,165,000 ac) receive pesticide treatments (Figure 9.1; M. Cyr, personal communications, 2005, 2006). Analogous statistics on fertilizer are not readily available, but presumably the fertilizer-treated area would be larger. Managed turfgrass systems usually are in closer proximity to population centers than row crop agriculture. Clearly, this is an economically important activity that has the potential for human and environmental exposure.

Mathematical Models for Turfgrass Chemicals: What Are They?

Definition

Mathematical models are assemblages of concepts in the forms of mathematical equations that portray understanding of natural phenomena (ASTM 1984; Cohen et al. 1995). In the current context, mathematical models simulate off-site transport of turfgrass chemicals to surface water and groundwater.

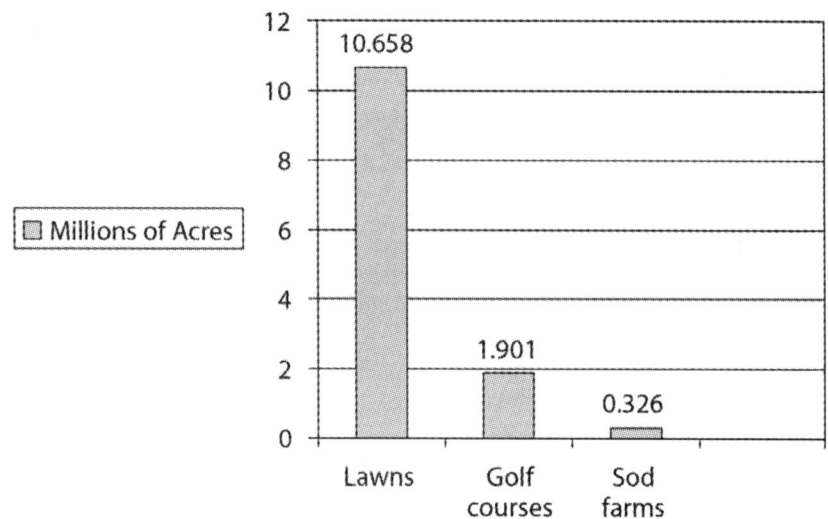

Figure 9.1. Pesticide applications to turf.
Source: Personal correspondence with M. Cyr of Kline & Co., Inc., 2005 and 2006. Derived from various Kline reports and internal data. (Data for 2004, and adapted by ETS.)

Uses of Mathematical Models for Turfgrass Chemicals

Many turfgrass chemical products, especially pesticides and fertilizers, have been developed to improve the vegetative vigor of turfgrass, maintain its health, and improve its ability to provide its intended uses, e.g., lawns, buffers, golf courses, and ball fields. It is important for scientists, regulators, the regulated community, and the general public to understand the extent to which these compounds may persist and move off-site to nontarget environments, particularly in concentrations that might adversely affect the environment.

It would be desirable to conduct controlled monitoring and dissipation studies in every possible combination of land use and environmental condition (climate, soil, and hydrology), but this is infeasible. Consequently, models have been developed with the goal of extrapolating from a relatively limited data set to many different points in time and space. Their uses include national pesticide registration decisions; environmental permitting at the local/watershed scale, particularly for golf courses; applied research on pesticides and nutrients; and—a subject area that bridges all the preceding uses—risk management and the design of Best Management Practices.

More information on the applications of models in a regulatory context is provided later in this section. There also is a discussion about the sensitivity of model output to the input data.

Unique Aspects of the Turfgrass System Relevant to Modeling Turfgrass Chemical Fate

Thatch is the organic layer of shoots, roots, and plant stems that exists between the verdure (the green vegetation) and the soil. The high organic carbon content of thatch enables it to bind pesticides (Lickfeldt and Branham 1995), retain water, and control erosion. Thus, this layer, which cannot be found in other crops, is nature's own pollution control measure.

Turfgrass has a dense bioactive root zone. For example, Beard (2000) reports turfgrass root system biomass is in the range of 11,000 to 16,100 kilograms ha^{-1}, with approximately 122,000 roots and 6.1 x 10^7 root hairs per liter (L) in the upper 150 millimeters of soil, plus a combined length of more than 74 kilometers and a surface area approximately 2.6 meter^2. Its root system is denser than that of most other crops. This almost certainly enhances the biodegradation of chemicals relative to other plant systems. There is some literature that demonstrates shorter dissipation half-lives relative to bare ground or row crops (Horst et al. 1996; Racke 1993).

Turfgrass receives frequent mowing, which dramatically changes the leaf area index, canopy cover, and surface roughness and significantly affects evapotranspiration and surface runoff. Mowing also affects water and chemical uptake and chemical residue distributions in turfgrass.

Most Commonly Used Models: Descriptions, Advantages and Disadvantages, Sucesses and Failures

This section briefly describes the mathematical models most often used to simulate the fate of turfgrass chemicals. Much more analysis is provided of the United States Environmental Protection Agency's (EPA) Pesticide Root Zone Model (PRZM) and Exposure Analysis Modeling System (EXAMS) models because of their heavy use by regulators and the regulated community. The last discussion of models focuses on nutrient modeling. A summary of some key algorithms employed in the three most frequently used models to simulate turfgrass chemical fate can be found in Table 9.1.

Table 9.1. Key model algorithms

	Model algorithm comparison		
	PRZM	RZWQM	TurfPQ
Runoff	Curve number	Infiltration-runoff model	Curve number
Infiltration-redistribution	Field capacity model (soil water accounting procedure)	Green-Ampt for infiltration, Richards' for redistribution	N/A
Pesticide adsorption	Equilibrium model	Equilibrium model and two-site equilibrium-kinetics model	Equilibrium model
Plant growth	Mechanistic growth model	Generic growth model, mechanistic growth for turfgrass	N/A
Evapotranspiration	Pan evaporation or Hamon's formula	Shuttleworth-Wallace model or pan evaporation	N/A

Use of Models in a Tiered Assessment

Three of the following models described (PRZM, Root Zone Water Quality Model [RZWQM], and TurfPQ) were designed to estimate edge-of-field pesticide concentrations in runoff water, and the first two models also estimate leachate concentrations leaving the root zone. (The first two models can also be used to simulate nutrient transport.) This application/level of prediction satisfies some of the model uses previously described, but it is insufficient for quantitative risk assessment. For the latter, the models must be linked to receiving waters, i.e., streams, lakes, and groundwater (as part of aquifers).

In risk assessment, this is done in a tiered concept, whereby the goal of relatively simple Tier I is to screen out chemical/scenario combinations unlikely to cause environmental impacts and pass to a higher tier those chemical/scenario combinations that either may truly cause a problem or are "false positives" (USEPA 2000). When models such as PRZM are used in this capacity, it is important to use input parameters appropriate for the tier level of the particular assessment. This summary of modeling context is important to keep in mind for the remainder of this section.

A turfgrass lawn enhances the appearance of a home. Photo courtesy of Jim Novak, Turfgrass Producers, International.

Water and turfgrass are important elements on a golf course. Photo courtesy of the U.S. Golf Association.

A lawn and play equipment provide enjoyment at a city park. Photo courtesy of Shelley Hart, Hart Arts.

More than half of the world's cool-season grass seed—some 500 million pounds—is produced in the U.S. Pacific Northwest. Turf in parks and golf courses is an important market for this seed. Photo by Jack Dykinga, Agricultural Research Service, U.S. Department of Agriculture.

At a Savannah, Georgia, golf course the newer TifEagle bermudagrass overseeded with *Poa trivialis* is mowed at 0.125 inches. Photo by Wayne Hanna, Agricultural Research Service, U.S. Department of Agriculture.

Carefully placed sprinkler heads cover dry spots in large turfgrass areas. Photo courtesy of John M. Gurke, CGS, Aurora Country Club, Aurora, Illinois.

Technicians collect water samples from monitor wells in a turfgrass field near Corvallis, Oregon. Photo by Brian Prechtel, Agricultural Research Service, U.S. Department of Agriculture.

An agricultural engineer examines discharge water from a turfgrass system as part of a research program designed to assess how land uses and management affect water quality. Photo by Peggy Greb, Agricultural Research Service, U.S. Department of Agriculture.

Scientists use the Root Zone Water Quality Model to examine nitrate distribution in a simulated soil profile. The model enables scientists to forecast potential environmental pollution, such as from excessive nitrate leaching. Photo by Scott Bauer, Agricultural Research Service, U.S. Department of Agriculture.

Landscaping with plants suited to a dry, rocky environment. Photo courtesy of Dale Layfield, Clemson University, Clemson, South Carolina.

Landscaping with flowers, plants, and turfgrass adapted to local conditions. Photo courtesy of Jim Novak, Turfgrass Producers International, Barrington Hills, Illinois.

A researcher evaluates new spray nozzles designed and configured to increase watering efficacy. Photo by Stephen Ausmus, Agricultural Research Service, U.S. Department of Agriculture.

A sprinkler irrigation system in use. Photo courtesy of Derek Settle, Director of Turfgrass Programs, Chicago District Golf Association.

A "Water Smart" column appears daily in the *Amarillo Globe News* and provides urban lawn watering guides based on data from the North Plains Evapotranspiration Network. Photo by Scott Bauer, Agricultural Research Service, U.S. Department of Agriculture.

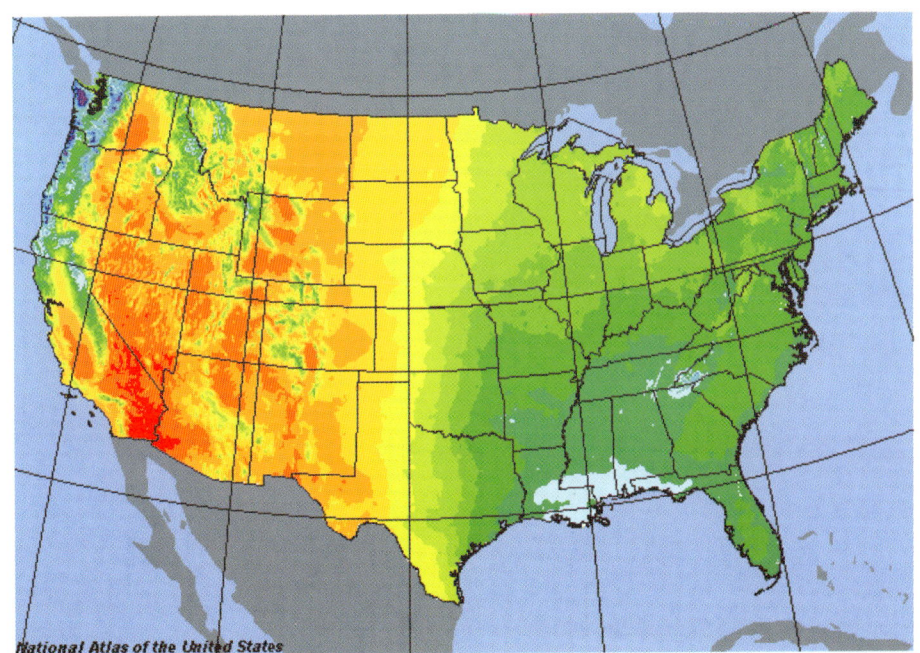

Figure 1.2. Precipitation patterns in the United States.

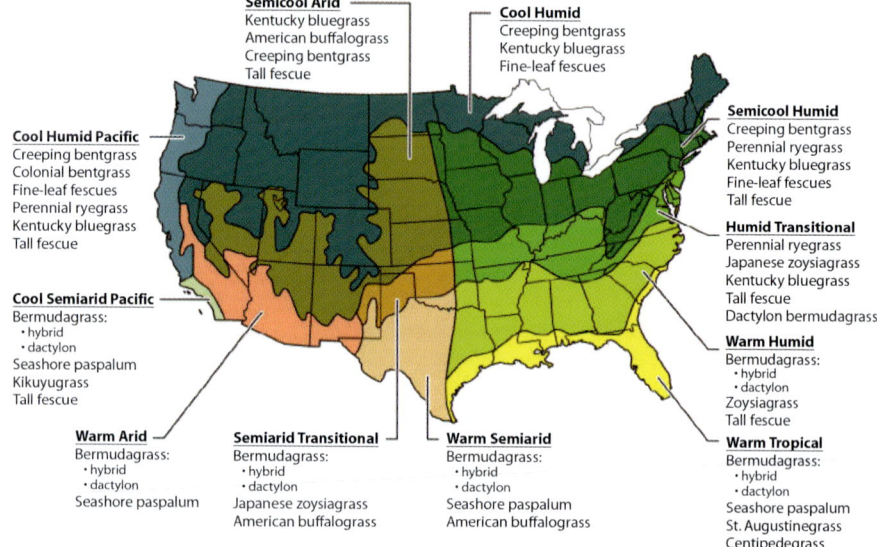

Figure 5.2. Major turfgrass climatic zones and geographic distribution of species in the United States (adapted from Beard 2002).

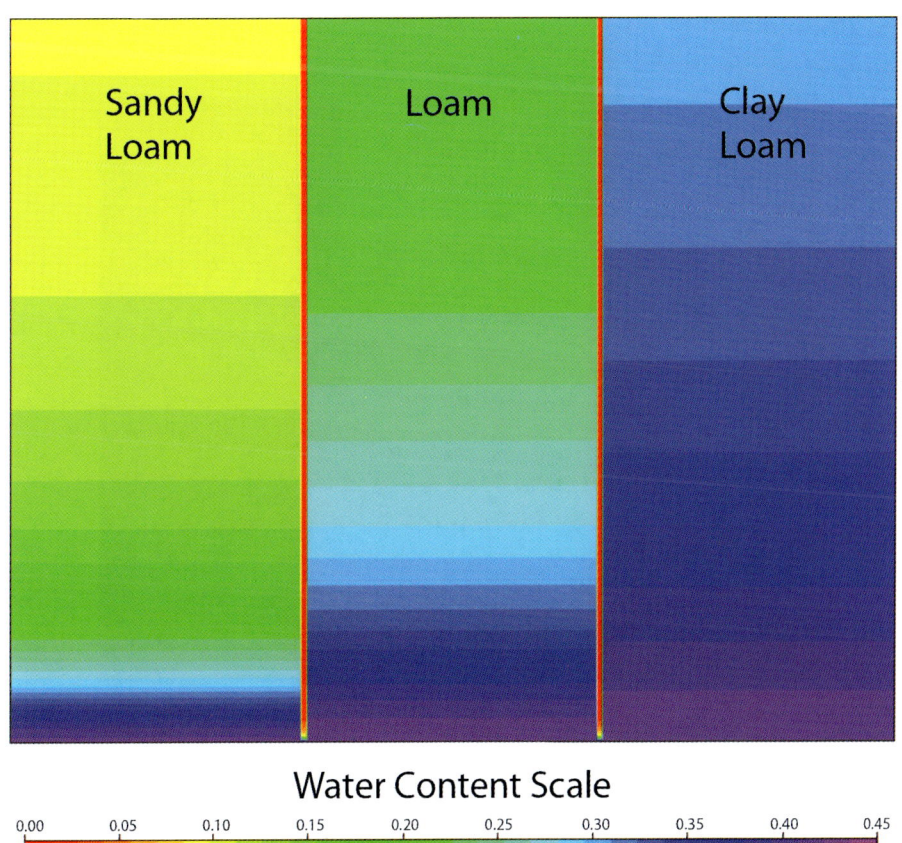

Figure 6.1. A color scale is used to show soil water content as a function of height above a water table for three soils. At the water table (bottom) all soils are saturated and show maximum water content values. Moving upward from the water table, water content values of all soils decline as the soils become drier. The relative dryness of the soils approaching the top, at 2-m height, reflects how different soil textures retain water. It is important to note, however, that this graphic shows an equilibrium state (a condition of no water flow) that rarely exists in nature.

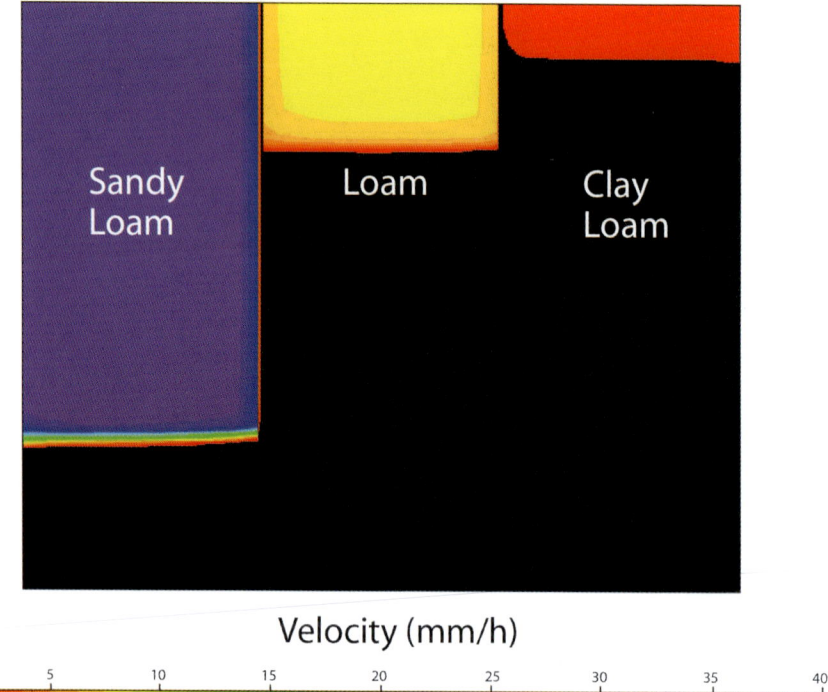

Figure 6.2. A color scale is used to show the velocity of water flow (mm h^{-1}) and depth of water penetration within 2-m columns of three different soils. This view is at 12 hours after establishing a shallow ponding of water at the soil surface (top). The coarse-textured sandy loam soil exhibits greater velocities and a deeper water penetration than the fine-textured clay loam. The medium-textured loam soil is intermediate in both velocity and depth after 12 hours.

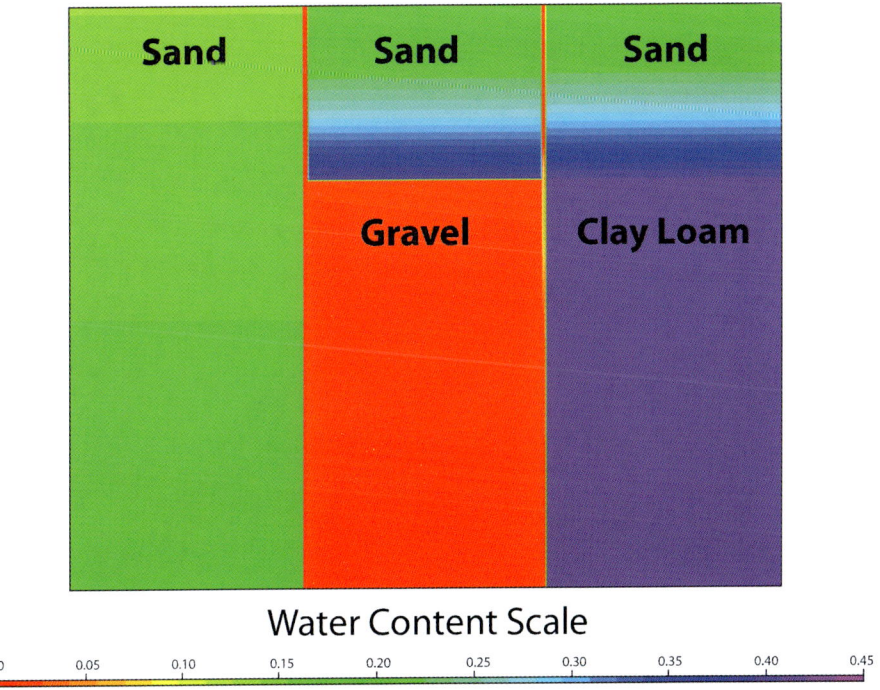

Figure 6.4. A color scale is used to show soil water content within 1-m-long soil columns of three soils. The column on the left is sand throughout, whereas the other columns are layered as indicated, with layer interfaces at 300 mm depth. This illustration is at 24 h after applying a 75-mm depth of water to the soil surface. In both layered soil columns, a capillary fringe has formed as shown by the near-saturated water contents above the layer interface. This fringe is absent in the sand column due to uninterrupted water flow.

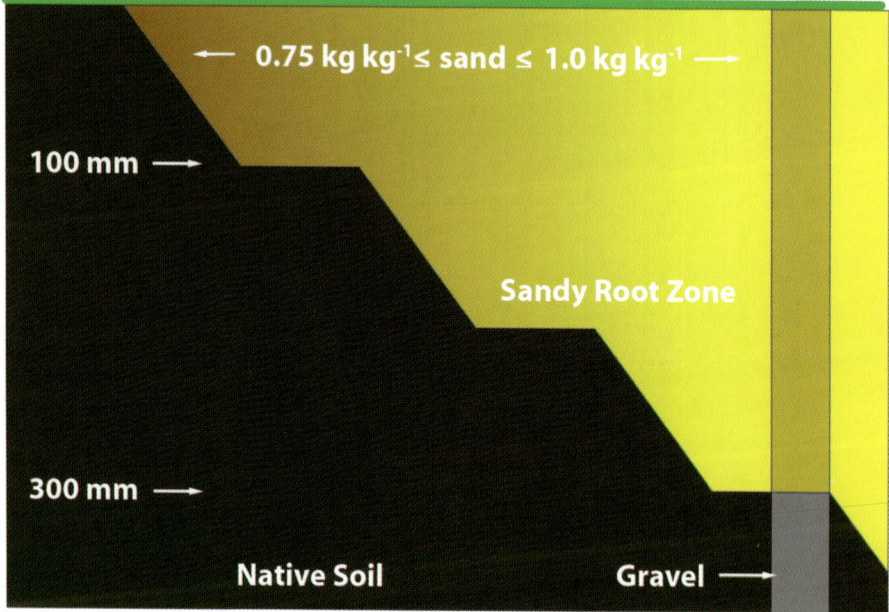

Figure 6.5. The range of root zones and their respective depths as commonly found in high-traffic turfgrass soils. There is a general tendency for higher sand content root zones to be associated with deeper root zone depths. The exception is when the root zone is placed over gravel. Layering of soil materials as shown in this figure offers the opportunity for water conservation where rainfall runoff may be greatly reduced yet water for evapotranspiration is retained in the root zone as a capillary fringe.

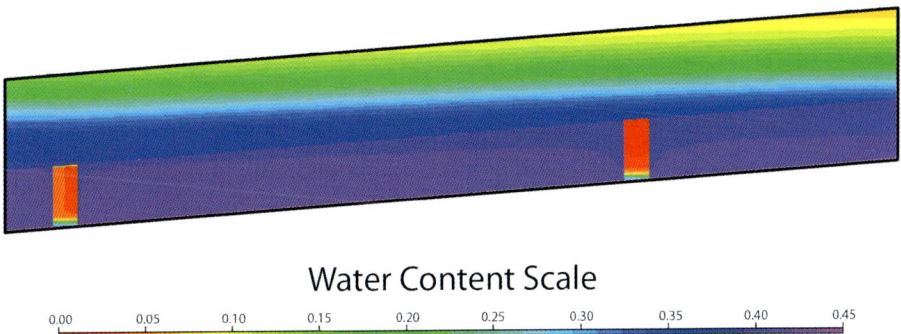

Figure 6.6. This illustration shows the impact of a 3% slope on lateral water flow in a layered soil profile. A color scale is used to show soil water content 48 h following a 64-mm rain, within a 7.2-m-long by 0.5-m-thick section (shown at 2X vertical exaggeration). The upper, 0.3-m layer is sand and the lower 0.2-m layer is a clay loam soil. The red rectangles are gravel drainage trenches 4.6 m apart, extending through the clay loam layer. Lateral water flow after 2 days results in a sizable difference in water contents within the sand layer.

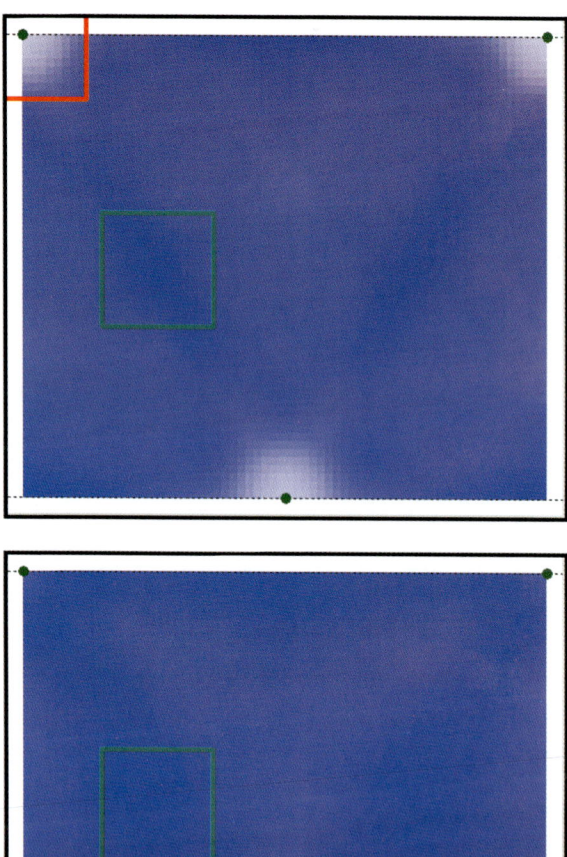

Figure 13.4a (top image) and b (bottom image). Two densograms graphically represent the coverage of two different nozzle combinations when triangularly spaced at the same distance. The densogram allows a visual assessment of the location(s) and size(s) of wet and/or dry areas of coverage between the sprinklers. Note in (a) that the lighter color density predicts dry areas developing immediately surrounding each sprinkler location, shown as a dot. In (b), the more uniform color density predicts more uniform coverage. The two boxes within each example represent the location of the wettest (green) and driest (red) contiguous areas.

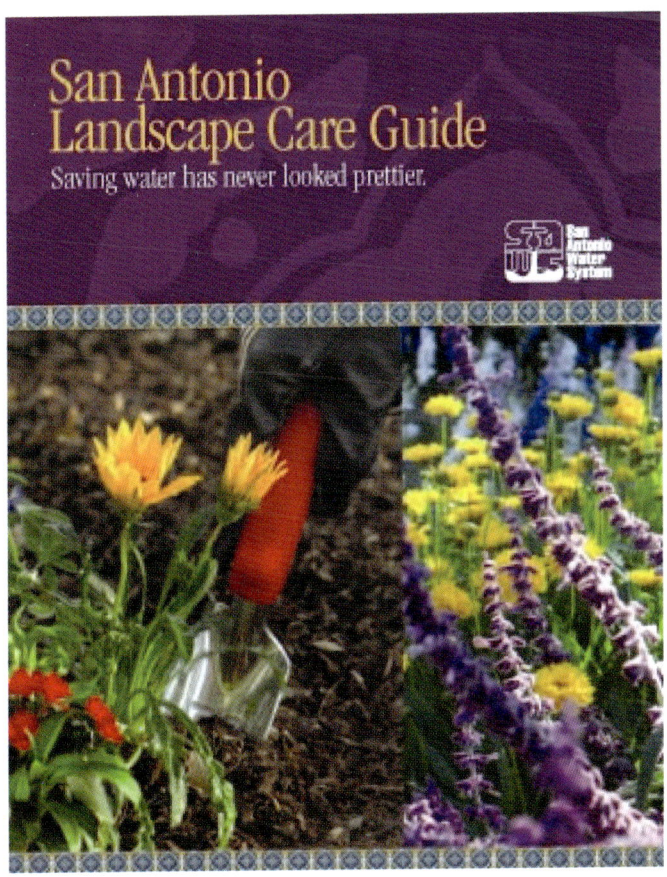

Figure 15.2. *San Antonio Landscape Care Guide.*

Figure 15.3. Planting turfgrass test plots to determine which turfgrass species/cultivars will survive 60 d of drought.

PRZM and PRZM/EXAMS

PRZM is a one-dimensional, dynamic, compartmental model that simulates chemical (pesticides and nitrogen) movement in unsaturated soil systems within and immediately below the plant root zone. It has two major components—hydrology (and hydraulics) and chemical transport. The hydrologic component for calculating runoff and erosion is based on the Soil Conservation Service/National Resources Conservation Service curve number technique and the Universal Soil Loss Equation. The chemical transport component can simulate pesticides or organic and inorganic nitrogen species. For pesticides, the transport component can simulate pesticide application on the soil or on the plant foliage. Biodegradation also can be considered in the root zone. Dissolved, adsorbed, and vapor-phase concentrations in the soil are estimated by simultaneously considering the processes of pesticide uptake by plants, surface runoff, erosion, decay, volatilization, foliar washoff, advection, dispersion, and retardation. For nitrogen, simulation of surface applications, atmospheric deposition, and septic effluent discharge all may be simulated. Predictions are made on a daily basis. Output can be summarized for a daily, monthly, or annual period. Daily time series values of various fluxes or storages can be written to sequential files during program execution for subsequent analysis.

Durborow and colleagues (2000) evaluated the ability of PRZM (as well as the U.S. Department of Agriculture's Groundwater Loading Effects of Agricultural Management Systems [GLEAMS] model) to accurately predict pesticide losses in runoff and leachate in small turfgrass test plots. They found that the PRZM 3.0 evaluation of the Penn State data did not initially simulate the runoff hydrology well, but that seemed to be because the PRZM does calculations for daily steps and the runoff results were on an hourly basis. For example, a 1-inch (2.54 centimeters) storm event in 1 hour generates far more runoff than the same amount of rain spread over 24 hours.

The PRZM 2.3 and 3.12 predictions of percolate volumes ranged from poor to fair for the Georgia study and very good for the Nebraska study. Predictions of pesticide leachate ranged from poor to good, but PRZM generally tended to overpredict pesticide mass.

The EPA usually conducts its pesticide runoff assessments by coupling PRZM with EXAMS. EXAMS is an aquatic ecosystem model that rapidly evaluates the fate, transport, and exposure concentrations of synthetic organic chemicals—pesticides, industrial materials, and leachates from disposal sites. EXAMS contains an integrated Database Management System specifically designed for storage and management of project databases required by the software. User interaction is provided by a full-featured Command Line Interface (CLI), context-sensitive help menus, an online data dictionary and CLI user's guide, and plotting capabilities for review of output data. EXAMS provides 20 output tables that document the input data sets and provide integrated results summaries to aid in ecological risk assessments.

The EPA uses coupled PRZM and EXAMS models for risk and exposure assessments for chemical registration and re-registration. The EPA has created 38 standard scenarios covering various crops and geographical regions for this purpose. Two of the scenarios were established for turf systems, within a golf course context, in Pennsylvania and Florida.

Comparison of PRZM/EXAMS Results with Monitoring Data

Recently, Jackson and colleagues (2005) provided a comprehensive comparison of the EPA's predictions of pesticide residues in EPA's standard "index" reservoir scenario with watershed-scale monitoring data. The scenarios analyzed included 42 combinations of pesticides and application sites, including one fungicide applied to turfgrass, following the EPA's guidance (Gallagher, Touart, and Lin 2001; Jones et al. 1999, 2000; USEPA 2000, 2001, 2002). The focus of this work was to evaluate the ability of two exposure models—the FQPA Index Reservoir Screening Tool (FIRST) and PRZM/EXAMS—to predict concentrations found in drinking water when compared with actual monitoring data from water systems. Once model runs were completed using FIRST, it was then possible to compare the predicted exposure estimates to actual monitoring data collected. Figure 9.2 is the comparison of Tier I predicted reservoir concentrations with actual measured values. Modeled versus monitored data are sorted from greatest overprediction to the least as paired comparisons. Results of the analysis indicated that modeling resulted in several orders of magnitude overprediction compared with actual water concentrations. To determine if there was any relationship between overprediction and model input, input factors were analyzed using a backward, stepwise regression. The backward, stepwise analysis was used to confirm and check that the authors had not overlooked any process that might contribute to predictions. It was determined that the total compound application rate (application rate x number of applications) was the best indicator of model overprediction.

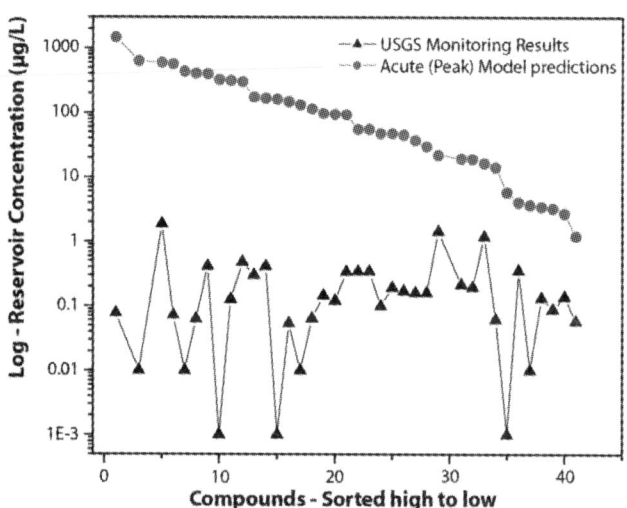

Figure 9.2. Comparison of Tier I acute model predictions versus monitoring. (Jackson et al. 2005. Reprinted with permission from the American Chemical Society.)

On the basis of the comparison in Figure 9.3, it is apparent that although some reduction in overprediction was obtained using Tier II modeling methodology, the reductions were not consistent nor a great improvement in accuracy. The regression developed in Figure 9.4 describes the relationship between total pounds (lb) applied and the Tier II acute PRZM/EXAMS model predictions

$$y = 2.156 + 1.03584\,x \tag{1}$$

where y = log of model overprediction and x = log of total active applied.

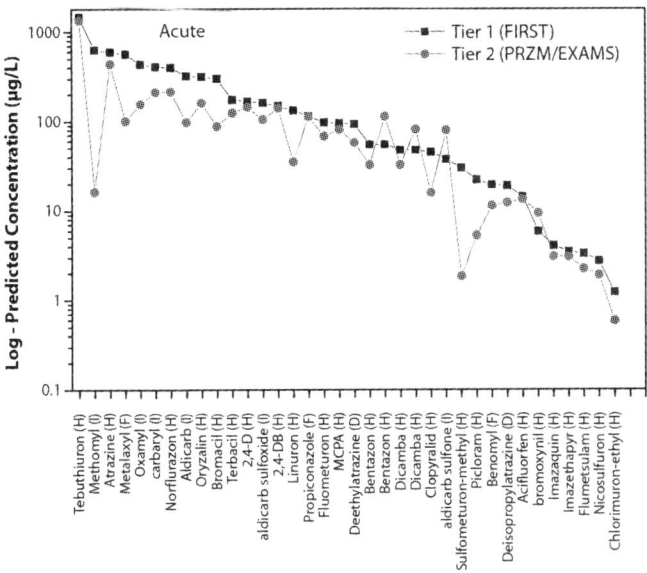

Figure 9.3. Comparison of Tier I with Tier II acute exposure predictions. (Jackson et al. 2005. Reprinted with permission from the American Chemical Society.)

As a computational example, the dicamba-use pattern has a maximum application rate of 0.5 lb/ac x two applications for a total seasonal active application rate of 1.0 lb/ac. Then,

log of 1.0 lb/ac = 0.0
x = log of 1.0 lb/ac or 0.0
y = 2.156 + 1.03584 X 0.0

Thus the overprediction factor (modeling/monitoring) = 10^y = $10^{2.156}$ ≈ 143.21. Residue correction = 1/143.21 x model prediction of 47.4 micrograms/liter (μg/L), or 1/143.21 x 47.4 μg/L = 0.33 μg/L.

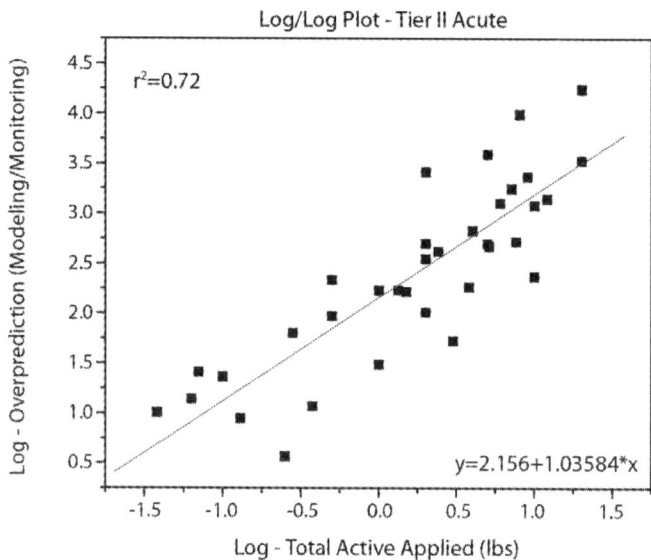

Figure 9.4. Relationship between total active applied and PRZM/EXAMS overprediction used to develop the overprediction factor. (Jackson et al. 2005. Reprinted with permission from the American Chemical Society.)

The PRZM/EXAMS scenario yielded an acute concentration of 47.4 µg/L. Taking into account the overprediction factor, a reasonable monitoring concentration value of 0.33 µg/L might be expected in larger bodies of water using the overprediction correction factor. The actual maximum concentration found in the U.S. Geological Survey (USGS) reservoir study was 0.19 µg/L. To address the concern that the 2-year monitoring study might not adequately describe the magnitude and frequency of residues occurring in water, available monitoring data were obtained from the USGS National Agricultural Water Quality Assessment (NAWQA) program data server. The data are from all available sources of surface water in the database and are presented as an indication that the data in the reservoir study are consistent with other data collected to describe pesticide residues and the analyses of the International Life Sciences Institute (1999) and RTI. The overprediction correction factor also was applied for each of the presented compounds summarized from the NAWQA program. Based on a paired t test, the predicted concentrations after application of the overprediction factor to modeled data and the NAWQA monitoring data were not significantly different ($p < 0.05$).

In a recent presentation, Ma, Cohen, and Barnes (2005) highlighted fundamental concerns with the way PRZM is used to simulate pesticide loss in runoff from turfgrass. One key concern is that erosion in turfgrass is handled the same as for a row crop, i.e., the thatch layer is actually allowed to "erode," which would lead to higher predicted runoff losses than would normally occur for chemicals that tend to bind to organic matter.

TurfPQ

The following description is primarily based on Haith (2001, 2002). TurfPQ is a mathematical model of pesticide runoff from turfgrass. It is based on experimental data. Compared with previous pesticide runoff models, which were developed for agricultural crops, TurfPQ is relatively simple in structure and requires a minimal number of inputs. A daily record of temperature and precipitation is needed, as well as the dates and quantities of pesticide applications. Model input parameters are runoff curve number, pesticide decay half-life, organic carbon partition coefficient, and the organic carbon in turfgrass vegetation. Default values have been developed for these parameters, and the model was tested for 52 pesticide runoff events involving six pesticides measured in plot studies in four states. TurfPQ typically produced conservative overpredictions of pesticide runoff, particularly with strongly adsorbed pesticides. Mean predicted pesticide runoff was 2.9% of application, compared with an observed mean of 2.1%. The model captured the dynamics of the pesticide runoff events well with $R^2 = 0.65$.

RZWQM

As described on the U.S. Department of Agriculture–Agricultural Research Service–Great Plains Systems Research Web site (USDA 2007), RZWQM simulates major physical, chemical, and biological processes in an agricultural crop production system. It is a one-dimensional (vertical in the soil profile), process-based model that simulates the growth of the plant and the movement of water, nutrients, and agro-chemicals in a cropping system under a range of common management practices. The model includes simulation of a tile drainage system and a simplified turfgrass component for simulating water and chemical behavior in turfgrass systems.

RZWQM consists of six major submodules/processes that define the simulation program, a Numerical Grid Generator, and an Output Report Generator. These six submodels are physical processes, plant growth processes, soil chemical processes, nutrient processes, pesticide processes, and management processes. Interactions between these programs are achieved through the use of seven input data files and three generated output files. The user can create and modify input files using a commercial editor. The model generates three general output files with 25 optional debugging output files that provide detailed results. The Output Report Generator uses model results to create summary tables and publication-quality graphical output in two- and three-dimensional formats. The most recent version of the model has a Microsoft Windows user interface.

In addition to commonly available climatic data, such as daily minimum and maximum air temperature, wind speed, radiation, and relative humidity, RZWQM also requires actual breakpoint rainfall data for accurate simulations.

There is a series of papers in *Agronomy Journal* (Vol. 91[2], 1999) and *Pest Management Science* (Vol. 60, 2004) discussing the model and validation of each component.

Generic Estimated Environmental Concentration (GENEEC)

The EPA GENEEC model is a screening level (Tier I) model that was designed by the EPA's Office of Pesticide Programs (OPP) to mimic the results of a Tier II model (i.e., the EPA's PRZM/EXAMS) (Parker, Jones, and Nelson 1995). The model conservatively assumes a 10-ha farm field is applied with a pesticide, and the pesticide runs off into a 1-ha pond with no renewable source of water. The model outputs the estimated environmental concentrations in the water column at peak, maximum 4-day, 21-day, 60-day, and 90-day intervals, based on easily available input parameters.

Screening Concentration in Ground Water (SCI-GROW)

SCI-GROW is a screening level model (Tier I) that OPP uses to calculate pesticide concentrations in vulnerable groundwater (Cohen 2000; USEPA 2007). These concentrations are approximately the upper 99th percentile of actual USGS NAWQA monitoring results. The model provides an exposure value that is used to determine an upper limit of the potential risk to the environment and to human health from drinking water contaminated with the pesticide(s) modeled. The SCI-GROW estimate is based on environmental fate properties of the pesticide(s) (aerobic soil degradation half-life and linear sorption coefficient normalized for soil organic carbon content), the maximum application rate, and existing data from small-scale prospective groundwater monitoring studies at sites with sandy soils and shallow groundwater. Pesticide concentrations estimated by SCI-GROW represent conservative or high-end exposure values because the model is based on groundwater monitoring studies that were conducted by applying pesticides at maximum allowed rates and frequency to vulnerable sites (i.e., shallow aquifers; sandy, permeable soils; and substantial rainfall and/or irrigation to maximize leaching).

Experience with golf course well-monitoring indicates that most pesticide analyses yield no detections, which complicates the ability to compare SCI-GROW results with monitoring results. The predicted concentration for one detected pesticide at a northeast U.S. golf course, however, compared well with actual monitoring results, i.e., the predicted SCI-GROW concentration for pesticide X = 0.2 parts per billion (ppb), the actual groundwater concentration ranged from non-detect to a high of 0.9 ppb, 70% of the detectable concentrations were 0.2 ppb. But, if Maximum Residue Limits were lower, i.e., < 0.1 ppb, then a better comparison could be made.

Soil and Water Assessment Tool (SWAT)

SWAT is a process-based, continual, daily, time-step model that evaluates land management decisions in large, ungauged rural watersheds. It is designed to predict long-term, nonpoint, source pollution impacts on water quality such as sediment, nutrient, and pesticide loads (Arnold et al. 1994). Model processes include calculations of water balance (i.e., surface runoff, return flow, percolation, evapotranspiration, and transmission losses), crop growth, nutrient cycling, and pesticide movement. Model outputs include subbasin and watershed values for surface flow; groundwater and lateral flow; crop yields; and sediment, nutrient, and pesticide yields. SWAT can analyze watersheds and river basins of 100 square miles by subdividing the area into homogenous parts.

Nutrient Modeling

Models that can simulate nutrient fate and transport include GLEAMS, Leaching Estimation and Chemistry Model for nutrient simulation (LEACHN), Nitrate Leaching and Economic Analysis Package (NLEAP), PRZM, RZWQM, SWAT, and others. These models were developed mainly for simulating nutrient behavior in agricultural systems. Their validity for turfgrass systems has not yet been thoroughly tested. Thus, the same issue of applicability exists as for the pesticide component included in these models. Of these models, NLEAP (Shaffer, Halverson, and Pierce 1991) is relatively simple and specifically designed for determining potential nitrate leaching associated with agricultural practices. NLEAP calculates potential nitrate leaching below the root zone and to groundwater supplies. It also provides three levels of analyses to determine leaching potential: an annual screening, a monthly screening, and an event-by-event analysis. PRZM and RZWQM can only simulate nitrate (N) behavior; GLEAMS, LEACHN, and SWAT can simulate both N and phosphorus behavior; and LEACHN can only simulate nutrient leaching.

Realistic Model Input Parameters: Their Impacts on Results and the Implications of Compounding Low-Probability Assumptions

Critical Assumptions

For pesticides with short to moderate persistence, one of the most sensitive categories of input parameters is the degradation rate constant (k). This constant is used to represent transformations from photolysis on surfaces and in water, aerobic metabolism in the thatch layer and root zone, foliar decay, and transformation below the root zone. A simplified form of the algorithm in which this usually occurs is equation 2

$$C_t = C_0 e^{-kt}, \quad (e=2.72) \tag{2}$$

where C_t = the pesticide concentration at time (t), and C_0 = the initial concentration.

Typically, the exponential term (-kt) includes a soil partitioning term as well. Thus k is part of an exponential term, and choice of an inappropriate k can have significant consequences for the model output and risk assessment (e = 2.72). For example, if a k is chosen based on a conservative estimate from a "look-up" table, which does not consider enhanced degradation in the thatch root system, a two-fold error in k can yield a 7.4-fold error in the predicted concentration.

The previous discussion is relevant to pesticide chemistry, which is only one category of model input. Two example sensitive field input parameters that are relevant to turfgrass system model simulations are the soil erodibility parameter (K), and the runoff factor that replaced the rainfall energy/erosivity factor (R) (Williams and Berndt 1977) in the original Universal Soil Loss Equation (Wischmeier and Smith 1978), which was based on agricultural systems. The OPP/EPA standard cool-season turfgrass scenario assigns a K value of 0.33 to the system. A K value of 0.05 may be more appropriate, because thatch does not erode. Such a change in this single parameter would generate a 6.6-fold reduction in soil lost to erosion (Figure 9.5). This could have implications for pesticides that are bound to soils.

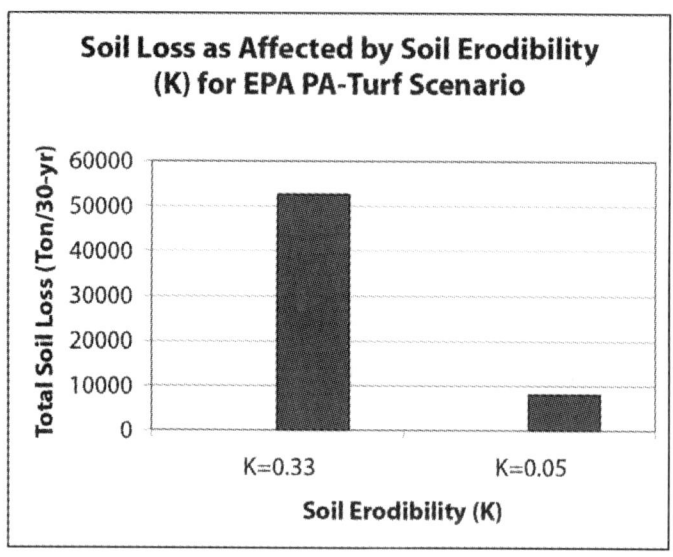

Figure 9.5. Implications of an erodibility parameter K = 0.33 for the EPA/OPP cool-season turf scenario.

A final example of sensitive modeling parameters is in the category of pesticide use—the assumptions made about treated turfgrass in a watershed. The OPP/EPA assumes that 100% of its standard pond and drinking water reservoir watersheds consist of turfgrass, which can be part of a golf course, a sod farm, or a series of home lawns. Research indicates this assumption errs typically by two orders of magnitude for golf courses and by a factor of two or more for lawns. Furthermore, the OPP/EPA assumes 100% of these turfgrass areas are treated with the subject pesticide at the maximum label rate. This latter combination of assumptions is not unreasonable for placing upper limits on risk assessments by a regulatory agency, but decreased levels of treatment areas and application rates should be explored as well. [Note: A recently reviewed confidential EPA risk assessment acknowledged, possibly for the first time, that these assumptions are likely too high.]

Low-Probability Assumptions

At a screening level of risk assessment (Tier I), it is appropriate to make a series of low probability assumptions to ensure that potentially risky chemicals do not escape further scrutiny. The collection of these assumptions yields highly improbable results. This can be illustrated via the following equation:

$$^1P_{low} \times {}^2P_{low} \times {}^3P_{low} \times {}^4P_{low} = P_{unreasonably\ low} \qquad (3)$$

1 = assume the upper 90th percentile pesticide transformation half-life

2 = the highest 90th percentile vulnerability approximation for the crop scenarios

3 = the highest 90th percentile for the metabolism rate constant

4 = assumptions of 100% turfed area, 100% treated.

This is inappropriate, however, for the Tier II or refined Tier II assessments that the OPP/EPA has used to justify significant regulatory actions against turf pesticides. For such actions, it is appropriate to regulate based on an estimate of the upper limit of exposure as well as a characterization of a range of exposures. Instead, when the OPP/EPA regulates based on a concept illustrated by equation 3, it goes beyond the extreme upper limit.

Summary

Turfgrass treated with pesticides and fertilizers is important to support various land uses in the United States, and it is an important commodity. Researchers who develop various approaches to turfgrass management, regulators and the regulated community concerned about off-site transport of pesticides and nutrients, and various scientists and engineers who design Best Management Practices for managed turf rely on mathematical models to predict the fate of turfgrass chemicals. Most of these models have not been designed for turfgrass, and the unique aspects of turf relative to row crops should be incorporated into model algorithms and input guidance. In addition, there can be fundamental questions about the overall model application scenarios regarding the extent to which they reliably predict reality. Models remain useful tools, but their content and application must be continually scrutinized and improved.

Conclusions

- Mathematical models are powerful tools that must be used to extrapolate from limited, expensive, and difficult-to-generate experimental data sets to different points in space and time.

- Some models can often give reasonable, edge-of-plot predictions of pesticides and N in leachate and runoff for turfgrass systems.

- There are fundamental, conceptual model and algorithm issues when evaluating chemical fate in turfgrass compared with row crop ag systems. There are special concerns about simulation of the thatch layer, degradation rate constants in the turfgrass system versus plain soil, and the portions of the modeled watersheds that consist of turfgrass. In a related context, there are fundamental issues when evaluating pesticide runoff into the standard OPP/EPA drinking water reservoir.

- In most instances, these suspected or documented technical issues tend to result in overpredictions of pesticide risks. This happens when a series of low probability assumptions is compounded to yield a highly unlikely scenario.

- It should be remembered that "Mother Nature rolls dice," and that rather than representing extreme scenarios, models are best used to represent a reasonable range of possible outcomes.

Recommendations

1. Institutions that create and maintain mathematical models to predict offsite transport of pesticides and fertilizers should continually evaluate the performance of these models against monitoring data. Therefore, model validation and evaluation should be an ongoing task.

2. Institutions that create and/or apply the constructed scenarios used to apply the models, e.g., the standard surface water drinking water scenario and the turfgrass runoff scenario, also should continually evaluate the extent to which these scenarios accurately represent reality and produce realistic results.

3. The following specific technical issues should be addressed when modeling the fate of turfgrass chemicals:

    - rainfall impacts on thatch vs. soil;

    - whether the models cause the thatch layer to erode;

    - the effective organic carbon content of thatch relevant to pesticide sorption; and

    - aerobic metabolism half-lives in the turfgrass root zone, which are presumably significantly more rapid than the kinetics derived from the standard soil test. Generating laboratory metabolism kinetics on bioactive turfgrass root zone soils will pose some technical challenges, but this should be done.

4. For risk assessment tiers beyond Tier I, the EPA should take care to include some realistic assumptions such that a range of probable outcomes is presented, not just an unreasonable, worst case.

Glossary

EXAMS (Exposure Analysis Modeling System) is an aquatic ecosystem model that evaluates the fate, transport, and exposure concentrations of synthetic organic chemicals — pesticides, industrial materials, and leachates from disposal sites.

GENEEC (GENeric Estimated Environmental Concentration) is an EPA Tier 1 screening model.

GLEAMS (Groundwater Loading Effects of Agricultural Management Systems) is a field-scale compartment model for water and chemical runoff and leaching.

LEACHM (Leaching Estimation and Chemistry Model) is a physically based, integrated process model for simulating water and chemical leaching based on the Richard's equation. LEACHN is a submodel of LEACHM for nutrient simulations.

NLEAP (Nitrate Leaching and Economic Analysis Package) is a simple, screening-level, nitrate-leaching model.

PRZM (Pesticide Root Zone Model) is a model for predicting pesticide and nitrogen fate in the crop root and unsaturated soil zones. It also can be used to predict runoff when linked with EXAMS.

RZWQM (Root Zone Water Quality Model) is a physically based, integrated process model for simulating major processes occurring in a typical agricultural system.

SCI-GROW (Screening Concentration In Ground Water) is a screening-level model that uses a regression equation based on the results of small-scale prospective groundwater studies of exceptionally vulnerable sites using the maximum allowable pesticide rates.

SWAT (Soil and Water Assessment Tool) is a watershed-scale model for simulating water and chemical runoff and leaching. It also contains a turfgrass component.

TurfPQ is a curve-number, method–based, regression model for pesticide runoff from turfgrass.

Literature Cited

American Society of Testing and Materials (ASTM). 1984. *Standard Practice for Evaluating Environmental Fate Models of Chemicals*. ASTM, Philadelphia, Pennsylvania.

Arnold, J. G., J. R. Williams, R. Srinivasan, K. W. King, and R. H. Griggs. 1994. *SWAT: Soil and Water Assessment Tool*. USDA, Agricultural Research Service, Grassland, Soil and Water Research Laboratory, Temple, Texas.

Beard, J. B. 2000. Turfgrass Benefits and the Golf Environment. Pp. 36–45. In J. M. Clark and M. P. Kenna (eds.). *Fate and Management of Turfgrass Chemicals*. ACS Series 743. American Chemical Society, Washington, D.C.

Cohen, S. Z. 2000. Recent examples of pesticide assessment and regulation under FQPA. *Ground Water Mon & Rem* 20(1):41–43.

Cohen, S. Z., R. D. Wauchope, A. W. Klein, C. V. Eadsforth, and R. Graney. 1995. Offsite transport of pesticides in water: Mathematical models of pesticide leaching and runoff. *Pure & Appl Chem* 67(12):2109–2148.

Durborow, T. E., N. L. Barnes, S. Z. Cohen, G. L. Horst, and A. E. Smith. 2000. Calibration and validation of runoff and leaching models for turf pesticides, and comparison with monitoring results. Pp. 195–227. In J. M. Clark and M. P. Kenna (eds.). *Fate and Management of Turfgrass Chemicals*. ACS Series 743. American Chemical Society, Washington, D.C.

Gallagher, K., L. Touart, and J. Lin. 2001. *A Probabilistic Model and Process to Assess Risks to Aquatic Organisms*. SAP, Crystal City, Virginia, May 13–16. Office of Pesticide Programs, Environmental Protection Agency, Washington, D.C., http://epa.gov/scipoly/sap/meetings/2001/march/aquatic.pdf (23 April 2007)

Haith, D. A. 2001. TurfPQ, a pesticide runoff model for turf. *J Environ Qual* 30(3):1033–1039.

Haith, D. A. 2002. TurfPQ, a pesticide runoff model for turf – Errata. *J Environ Qual* 31(2):701–702.

Horst, G. L., P. J. Shea, N. Christians, D. R. Miller, C. Stueffer-Powell, and S. K. Starrett. 1996. Pesticide dissipation under golf course fairway conditions. *Crop Sci* 36:362–370.

International Life Sciences Institute (ILSI). 1999. *A Framework for Estimating Pesticide Concentrations in Drinking Water for Aggregate Exposure Assessments*. ILSI, Washington, D.C., http://rsi.ilsi.org/NR/rdonlyres/ACB8DC87-FE21-44CA-8D00-348AFE8FBF2B/0/dwpests.pdf (17 January 2007)

Jackson, S., P. Hendley, R. Jones, N. Poletika, and M. Russell. 2005. Comparison of regulatory method estimated drinking water exposure concentrations with monitoring results from surface drinking water supplies. *J Agric Food Chem* 53:8840–8847.

Jones, R. D., S. Abel, W. Effland, R. Matzner, and R. Parker. 1999. *An Index Reservoir for Use in Assessing Drinking Water Exposure*. SAP, Crystal City, Virginia, July 20–23. Office of Pesticide Programs, Environmental Protection Agency, Washington, D.C.

Jones, R. D., J. Breithaupt, J. Carleton, L. Libelo, J. Lin, R. Matzner, R. Parker, W. Effland, N. Thurman, and I. Kennedy. 2000. *Guidance for Use of the Index Reservoir and Percent Crop Area Factor in Drinking Water Exposure Assessments*. Office of Pesticide Programs, Environmental Protection Agency, Washington, D.C.

Lickfeldt, D. W. and B. E. Branham. 1995. Organic chemicals in the environment; Sorption of nonionic organic compounds by Kentucky bluegrass leaves and thatch. *J Env Qual* 24:980–985.

Ma, Q., S. Z. Cohen, and N. L. Barnes. 2005. Modeling offsite transport of turf-applied pesticides: Model and data needs. Presentation before the 230th American Chemical Society National Meeting, August 31, Washington, D.C.

Parker, R. D., R. D. Jones and H. P. Nelson. 1995. GENEEC: A screening model for pesticide environmental exposure assessment. Pp. 485–490. In *Proceedings of the International Exposure Symposium on Water Quality Modeling*. American Society of Agricultural Engineers, Orlando, Florida.

Racke, K. D. 1993. Environmental fate of chlorpyrifos. *Rev Environ Contam Toxicol* 131:1–151.

Shaffer, M. J., A. D. Halverson, and F. J. Pierce. 1991. Nitrate leaching and economic analysis package (NLEAP): Model description and application. Pp. 285–322. In R. F. Follett, D. R. Keeney, and R. M. Cruse (eds.). *Managing Nitrogen for Groundwater Quality and Farm Profitability*. Soil Science Society of America, Madison, Wisconsin.

U. S. Department of Agriculture (USDA). 2007. *RZWQM—Root Zone Water Quality Model*, Great Plains System Research Unit page, http://gpsr.ars.usda.gov/ products/rzwqm.htm (17 January 2007)

U. S. Environmental Protection Agency (USEPA). 2000. *Drinking Water Screening Level Assessment, Part A: Guidance for Use of the Index Reservoir in Drinking Water Exposure Assessments*. Pest. Science Policy. Office of Pesticide Programs, Environmental Protection Agency, Washington, D.C., http://www.epa.gov/oppfod01/trac/science/ reservoir.pdf (17 January 2007)

U. S. Environmental Protection Agency (USEPA). 2001. *Interim Guidance for Developing Input Parameters in Modeling the Environmental Fate and Transport of Pesticides: Version I*. Water Quality Tech Team, Environmental Fate and Effects Division, Office of Pesticide Programs, Environmental Protection Agency, Washington, D.C.

U. S. Environmental Protection Agency (USEPA). 2002. *Guidance for Selecting Input Parameters in Modeling the Environmental Fate and Transport of Pesticides, Version II*. Office of Pesticide Programs, Environmental Protection Agency, Washington, D.C., http://www.epa.gov/oppefed1/models/water/input_guidance2_28_02.htm (17 January 2007)

U.S. Environmental Protection Agency (USEPA). 2007. Water Models page, http://www.epa.gov/oppefed1/models/water/index.htm#scigrow (17 January 2007)

Williams, J. R. and H. D. Berndt. 1977. Sediment yield prediction based on watershed hydrology. *Transactions ASAE* 44(6):1100–1104.

Wischmeier, W. H. and D. D. Smith. 1978. *Predicting Rainfall Erosion Losses – A Guide to Conservation Planning*. USDA Handbook 537. Washington, D.C. 58 pp.

10

Urban Landscape Water Conservation and the Species Effect

Dale A. Devitt and Robert L. Morris

Introduction

Populations in urban areas throughout the world are expected to double over the next 50 to 75 years (Jury and Vaux 2005). California's population by the year 2020 is forecast to increase by 15 million people over the 1995 estimate, much of this occurring in the south coast region (California Department of Water Resources 1998). Nevada's population in the 1990s increased by more than 60%, mostly in and around the city of Las Vegas (U.S. Census Bureau 2000). In many of these areas, growth has been associated with urban sprawl, such as the 354-square-mile increase reported for Phoenix, Arizona, from 1970 to 1990, and the doubling of Albuquerque, New Mexico's geographic area during the same time period (Western Resource Advocates 2004). Such growth creates complex problems for cities, counties, states, and federal agencies.

One such problem is the acquisition, treatment, and delivery of enough water to meet residential indoor and outdoor needs. Long before water demands approach the limitations of water resource availability, it is critical that well-thought-out, sustainable water management plans be put into place. Although water savings can and should occur indoors, outdoor water savings can be significantly higher, especially in arid environments such as the southwestern United States. As urban development increases, total residential landscape water use also increases. A single home is somewhat insignificant when one looks at the water balance for a large community, but when 6,000 people move into a city like Las Vegas each month for 20 years, it translates into significant increases in the amount of irrigated landscapes.

Many of these new residents come from wetter regions of the country and bring with them their own "sense of place." This sense of place is translated into their expectation of what an urban landscape should become, which impacts plant selection, density of plantings, and sense of landscape scale (mature size of plant material used). Along with a sense of place, society as a whole expects a level of plant quality that differs significantly from the quality of that same plant observed growing in its native environment.

Besides the end-user expectations, we also have municipalities and homeowner associations dictating landscape design, level of plant quality, plant selection, size of plant material used, and plant densities. Easterbrook (1999) of the Fannie Mae Foundation stated, "Smart growth threatens to derail one of the key engines of the national economy: suburban sprawl. Despite its negative image, sprawl is efficient and reflects consumer preference. In a nation where so much developable land remains, sprawl is hardly the environmental threat it is made out to be. The real threat is that the nation might adopt policies that halt development and frustrate the millions of people who seek their share of the suburban dream."

Such a dream must be in balance with nature when it comes to the design and maintenance of living spaces and landscapes. John Lyle (1993) argues, "For thousands of years, cities have existed apart from nature. They rush the water falling upon their roofs and streets as rain out through concrete pipes and channels into the nearest bay or river and at the same time, bring water in from distant landscapes through similar concrete channels."

Urban Landscapes

Urban landscapes can provide beauty, decrease runoff from storm events, provide cooling, and remove environmental pollutants, to name just a few good reasons they are valuable. Unfortunately, urban landscapes also can require significant amounts of water, and this water often is applied inappropriately. In some regions of the United States, water issues are not as critical as in other regions. For example, the Colorado River Basin is currently in an extended 5-year hydrologic drought. During the twentieth century, at least six major droughts impacted North America. Unfortunately, it is difficult to predict when droughts may occur, where they may occur, and for how long they may last. It is essential that communities develop and implement sustainable water management plans. Even the best-laid plans, however, will be severely tested by unpredictable growth in populations. In times of drought when municipalities look to other sectors of society for additional water, it is essential they first demonstrate good stewardship of the water they use. The focus on water conservation programs for urban landscapes in many communities is appropriate and can lead to significant water savings.

Urban landscapes vary not only in size, composition, functionality, microenvironments, and edaphic factors, but also in the cultural management practices imposed. As such, the amount of water applied and the potential to conserve water varies with each landscape setting. But it is clearly the species, size, and number of plants (density) used in a landscape that drive its water requirement. Irrigations must satisfy transpiration and evaporation losses, irrigation inefficiencies, and any leaching requirements, which in turn dictate the total amount needed. Except under extreme drought conditions, the goal of conservation efforts in urban landscapes is to apply water more closely in parallel with landscape requirements while meeting the quality expectations of the end user.

Because of water restrictions, regulations, and pricing, landscape water managers in the southwestern United States (whether they are homeowners or professionals) have begun the process of decreasing total irrigation amounts, improving irrigation systems, and redesigning to decrease the area of what is thought to be high-water-use landscape plantings (encouraged and directed through educational programs). Movement toward more xeric landscapes has gained the interest of the general public and therefore the nursery and landscape industry.

The Green Industry and end users responsible for water management have been slow to understand and adopt the changes necessary to convert, manage, and sustain these xeric landscapes. Depending on the region of the United States, landscapes that can survive solely on precipitation are perhaps possible, but typically not acceptable to the end user or the community. In the more arid regions such as the Mojave Desert, landscapes that can survive on 4 inches of annual rainfall create a sparsely vegetated landscape dominated by creosote bush and other desert species. Such a landscape represents an extreme case in plant selection, total plant numbers, total plant biomass, and xeric design, and unfortunately (because of current housing design) results in higher residential energy use (McPherson, Simpson, and Livingston 1989). The challenge is to win over the general public in support of landscapes that are aesthetically acceptable, require less water, and do not increase energy consumption unnecessarily. Education and research must be provided so water savings can be calculated during the design phase and actually captured for the community and not traded for higher energy costs.

It is one thing to know that certain plants require less water and another thing actually to apply the right amount. Water use of various turfgrass species has been quantified by many scientists under many different growing conditions (Beard 1989; Feldhake, Danielson, and Butler 1983; Huang and Fry 1999). Far less is known, however, about the actual water use of ornamental plants, especially large trees, and even less about mixed landscapes. Much of this discrepancy is related to the number of ornamental species that need to be studied.

There are perhaps only 12 major turfgrass species used extensively in urban landscapes throughout the United States, whereas the number of ornamental species safely exceeds several thousand. It may be this paucity of research on ornamentals and total landscape water use, compared with research that has enabled the precision irrigation of turfgrass, that has led to turfgrass restrictions or its removal in many water conservation programs. This has resulted in the substitution of turfgrass, with a precise water use, for ornamentals of unknown but gradually increasing water use. Frequently, this is done without regard to maximum planting densities.

Plant Water Use/Landscape Water Use

A fundamental axiom in plant biology is that as total biomass increases (canopy size, leaf area), total transpiration increases (de Wit 1958; Heilmeier et al. 2002). Any factor that accelerates plant growth has the potential to accelerate plant water use in both turfgrass and ornamental species (e.g., increased irrigation, nitrogen applications, and other plant stimulation through cultural management). Stabler and Martin (2000, 2004) found that increased irrigation volumes stimulated shoot growth for oleander, Texas sage, red bird of paradise, and blue palo verde. In a study of 20 tall fescue cultivars, Bowman and Macaulay (1991) found evapotranspiration (ET) to be correlated strongly with dry yield.

All plants are not the same in regard to growth and transpiration. Transpiration on a leaf surface area basis can be quite different for many species. Thus, plants of similar size may have entirely different water use rates (Wullschleger, Meinzer, and Vertessy 1998). On the other hand, many species may have similar transpiration rates on a leaf surface area basis (Sala, Devitt, and Smith 1996) but have an entirely different canopy size. A disregard for plant size (especially at maturity) can result in significant irrigation errors over time at both the plant and total landscape level.

Most urban irrigated landscapes in arid and semiarid climates will need to be redesigned into lower-water-use landscapes as water becomes scarcer. Communities will need to implement strategies that will drive total landscape water use down. An excellent example of this can be found in the community of Civano (Tucson, Arizona), which reports water use of only 77 gallons per capita per day. This low water consumption is based on a development that has (1) small residential lot sizes, (2) Xeriscape™ landscapes on private lots and common areas, and (3) 35% of the area dedicated as Sonoran Desert open space (Western Resource Advocates 2004). It is not critical that a universal approach be taken by all communities, but rather, that the final outcome of less water use outdoors is attained.

Xeriscape™ landscaping is founded on solid principles of planning and design, soil analysis, selection of suitable plants, practical turfgrass areas, efficient irrigation, use of mulches, and appropriate maintenance (Clewis 1991). Unfortunately, few homeowners possess the knowledge necessary to convert traditional landscapes to Xeriscape™ successfully. Even if such landscapes were installed correctly, information still is lacking on proper irrigation of most landscape ornamentals and on total landscape water use. Additionally, one cannot assume that changes in behavioral practices related to irrigating and managing landscapes would occur. Peterson, McDowell, and Martin (1999) reported that irrigation management by homeowners in the desert Southwest did not change substantially after landscapes were converted from "traditional" to more xeric designs, and no decrease in total landscape water use was realized.

Landscapes will need to be well-thought-out in terms of size, use and appropriate design, species composition, and the kind of irrigation and cultural management practices required. This practice will require a significant change in how most landscapes are planned, designed, developed, and maintained.

Financial constraints have been shown to be a barrier in making changes to more xeric-type landscapes. Spinti, St. Hilaire, and VanLeeuwen (2004) found that in Las Cruces, New Mexico, a significantly higher percentage of participants in the lower and lower-middle income categories, versus the middle-high and high income categories, indicated that they would like to alter their landscapes, but that financial constraints were barriers to making those changes. Martin (2003) reported that in Arizona most homeowners preferred an "oasis-type" landscape design to either a desert or mesic design. Oasis designs have a designated high-water-use area called an "oasis" or "mini-oasis," but overall landscape water savings occur because of lower water use elsewhere in the landscape. Martin further suggested that plant spacing, vegetation coverage, plant size, and growth rate can be more important in determining landscape water use than actual plant selection. In a review of water-efficient landscape ordinances, California's Department of Water Resources concluded that "there is very little information, guidance, or training available related to maintaining landscapes with water conservation in mind" (California Urban Water 2005).

As water availability decreases, whether because of regulations, pricing, or personal preference, landscape managers and homeowners need to recognize that they control when and how much water is applied to a landscape. To conserve water in the landscape, it is critical to apply irrigations in parallel to the environmental demand and to adjust the irrigation volume based on the species and its level of development. When water is not a limiting resource, homeowners and landscape water managers typically give little consideration to water management, often changing irrigation clocks only once or twice a year. But as water pricing increases, or restrictions are imposed, homeowners and landscape water managers will need to become better irrigators. More sophisticated (but also more user-friendly) irrigation clocks will need to be installed; irrigation systems will need to be designed or redesigned in a way that groups plants of similar watering frequency on the same irrigation circuit; and plants known to have high water requirements will need to be eliminated, decreased in numbers or area, or placed in designated, functional, high-water-use areas of the landscape (oases). But most importantly, irrigation managers must become better educated as research-based information becomes available, because even a landscape designed to be water efficient can be overwatered.

All plants use water, and they use more water as leaf surface area increases, as environmental demand increases, as water becomes more available, and as growing conditions are made more favorable. Physiological characteristics such as control of stomata (Cowan 1986), location and position of stomata, leaf reflectance, canopy density, trichome density, cuticle thickness, cuticular wax content, leaf thickness, canopy aerodynamics, growth habit and orientation, potential for rapid growth, and rooting depth generally will separate water use of one species from another (Balok and St. Hilaire 2002; Beard 1973; Kirkham 1999; Tipton and White 1995). Growing conditions, however, will have a large impact on the amount of water a given species will use (resulting in water use "overlap" when species are compared).

Disregarding plant quality (end-user acceptance), how a plant is grown and under what conditions dictate where in the general plant water-use range it will be found. Montague, Kjelgren, and Rupp (1998) investigated the surface energy balance effects on gas exchange of three shrub species. Shrubs growing in mulch had higher leaf temperatures and greater leaf-to-air vapor pressure differences than shrubs growing in turfgrass. Skunkbush sumac had greater stomatal conductance and water loss in turfgrass than in mulch. Euonymus and dogwood, however, showed no differences based on treatments. Zajicek and Heilman (1991) reported greater morning stomatal conductance for crape myrtle over mulch than over bare soil or turfgrass, which led to greater daily water use. Typically, bigger trees use more water than smaller trees of the same species (Devitt, Morris, and Neuman 1994), plants under stress use less water than plants not under stress (Brown, Devitt, and Morris 2004; Carmen Garcia-Navarro and Evans 1999), plants in protected microenvironments use less water than plants subjected to advection and increased radiation loading (Kjelgren and Rupp 1997), and most importantly, larger areas of mixed landscaping use greater amounts of water than smaller areas of mixed landscaping (Devitt and Morris 2006). In Las Vegas, when mixed landscapes were compared from 1968 to 1994, a good correlation existed between residential lot size and residential water consumption (Las Vegas Valley Water District. Personal communication). A general observation was made that low-density development uses more water than high-density development (Western Resource Advocates 2004).

Water in the Southwest historically has been inexpensive and abundant. Peterson, McDowell, and Martin (1999) pointed out that urban landscapes across the Phoenix metropolitan area were developed in such a way that native plants were being replaced community-wide with imported, exotic species, creating a large urban oasis. Hope and colleagues (2003) indicated that such increases in nonnative plant diversity in the greater Phoenix area were positively associated with family incomes. In neighborhoods where family income was above the median value, plant diversity increased two-fold over neighborhoods where family income was below the median value. Making irrigation decisions for landscapes dominated by turfgrass is fairly straightforward. In mixed landscapes, however, Kjelgren, Rupp, and Kjelgren (2000) pointed out that such recommendations are complicated by the diversity of species and their water-use characteristics, making irrigation decisions less precise.

Is There a Species Effect on Landscape Water Use?

Studies have been published during the last few decades quantifying water use of turfgrass, shrubs, trees, and groundcovers, clearly indicating that a real and significant difference in the water use of many species exists (Ayars, Johnson, and Phene 2003; Beeson 2005; Brown, Devitt, and Morris 2004; Carmen Garcia-Navarro, Evans, and Montserrat 2004; Devitt, Morris, and Neuman 1994; Devitt et al. 1998; Garrot and Mancino 1994; Kopec, Shearman, and Riordan 1988; Levitt, Simpson, and Tipton 1995). Unfortunately, all these studies have been conducted at different sites under different conditions, including plant size (seedlings in containers, young transplants, mature trees, trees in forests), level of fertility (native soils, potting soils, with and without nutrient additions), planting densities, climatic conditions (short summer monitoring, multiple year monitoring), and irrigation regimes (deficit irrigating, irrigating to avoid stress, irrigating to replace ET, irrigating with a leaching fraction [LF] included).

At the landscape level, no studies have looked at quantifying plant water use based on substituting one or more species while holding all other variables constant. Yet, even if exhaustive substitution experiments could be conducted, how transferable and important such results would be to a wide range of urban landscapes (different climatic, nutrient, edaphic, and irrigation conditions) is unknown. More importantly, where should the research emphasis be placed—species, landscape size, plant size, plant density, and/or cultural management? Many of these appropriate variables have been researched for turfgrass as noted previously.

The species effect on water use of urban landscapes can be a very important variable, but its impact will be dependent on which species are compared and under what conditions. This is probably why such contrasting results have been reported in the literature. Danielson, Feldhake, and Hart (1981); Kjelgren, Rupp, and Kjelgren (2000); Costello, Matheny, and Clark (2000); and Devitt et al. (1995a) all have reported different results on the impact turfgrass and ornamental species have on the water use of landscapes. All scientists (not just those previously mentioned) have different backgrounds and analytical skills, leading to differences in how data are interpreted and what conclusions are made. Danielson, Feldhake, and Hart (1981) stated, "Trees and shrubs are in a position to use more water than when grass only covers the same area. A water-conserving landscape should have a minimum of woody plants, and those should be species adapted to dry areas. Woody plants need some irrigation to survive drought even though some grasses may not." Kjelgren, Rupp, and Kjelgren (2000) stated, "Even a uniform sprinkler-irrigated landscape of herbaceous perennials or woody plants can potentially use less water than a turf landscape." Devitt and colleagues (1995a) stated, "Replacement of turfgrass with woody ornamental plants would not necessarily ensure large, long-lasting water savings. The results suggest that low-fertility bermudagrass/ryegrass offers low water usage in an arid environment, and that for even small immature trees, water use ratios would favor bermudagrass on a basal canopy area basis."

Costello, Matheny, and Clark (2000) stated, "The water needs of most tree species planted in turf are generally met by the relatively high water needs of turf. Turfgrass crop coefficients (Kc) range from 0.6 for warm-season species to 0.8 for cool-season species. This range is sufficient to satisfy the needs of all trees in the moderate, low, and very low water-use classification of landscape species (WUCOL) categories. Trees in the high category may need supplemental water, particularly if they are planted in warm-season turfgrass. Trees in cool-season turfgrass are not likely to need supplemental water." Devitt and colleagues (1995a) reported entirely different results if comparisons were made between (1) trees under high irrigation versus low fertility bermudagrass and (2) trees under low irrigation versus tall fescue under high irrigation. Tree-to-grass water-use ratios were as high as 2:4 during the growing season when low fertility bermudagrass was used for comparison. In the field, however, canopy volume-to-basal canopy area was as much as 4.5 times greater, and canopy volume-to-trunk diameter was 94 times greater, suggesting that water-use ratios may be significantly higher with mature trees.

Turfgrass Water Use

Water use of turfgrass has been shown by many investigators to vary by species, cultivar, climatic conditions, and cultural management (Beard 1989; Feldhake, Danielson, and Butler 1983; Huang and Fry 1999; Shearman 1985). Danielson, Feldhake, and Hart (1981) conducted research on urban lawn irrigation and management practices in Colorado. Their research indicated that maximum water use was influenced by mowing height, nitrogen fertility, shade level, grass species, and, to a slight degree, soil properties. They further indicated that adequately fertilized grass had minimal reduction in quality when irrigation was decreased to 70% of that required for maximum ET. Significant differences in water use by cultivar have been reported for tall fescue (18% range) (Bowman and Macaulay 1991; Kopec, Shearman, and Riordan 1988); creeping bentgrass (39 to 84% variation on different days) (Salaiz et al. 1991); Kentucky bluegrass (Ebdon and Kopp 2004); and buffalograss (Bowman et al. 1998). Cool-season grasses typically have been reported to use more water than warm-season types (Kneebone and Pepper 1979; Marsh et al. 1980; Shearman and Beard 1973). Feldhake, Danielson, and Butler (1983) reported that bermudagrass in Colorado used 24% less water than Kentucky bluegrass under identical management and microenvironmental conditions. Cool-season tall fescue was reported by Brown, Devitt, and Morris (2004) to have lower dry yields at lower nitrogen (N) rates, which they believed would correlate with lower ET rates, suggesting that further water savings could be achieved by managing N as a water conservation tool. Devitt and Morris (2006) reported a 20% decrease in ET of tall fescue when N was lowered from 1 pound per 1,000 square feet per month to 0.25 pound per 1,000 square feet per month in southern Nevada. Feldhake, Danielson, and Butler (1983) reported a 13% higher ET rate for Kentucky bluegrass in Colorado when 4 kilograms(kg)/1,000 meters(m)^2 of N (0.8 pounds N per 1,000 square feet) was applied each month during spring and summer compared with only one application for the season, applied in spring.

Nitrogen also has been reported to significantly influence ET of bermudagrass (golf courses having a 29% higher ET rate than a park) with differences attributed to cultural management and N fertilization in particular (Devitt, Morris, and Bowman 1992). Intensively managed bermudagrasses, growing under fairway conditions in an arid environment, were shown to have acceptable quality even when irrigated at 60% of potential evapotranspiration (ETo) (Garrot and Mancino 1994).

Much emphasis has been placed on turfgrass, and turfgrass removal, as a means to decrease urban landscape water use. In some instances, turfgrass removal has become an emotional rather than a rational issue. The role of turfgrass in residential design in the past, when water use was not an issue, was relegated to filling landscape design voids—in design terms called "negative space"—and perhaps it has been given a less valuable placement in the mixed landscape, collectively called the "yard" or the "lawn." This value placement in the hierarchy of landscape planting materials may have been transferred to landscapes dominated by turfgrasses where its functional use clearly has economic and social benefits (golf courses, athletic fields, school grounds, parks). Clearly, the removal of any irrigated vegetation should translate into water savings, and those landscapes with significant turfgrass areas would potentially have high water savings based on the percentage of removal. From a scientific point of view, however, should all turfgrass be lumped into one category? (See Beard [1989] for a full review.) Are the water savings the same for tall fescue and bermudagrass (contrasting CO_2 fixation pathways), and are there significant differences in water use based on cultural management (nitrogen application, cutting height, etc.)?

Many turfgrass species demonstrate a significant ability to recover from long-term water deficits (Brown, Devitt, and Morris 2004; White et al. 2001). Although tall fescue is a high water user (Kc >1.2 July) (Devitt and Morris 2006), it has been reported to handle significant levels of drought. In a study by Brown, Devitt, and Morris (2004), water savings of 60 centimeters during summer months was reported for tall fescue when deficit irrigated. Although plant stress increased, color and cover returned to pre-experimental values after a 28-day recovery period, with a well-defined I/ETo (irrigation volume divided by the potential evapotranspiration) threshold of 0.8 for both color and cover. Trees and shrubs, however, can suffer catastrophic cavitations (Sperry and Tyree 1990; Tyree and Cochard 1996; Tyree and Dixon 1986) when soil moisture reaches minimum threshold levels. This condition can lead to significant canopy dieback and/or mortality, which in an urban landscape can lead to removal from the landscape at a significant cost to the end user. Turfgrass can be removed from urban landscapes and replanted into more xeric designs at much less expense to the end user.

Many turfgrass species possess relatively good salt tolerance (Marcum and Murdoch 1994) and demonstrate little foliar damage when poor-quality water—such as reuse water—is used, in contrast to findings for many ornamental trees and flowering annuals (Devitt et al. 2005; Jordan et al. 2001).

Water Use of Trees and Ornamentals

The water use of trees has been shown to vary by species (Buwalda and Lenz 1993; Kjelgren and Montague 1996; Wullschleger, Meinzer, and Vertessy 1998); irrigation frequency (Renquist 1987); plant density (Mitchell et al. 1991; Natali, Xiloyannis, and Barbieri 1985); and the surrounding surface area and mulching material used (Singh, Kumar, and Prasad 1991; Zajicek and Heilman 1991). In a survey of whole-tree water use in forests and plantations (Wullschleger, Meinzer, and Vertessy 1998) that included 67 species in 35 genera, almost 90% of the observations indicated a range in maximum daily water use rates between 10 and 200 kg day^{-1} (species included only if >10 kg day^{-1}) for trees averaging 21 m in height. The highest rate was for *Euperua purpurea* Bth., a tree growing in the Amazonian rainforest, at 1,180 kg day^{-1}. Such values may not be typical of immature urban landscapes, but they do give a perspective to what large trees use in their natural setting. But because these values are for individual trees in larger stands, isolated trees in arid environments that have entirely different boundary layer conditions could have high water use even though the trees are significantly smaller in size.

Devitt, Morris, and Neuman (1994) showed that the yearly ET of oak was significantly influenced by planting size and LF imposed, with 92% of the variability in ET accounted for by the LF, trunk diameter, and canopy volume. The fact that these trees consumed greater amounts of water at the higher LFs would indicate that plant selection alone cannot necessarily lead to maximum water savings, but that water management still must play a critical role in developing low-water-use landscapes. In the same study, Chilean mesquite (*Prosopsis chilensis* [Molina] Stuntz.), a plant commonly used in southwestern xeric landscapes, typically used more water than a seedling selection of live oak (*Quercus virginiana* Mill.) called "Heritage" and categorized at The Arboretum at Arizona State University as "thriving on water" (Arizona 2006). A similar finding was reported by Levitt, Simpson, and Tipton (1995). Ansley and colleagues (1992) suggested that water use by mesquite increases with water availability. The increase in water use by planting size and LF treatment imposed suggested that as woody ornamentals mature, even greater amounts of water will be required relative to the total landscape water needs (Devitt, Morris, and Neuman 1994).

Schuch and Burger (1997) reported water use of 12 different container-grown woody ornamentals in Riverside and Davis, California. Relative rankings of water use by species at each location changed very little during the study period. Of the five highest and five lowest water users, four out of five species were the same for both sites. Balok and St. Hilaire (2002) quantified the drought response of seven southwestern landscape tree taxa. Drought had little impact on Texas red oak (*Quercus buckleyi* Buckl.) based on tissue water relations, large root/shoot ratios, and the presence of high stomatal density and waxy leaves. Arizona ash (*Fraxinus velutina* Torr.), however, had the most negative predawn and midday leaf water potentials, decreased stomatal conductances, and lowest root/shoot ratios. Montague, Kjelgren, and Rupp (1998) reported that skunkbush sumac (*Rhus trilobata* Nutt.), a native to arid habitats and considered drought tolerant, used more water under nonlimiting soil water conditions than did other less-drought-tolerant species.

Although xeric plant species possess many morphological and physiological characteristics that enable them to grow in water-limiting, arid environments, Stabler and Martin (2000) concluded that there is probably no advantage for these plants to conserve water when it is available readily in irrigated urban landscapes. Irrigations should be applied to maintain aesthetic value rather than to encourage excessive growth. To accomplish this, however, precision irrigation and solid knowledge of the plant's growth and water use rates may be required.

Shaw and Pittenger (2004) estimated ornamental plant water needs for acceptable aesthetic appearance. Irrigations were set at 0.36, 0.18, or 0.00 of ETo in Encinitas, California (coastal Mediterranean climate, irrigation plus rain accounted for 32 to 55 % of ETo). Sixteen species had decreased aesthetic quality at lowered irrigation amounts, whereas quality was not affected in 11 of the species studied. Shaw and Pittenger concluded that additional studies need to be performed to verify these findings in climates with higher ETo. Others (Araujo-Alves et al. 2000;

Carmen Garcia-Navarro and Evans 1999) also have evaluated the ability of plants to maintain ornamental value under stress conditions. Carmen Garcia-Navarro and Evans (1999) subjected four species to water stress and quantified a 65 to 70% lower water use than was measured in well-watered plants. In the instance of *Leucophyllum* and *Spiraea* species, however, such conditions led to early senescence and leaf drop. *Arctostaphylos* and *Viburnum* species were classified as low water users, and *Spiraea* and *Leucophyllum* species as moderate and heavy water users, respectively. Such results led the researchers to question possible results in mixed landscapes where competition for water among neighboring plants of different species and genera confound the response. They stated, "One cannot know whether a plant that thrives under minimal irrigation in such a garden does so because it consumes very little water or because it competes better for water than its neighbors."

A few studies have attempted to compare the water use of woody species with turfgrass species (Devitt et al. 1995a; Lownds and Berghage 1991) to assess possible water use trade-offs in landscape design. Staats and Klett (1995) compared three nonturfgrass groundcovers with Kentucky bluegrass (canopy temperatures, visual ratings). Plants were irrigated at 100, 75, 50, 25, or 0% of potential ET. Optimum irrigation for Kentucky bluegrass was found to be 50% of ETo, whereas snow-in-summer (*Cerastium tomentosum* L.) required irrigation at 50–75% ETo. Creeping potentilla (*Potentilla tabernaemontani* Asch.) required irrigation at 75% ETo, whereas goldmoss (*Sedum acre* L.) maintained good aesthetic appearance at rates as low as 25% ETo.

Tree-to-turfgrass water-use ratios reported by Devitt and colleagues (1995a) would confirm cool-season tall fescue as a high water user. Tree-to-grass ratios indicated that water use for tall fescue exceeded the water use for several ornamental trees on a basal canopy area basis. But that was not the case for low-fertility bermudagrass. Such results suggest that turfgrass selection should be given as much consideration as the selection of landscape ornamentals. Several cities have recognized this fact and developed regulations restricting not only turfgrass (turfgrass per lot size, lawn permits required), but also suspected high-water-use woody plants (Tucson and Mesa, Arizona) or the actual number of trees (spacing of trees, selection of tree species and tree equivalents) that a residential landscape can have (Aurora, Colorado).

Data on the water requirements of most ornamental shrubs, trees, and groundcovers are not very extensive and therefore have led to different approaches in developing irrigation plans for mixed landscapes. One such approach, known as the landscape coefficient method (Costello, Matheny, and Clark 2000), is based on a subjective classification of species into low-, medium-, and high-water-use categories (irrespective of plant size) and the combination of coefficients to adjust potential ET values based on species, density, and microenvironment effects. The end user must be able to classify these coefficients for the landscape successfully, have a sophisticated irrigation clock to make such adjustments, and detect declining plant vigor so incremental adjustments can be made before significant stress is magnified and the overall aesthetics of the landscape is threatened.

Considering the alternative, which is to take no action at all, this approach has merit; however, it is not based on enough hard science to cite it as an approach that gives quantifiable landscape water-use numbers that can be used as a means to judge individual or community-wide water savings associated with landscape changes and/or reductions. In almost all instances, the coefficients generated by this method are less than published crop coefficients for turfgrass. Clearly, if one plant species is removed and not replaced, permanent water savings likely will be realized. But if landscapes containing certain species are replaced by other species, evaluation of long-term savings must wait until the replacement species have fully matured.

Water conservation plans need to be based on an integrated planning process that considers the merits of all available options (landscape area, species, irrigation management, pricing) and encourages natural tradeoffs that lead to public acceptance and the net savings of water.

Crop Coefficients

Crop coefficients used in landscape irrigation scheduling are the ratio of actual evapotranspiration (ETa) to ETo. Once such coefficients have been generated, only estimates of potential ET are required to estimate ETa needed for scheduling irrigations. Crop coefficients only are valid and transferable, however, when plants are compared under similar growing conditions. These conditions generally are defined as nonlimiting water conditions. Snyder and Eching (2005) suggested incorporating a stress coefficient with the density, microclimate, and vegetation coefficients proposed by the WUCOL method (Costello, Matheny, and Clark 2000). Such coefficients would be valid only if the exact kind of stress, duration, and intensity were replicated. But such a coefficient can be used as another subjective evaluation of the landscape, as proposed by Costello, Matheny, and Clark (2000), to define irrigation needs.

Allen and colleagues (2005) stated, "There can be considerable uncertainty in Kc-based ET predictions due to uncertainty in quality and representativeness of weather data for the ETo estimate and uncertainty regarding similarity in physiology and morphology between specific crops and varieties in an area and the crop for which the Kc was originally derived." Crop coefficients have been reported for both pan evaporation (Chalmers, Andrews, and Harris 1992; Sivyer et al. 1997) and empirical-based equations such as the Penman Combination method (Devitt et al. 1995b). The Food and Agricultural Organization of the United Nations now recommends the use of the Penman-Monteith equation (Allen et al. 1998) as the sole method for determining reference ETo.

It is important to note how ETa values are reported. Some authors have reported it based on surface planting area, container area, or shaded area (Beeson 2005; Burger et al. 1987; Devitt, Morris, and Bowman 1992), and others have reported it based on leaf surface area (Levitt, Simpson, and Tipton 1995; Montague et al. 2004). From a scientific perspective, all are valid, but they may have meaning only to specific audiences (such as container area for the nursery industry). Because few end users possess equipment to assess leaf surface area, crop coefficients based on planting area or shaded area may have greater meaning.

Levitt, Simpson, and Tipton (1995) estimated water-use coefficients on a total leaf area basis for oak and mesquite at 0.5 and 1.0, but also on a projected canopy area basis as 1.2 and 1.6. Montague and colleagues (2004) estimated Kc values for various tree species during short summer periods based on leaf surface area. Norway maple (*Acer platanoides* L.) was rated as a low water user (Kc 0.19); plane tree (*Platanus occidentalis* L.) and green ash (*Fraxinus pennsylvanica* Marsh.) as moderate water users (Kc 0.52 and 0.54, respectively); and littleleaf linden (*Tillia cordata* L.) and corkscrew willow (*Salix matsudana* Koidz. f. *tortuosa* Rehd.) as high water users (Kc 0.83 and 1.05, respectively).

Crop coefficients historically have been used in agriculture where they have been successful in estimating crop water requirements (Doorenbos and Pruitt 1975). Allen and colleagues (2005) reported that ETo predicted by the Kc method averaged 8% higher than ET determined by a water balance for the Imperial Irrigation District in southern California, with similar month-to-month and year-to-year trends. Doorenbos and Pruitt (1975) estimated Kc values for nine different deciduous fruit and nut trees, with values varying during the active growing period and whether the orchards were with or without groundcover. Kc values were approximately 0.2 units higher with groundcover, higher during peak summer, and higher under dry, windy conditions. All nine species grown with groundcover were assigned Kc values >1.1 for summer months. Klein (1983) estimated Kc values for 4-year-old peach trees as 0.68 (5), whereas Natali, Xiloyannis, and Barbieri (1985) estimated Kc values as high as 1.25 for peach under higher density. Research on peach by Ayars, Johnson, and Phene (2003) indicated that Kc increased as the midday canopy light interception increased ($r^2 = 0.86$), with maximum values of 1.06.

Crop coefficients have greater validity when derived and applied to large monospecific stands of agronomic crops such as alfalfa, fruit orchards, or even large turfgrass areas. Applying Kc values to mixed landscapes is still problematic and needs additional research to address the many shortcomings of this approach.

Turfgrass and the Mixed Landscape

Most end users maintain higher standards of acceptability for turfgrass than for other landscape species. Unfortunately, turfgrass in residential landscapes typically is irrigated via sprinkler systems that are operating with low distribution uniformities. To compensate for this poor uniformity, many end users who manage small turfgrass areas tend to over-irrigate. As the Christiansen uniformity coefficient goes down from 0.95 to 0.85, the field-based LF has to increase from 0.15 to 0.34 to maintain the area receiving the least amount of water with a LF of 0.05 (Jensen 1975). This over-irrigation is linked directly to system deficiencies. This response is not a species effect, but rather an irrigation effect.

Trees and shrubs frequently are placed on the perimeter of turfgrass areas where a significant portion of the root water uptake comes from irrigated turfgrass. Turfgrass often acts as a safety cushion in most mixed landscapes as trees and shrubs quickly develop lateral root systems below turfgrass areas to access supplemental water. The extent to which trees and shrubs drive the turfgrass irrigation requirement depends on the species and size of the trees and shrubs and the level of expectation by the end user. Water availability for other plants associated with turfgrass irrigations might be excessive for species such as desert willow (a smaller tree native to arid climates) but inadequate for 40-foot Mondell pines. How much of the irrigation water applied to turfgrass would be taken up by shrubs and trees depends on environmental demand and available soil moisture in the rootzones of the trees/shrubs versus turfgrass.

How dependent trees and shrubs are on turfgrass irrigation is revealed clearly when turfgrass is removed. If the water needs of trees are met by the water needs of turfgrass as suggested by Costello, Matheny, and Clark (2000), they must be altering the water balance of the turfgrass system, unless one assumes that such water extraction would come entirely from soil water otherwise destined for deep drainage. (It should be noted that the authors of this section have observed that tree root systems in irrigated urban landscapes in the desert Southwest often are quite shallow and exist extensively under adjacent turfgrass areas.)

Trees and the Mixed Landscape

Trees in landscapes often are more isolated, with energy balances entirely different from orchard, forest, or plantation conditions having higher water use fueled by advective energy (Oke 1987). When air mixing is good, transpiration should be controlled to a greater extent by stomata and vapor pressure deficits (Jarvis and McNaughton 1986). When plant stands have a more closed canopy such as in older orchards, however, water use has been reported to go down (Mitchell et al. 1991). Kjelgren and Rupp (1997) reported significantly lower water use of Norway maple and green ash seedlings grown in shelters compared with those not grown in shelters. They suggested that the lower values in the shelters were because of restricted air movement that severely decreased boundary layer conductance, essentially disconnecting stomatal conductance from the outside atmosphere.

Although trees and turfgrass can use significant amounts of water, turfgrass and appropriate, well-placed ornamentals can lower the cooling requirements of a home substantially. Clearly, 100% rock mulch for a landscape would represent the ultimate in landscape water conservation, but it provides little recreational value and elevates the exterior temperature of a home and its cooling costs (McPherson and Dougherty 1989). Well-designed landscapes can enhance the microclimate surrounding a home through wind channeling and shading, which leads to increases or decreases in humidity (or the perception of humidity) through adjustment in air movement (Rodie and Streich 2000).

The orientation of landscapes and the positioning of trees need to be given careful consideration, such as "peak load landscaping" suggested by Parker (1983). Sound economic analysis will be needed, however, to justify increasing the number of trees based on the availability of water and increased expenditure of water versus the amount of energy saved.

McPherson and Dougherty (1989) found that the annual water costs for certain tree species can be twice as much as the energy savings from their shade (this would vary based on cost of water, tree species, and environmental conditions). McPherson, Simpson, and Livingston (1989) suggested that landscape mini-oases containing plants that are in scale with the site, shade the walls of buildings, and provide ET cooling can balance the need to conserve energy and water effectively. McPherson also suggested that decreases in vegetation partly may be responsible for the growing urban heat islands in Phoenix and Tucson. Givoni (1981) stated that energy-efficient landscapes in hot climates should be more open than in cold climates to allow air flow to cool buildings, and that trees should be used to shade sidewalks, parking lots, streets, and other paved heat sinks. How this relates to water conservation is unknown.

Conclusions

Plant selection should be given serious consideration in the development of low water-using landscapes. Placing too much emphasis on plant selection as a means to conserve water, however, may provide a false sense of security to the homeowner and community and may mislead them in terms of true savings that will occur over time.

Currently there is a lack of scientific data on the water use of trees, shrubs, and groundcovers, as well as how this water use is influenced by climate, growing conditions, and irrigation. Therefore, we believe that in arid environments, emphasis instead should be placed on the following factors:

1. Price water based on its true societal value as a scarce resource.

2. Decrease irrigated landscape areas (especially those with high plant densities, large plants, and high irrigation volumes). Smaller, over-irrigated landscapes almost always use less water than larger, over-irrigated landscapes.

3. Track irrigations and adjust for changes in the seasonal demand of water. Irrigating based on seasonal demand will almost always use less water than irrigating based on guesswork. But when using poor-quality water—such as reuse water—more water will be needed to accomplish adequate leaching of soluble salts.

4. Adjust landscape expectations down whenever possible and be more flexible in plant selection (especially with those plants known to be high water users). Lower growth rates by decreasing fertilization and irrigations to achieve judicious size control. Accept smaller, slower-growing plants with more horizontal growth habits that might show some signs of previous stress. Many ornamental trees and shrubs will increase water use as they increase in size, as more water is applied, and as nitrogen fertility increases (additional research needed).

5. Purveyors. Use water allotments and water credits rather than restrict plant selection. Force end users and landscape designers/architects to make tough decisions about landscape composition and size based on accelerated pricing as allotments are exceeded or credits as savings occur.

6. Emphasize Xeriscape™ and other landscape design concepts that help to achieve water conservation. If "plant lists" are to be used, use plants that are documented to have lower water requirements. These plant lists must, however, be tempered by the lack of information on the influence that growing conditions can have on many tree and shrub species, moving them from low- to high-water-use classifications (additional research needed).

7. Emphasize functional landscapes and avoid banning entire plant categories without justification. Use turfgrass where appropriate, and use those with lower water requirements whenever possible. In warmer climates, turfgrass restrictions should focus on cool-season grasses.

Literature Cited

Allen, R. G., L. S. Pereira, D. Raes, and M. Smith. 1998. *Crop Evapotranspiration: Guidelines for Computing Crop Water Requirements*. Irrigation and Drainage Paper 56. Food and Agriculture Organization of the United Nations, Rome. 300 pp.

Allen, R. G., A. J. Clemmens, C. M. Burt, K . Solomon, and T. O'Halloran. 2005. Prediction accuracy for project wide evapotranspiration using crop coefficients and reference evapotranspiration. *J Irrig Drain Engin* 131(11):24–36.

Ansley, R. J., P. W. Jacoby, C. H. Meadors, and K. K. Lawerence. 1992. Soil and leaf water relations of differentially moisture-stressed honey mesquite (*Prosopis glandulosa* Torr). *J Arid Environ* 22:147–159.

Araujo-Alves, J. M. L., J. M. Torres-Pereira, C. Biel, F. de Herralde, and R. Save. 2000. Effects of minimum irrigation technique on ornamental parameters of two Mediterranean species used in xerigardening and landscaping. *Acta Hort* 541:353–358.

Arizona State University Arboretum. 2006. Plant list, http://www.azarboretum.org/plantlist/southernliveoak.htm (6 December 2006)

Ayars, J. E, R. S. Johnson, and C. J. Phene. 2003. Water use by drip-irrigated late-season peaches. *Irrig Sci* 22(3–4):187–194.

Balok, C. A. and R. St. Hilaire. 2002. Drought responses among seven southwestern landscape tree taxa. *J Amer Soc Hort Sci* 127:211–218.

Beard, J. B. 1973. *Turfgrass: Science and Culture*. Prentice Hall, Upper Saddle River, New Jersey.

Beard, J. B. 1989. Turfgrass water stress: Drought resistance components, physiological mechanisms, and species-genotypes diversity. Pp. 23–28. In H. Takatoh (ed.). *Proceedings of the Sixth International Turf Research Conference,* July 31–August 5, 1989. Japan Society of Turf Science, Tokyo.

Beeson, R. C., Jr. 2005. Modeling irrigation requirements for landscape ornamentals. *HortTechnol* 15(1):18–22.

Bowman, D. and L. Macaulay. 1991. Comparative evapotranspiration rates of tall fescue cultivars. *HortSci* 26(2):122–123.

Bowman, D. C., D. A. Devitt, D. R. Huff, and W. W. Miller. 1998. Comparative evapotranspiration of seventeen buffalograss (*Buchloedactyloides* [Nutt.] Engelm.) genotypes. *J Turfgrass Mgmt* 2: 1–10.

Brown, C. A., D. A. Devitt, and R. L. Morris. 2004. Water use and physiological response of tall fescue turf to water deficit irrigation in an arid environment. *HortSci* 39(2):388–393.

Burger, D. W., J. S. Harttin, D. R. Hodel, T. A. Lukaszewski, S. A. Tjosvoid, and S. A. Wagner. 1987. Water use in California's ornamental nurseries. *Calif Agr* 41(9–10):7–8.

Buwalda, J. G. and F. Lenz. 1993. Water use by European pear trees growing in drainage lysimeters. *J Amer Soc Hort Sci* 790:531–540.

California Department of Water Resources. 1998. *California Water Plan Update Bulletin 160–98*. Department of Water Resources, Sacramento, California.

California Urban Water Conservation Council. 2005. Findings, recommendations and proposed actions from work group three: Landscape design, plants, turfgrass and soils, http://www.cuwcc.org/Uploads/committee/AB2717_WG3_Draft_Recos_05-08-13.pdf (11 December 2006)

Carmen Garcia-Navarro, M. and R. Y. Evans. 1999. Evaluation of a method for classifying landscape plants by relative water use. *Slosson Report* 98–99:1–7.

Carmen Garcia-Navarro, M., R. Y. Evans, and R. S. Montserrat. 2004. Estimation of relative water use among ornamental landscape species. *Sci Hort* 99(2):163–174.

Chalmers, D. J., P. K. Andrews, and K. M. Harris. 1992. Performance of drainage lysimeters for the evaluation of water use by Asian pears. *HortSci* 27(3):263–265.

Clewis, B. 1991. Xeriscape™ landscaping: Gardening to conserve water. *Sci Technol Libr* 12: 133-138.

Costello, L. R., N. P. Matheny, and J. R. Clark. 2000. A guide to estimating irrigation water needs of landscape planting in California. The Landscape Coefficient Method and WUCOLS III. University of California Cooperative Extension and California Department of Water Resources, Sacramento.

Cowan, I. R. 1986. Economics of carbon fixation in higher plants. Pp. 133-170. In T. J. Givnish (ed.). *On the Economy of Plant Form*. Cambridge University Press, New York.

Danielson, R. E., C. M. Feldhake, and W. E. Hart. 1981. Urban lawn irrigation and management practices for water saving with minimum effect on lawn quality. Completion Report No. 106. Colorado Water Resources Research Institute, Fort Collins.

Devitt, D. A. and R. L. Morris. 2006. *Final Report. Water Savings Associated with ET Controllers on Residential Landscapes*. Southern Nevada Water Authority, Las Vegas.

Devitt, D. A., R. L. Morris, and D. C. Bowman. 1992. Evapotranspiration, crop coefficients, and leaching fractions of Irrigated Desert Turfgrass Systems. *Agron J* 84(4):717-723.

Devitt, D. A., R. L. Morris, and D. S. Neuman. 1994. Evapotranspiration and growth response to three woody ornamental species placed under varying irrigation regimes. *J Amer Soc Hort Sci* 119(3):452-457.

Devitt, D. A., D. S. Neuman, D. C. Bowman, and R. L. Morris. 1995a. Comparative water use of turfgrasses and ornamental trees in an arid environment. *J Turf Mgmt* 1(2):48-62.

Devitt, D. A., D. Kopec, M. J. Robey, R. L. Morris, P. Brown, V. Gibeault, and D. C. Bowman. 1995b. Climatic assessment of the arid southwestern United States for use in predicting evapotranspiration of turfgrass. *J Turf Mgmt* 1:65-81.

Devitt, D. A., A. Sala, S. D. Smith, J. Cleverly, L. K. Shaulis, and R. Hammett. 1998. Bowen ratio estimates of evapotranspiration for *Tamarix ramosissima* stands on the Virgin River in southern Nevada. *Water Resources Res* 34:2407-2414.

Devitt, D. A., R. L. Morris, M. Baghzouz, and D. S. Neuman. 2005. Foliar damage and flower production of landscape plants irrigated with reuse water. *HortSci* 40(6):1871-1878.

de Wit, C. T. 1958. Transpiration and crop yields. Versl Landbouwk Onderz, 64.6, IBS, Wageningen, The Netherlands.

Doorenbos, J. and W. O. Pruitt. 1975. *Crop Water Requirements. Irrigation and Drainage Paper FAO 24*. 179 pp.

Easterbrook, G. 1999. Comment on Karen A. Danielsen, Robert E. Lang, and William Fulton's "Retracing suburbia: Smart growth and the future of housing." *Housing Policy Debate* 10(3). Fannie Mae Foundation, Washington, D.C.

Ebdon, J. S. and K. L. Kopp. 2004. Relationships between water use efficiency, carbon isotope discrimination, and turf performance in genotypes of Kentucky bluegrass during drought. *Crop Sci* 44(5):1754–1762.

Feldhake, C. M., R. E. Danielson, and J. D. Butler. 1983. Turfgrass evapotranspiration. 1. Factors influencing rate in urban environments. *Agron J* 75:824–830.

Garrot, D. J., Jr. and C. F. Mancino. 1994. Consumptive water use of three intensively managed bermudagrasses growing under arid conditions. *Crop Sci* 34(1):215–221.

Givoni, B. 1981. *Man, Climate, Architecture*. 2nd ed. Van Nostrand Reinhold, New York. 483 pp.

Heilmeier, H., A. Wartinger, M. Erhard, R. Zimmermann, R. Horn, and E. D. Schulze. 2002. Soil drought increases leaf and whole-plant water use of *Prunus dulcis* grown in the Negev Desert. *Oecologia* 130:329–336.

Hope, D., G. Corinna, W. Zhu, W. F. Fagan, C. L. Redman, N. B. Grimm, A. L. Nelson, C. Martin, and A. Kinzig. 2003. Socioeconomics drive urban plant diversity. *Proc Natl Acad Sci* 100(15):8788–8792.

Huang, B. and J. D. Fry. 1999. Turfgrass evapotranspiration. Pp. 317–334. In M.B. Kirkham (ed.). *Water Use in Crop Production*. Food Products Press, Binghamton, New York.

Jarvis, P. and K. McNaughton. 1986. Stomatal control of transpiration: Scaling up from leaf to region. *Adv Ecol Res* 15:1–49.

Jensen, M. E. 1975. *Scientific Irrigation Scheduling for Salinity Control of Irrigation Return Flows*. Environmental Protection Technology Series, EPA-600/2-75—064. U.S. Environmental Protection Agency, Ada, Oklahoma.

Jordan, L. A., D. A. Devitt, R. L. Morris, and D. S. Neuman. 2001. Foliar damage to ornamental trees sprinkler irrigated with reuse water. *Irrig Sci* 21:17–25.

Jury, W. A. and H. Vaux. 2005. The role of science in solving the world's emerging water problems. *Proc Nat Acad Sci* 102(44):15715–15720.

Kirkham, M. B. 1999. *Water Use in Crop Production*. Food Products Press, Binghamton, New York. 385 pp.

Kjelgren, R. and T. Montague. 1996. Isolated tree water use in various urban surfaces. *ASAE International Conference on Evapotranspiration and Irrigation Scheduling*. November 3–6, 1996. San Antonio, Texas.

Kjelgren, R. and L. Rupp. 1997. Establishment in tree shelters I: Shelters reduce growth, water use, and hardiness, but not drought avoidance. *HortSci* 32(7):1281–1283.

Kjelgren, R., L. Rupp, and D. Kjelgren. 2000. Water conservation in urban landscapes. *HortSci* 35(6):1037–1040.

Klein, I. 1983. Drip irrigation in peach and grape based on soil matric potential conserves water. *HortSci* 18:942–944.

Kneebone, W. R. and I. L. Pepper. 1979. *Water Requirements of Urban Lawns: Arizona.* Project B-035-WYO, Office of Water Research and Technology, U.S. Department of the Interior, Washington, D.C.

Kopec, D. M., R. C. Shearman, and T. P. Riordan. 1988. Evapotranspiration of tall fescue turf. *HortSci* 23(2):300–301.

Levitt, D. G., J. R. Simpson, and J. L. Tipton. 1995. Water use of two landscape tree species in Tucson, Arizona. *J Amer Soc Hort Sci* 120(3):409–416.

Lownds, N. K. and R. D. Berghage. 1991. Relationships between water use of container grown landscape plants and turf evapotranspiration rates. *HortSci* 26(6):Abstract.

Lyle, J. T. 1993. *Urban Ecosystems.* Context Institute, Langley, Washington.

Marcum, K. B. and C. L. Murdoch. 1994. Salinity tolerance mechanisms of six C_4 turfgrasses. *J Amer Soc Hort Sci* 119:779–784.

Marsh, A. W., R. A. Strohman, S. Spaulding, V. Younger, and V. Gibeault. 1980. Determining the water requirements of various turfgrass species. *Irr Assoc Ann Tech Conf Proc* 1:34–45.

Martin, C. 2003. *Landscape Water Use in a Desert Metropolis: Phoenix, Arizona.* Research Vignette No. 1, Consortium for the Study of Rapidly Urbanizing Regions, Phoenix, Arizona.

McPherson, E. G. and E. Dougherty. 1989. Selecting trees for shade in the southwest. *J Abor* 15:35–43.

McPherson, E. G., J. R. Simpson, and M. Livingston. 1989. Effects of three landscape treatments on residential energy and water use in Tucson, Arizona. *Energy and Buildings* 13:127–138.

Mitchell, P. D., A. M. Boland, J. L. Irvine, and P. H. Jerie. 1991. Growth and water use of young closely planted peach trees. *Sci Hort* (Amsterdam) 47(3–4):283–294.

Montague, T., R. Kjelgren, and L. Rupp. 1998. Surface energy balance affects gas exchange of three shrub species. *J Arbor* 24(5):254–262.

Montague, T., R. Kjelgren, R. Allen, and D. Wester. 2004. Water loss estimates for five recently transplanted landscape tree species in a semi-arid climate. *J Environ Hort* 22(4):189–196.

Natali, S., C. Xiloyannis, and A. Barbieri. 1985. Water consumption of peach trees grafted on four different rootstocks. *Acta Hort* 173:355–362.

Oke, T. R. 1987. *Boundary Climates*. Routledge Taylor and Francis Group, Oxford, U.K. 435 pp.

Parker, J. H. 1983. Landscaping to reduce the energy used in cooling buildings. *J Forest* 81: 82–84.

Peterson, K. A., L. B. McDowell, and C. A. Martin. 1999. Plant life form frequency, diversity and irrigation application in urban residential landscapes. *HortSci* 34:491.

Renquist, R. 1987. Evapotranspiration calculations for young peach trees and growth responses to irrigation amount and frequency. *HortSci* 22:221–223.

Rodie, S. and A. Streich. 2000. *Landscape Sustainability*. Nebraska Cooperative Extension, G1405, http://ianrpubs.unl.edu/sendlt/g1405.pdf (7 December 2006)

Sala, A., D. A. Devitt, and S. D. Smith. 1996. Water use by *Tamarix ramosissima* and associated phreatophytes in a Mojave Desert floodplain. *Eco Appl* 6:888–898.

Salaiz, T. A., R. C. Shearman, T. P. Riordan, and E. J. Kinbacher. 1991. Creeping bentgrass cultivar water use and rooting responses. *Crop Sci* 31(5):1331–1334.

Schuch, U. K. and D. W. Burger. 1997. Water use and crop coefficients of woody ornamentals in containers. *J Amer Soc Hort Sci* 122:727–734.

Shaw, D. A. and D. R. Pittenger. 2004. Performance of landscape ornamentals given irrigation treatments based on reference evapotranspiration. *Acta Hort* 664:607.

Shearman, R. C. 1985. Turfgrass culture and water use. Pp. 61–70. In *Turfgrass Water Conservation*. University of California Publication 21405, Oakland.

Shearman, R. C. and J. B. Beard. 1973. Environmental and cultural preconditioning effects on the water use rate of *Agrostis palustris* Huds., cultivar Penncross. *Crop Sci* 13:424–427.

Singh, S. B., P. Kumar, and K. G. Prasad. 1991. Response of various tree species to irrigation regimes and leaf residue management. *J Indian Soc Soil Sci* 39:229–232.

Sivyer, D. J., J. R. Harris, N. Persaud, and B. Appleton. 1997. Evaluation of a pan evaporation model for estimating post-planting street tree irrigation requirements. *J Arbor* 23(6):250–256.

Snyder, R. L. and S. Eching. 2005. *Estimating Urban Landscape Evapotranspiration*. University of California, Atmospheric Science, Davis; California Department of Water Resources, Office of Water Use Efficiency, Sacramento.

Sperry, J. S. and M. T. Tyree. 1990. Water stress induced xylem embolism in three species of conifers. *Plant Cell Environ* 13:427–436.

Spinti, J. E., R. St. Hilaire, and D. VanLeeuwen. 2004. Balancing landscape preferences and water conservation in a desert community. *HortTechnol* 14:72.

Staats, D. and J. E. Klett. 1995. Water conservation potential and quality of non-turf groundcovers versus Kentucky bluegrass under increasing levels of drought stress. *J Environ Hort* 13:181–185.

Stabler, L. B. and C. A. Martin. 2000. Irrigation regimens differentially affect growth and water use efficiency of two southwest landscape plants. *J Arid Environ* 18:66–70.

Stabler, L. B. and C. A. Martin. 2004. Irrigation and pruning affect growth and water use efficiency of two desert-adapted shrubs. *Acta Hort* 638:255–258.

Tipton, J. L. and M. White. 1995. Differences in leaf cuticle structure and efficacy among eastern redbud and Mexican redbud phenotypes. *J Amer Soc Hort Sci* 120:59–64.

Tyree, M. T. and H. Cochard. 1996. Summer and winter embolism in oak: Impact on water relations. *Ann Sci For* 53:173–180.

Tyree, M. T. and M. A. Dixon. 1986. Water stress induced cavitation and embolism in some woody plants. *Physiol Plant* 66:397–405.

U. S. Census Bureau. 2000. *2000 Census*, http://www.census.gov (2 February 2007)

Western Resource Advocates. 2004. Urban sprawl: Impacts on urban water use. Pp. 93–104. In *Smart Water: A Comparative Study of Urban Water Use across the Southwest*. Western Resource Advocates, Boulder, Colorado.

White R. H., M. C. Engelke, S. J. Anderson, B. A. Ruemmele, K. B. Marcum, and G. R. Taylor, II. 2001. Zoysiagrass water relations. *Crop Sci* 41:133–138.

Wullschleger, S. D., F. C. Meinzer, and R. A. Vertessy. 1998. A review of whole-plant water use studies in trees. *Tree Physiol* 18:499–512.

Zajicek, J. M. and J. L. Heilman. 1991. Transpiration by crape myrtle cultivars surrounded by mulch, soil, and turfgrass surfaces. *HortSci* 26:1207–1210.

11

Turfgrass Water Requirements and Factors Affecting Water Usage

Bingru Huang

Introduction

Turfgrasses, like other agronomic, horticultural, and landscape vegetation, requires water for growth and survival. Without adequate water, turfgrass becomes brown, desiccated, and may die in severe instances. Loss of ground cover by turfgrass can have significant negative impacts on the aesthetics and functionality of our environment because healthy turfgrass provides many important benefits (see Section 2). Therefore, irrigation is desirable to maintain the functional benefits of healthy, actively growing turfgrass in areas where rainfall cannot meet the water demand of plants. As water availability is becoming increasingly limited and more costly, water conservation in turfgrass culture has become extremely important.

Water use of turfgrasses is evaluated based on the total amount of water required for growth and transpiration (water loss from the leaf) plus the amount of water lost from the soil surface (evaporation). Transpirational water consumption accounts for over 90% of the total amount of water transported into the plants, with 1 to 3% actually used for metabolic processes (Beard 1973; Hopkins 1999). Water usage rates vary with species and cultivars and are affected by many external factors, especially environmental conditions. This article reviews the water-use characteristics of different turfgrass species and examines environmental factors affecting turfgrass water use.

The amount of water used under deficit irrigation may be calculated based on the actual evapotranspiration (ET) rate. One of the simplest and oldest methods to estimate ET is to measure the evaporation from a large standardized pan. Actual ET may be measured more accurately using weighing lysimeters. More recently, researchers have developed mathematical models (or modified Penman equations) to estimate potential evapotranspiration (PET) using climatic data of solar radiation, wind speed, relative humidity, and temperature. PET values represent a nonlimiting plant water status and an extended cover of short, green vegetation over the soil surface. Reference ET is a starting point for estimating irrigation needed for turfgrass areas by multiplying with the crop coefficient (Kc). The Kc for turfgrass depends on the type of grass (warm- or cool-season), cutting height, and desired turfgrass quality.

Water Use Characteristics of Cool-Season and Warm-Season Turfgrass Species

Turfgrasses are classified into two groups based on their climatic adaptation: cool-season and warm-season. Cool-season grasses mainly grow in temperate and subarctic climates, whereas warm-season grasses are adapted to tropical and subtropical areas (see U.S. climatic regions in Section 5). These two groups of turfgrass species have different water requirements (Table 11.1) and vary in water-use characteristics. Most cool-season grasses generally are higher water users than warm-season grasses. Typical ET rates range from 3 to 8 millimeters (mm) per day for cool-season grasses and from 2 to 5 mm per day for warm-season grasses (Beard 1994). Declining soil moisture levels will progressively lower the water-use rate by up to 80% (Figure 11.1). In addition, the comparative water-use rankings for different species and cultivars may change across different climatic conditions and cultural regimes, and also depends on individual species and cultivar adaptation.

Table 11.1. The relative maximum evapotranspiration rates of 24 turfgrass species (Modified from Beard and Beard 2004)

Relative ranking	ET rate (mm d-1)[a]	Turfgrass species[b]
Very low	<6	*American buffalograss
Low	6 – 7	*Hybrid bermudagrass Centipedegrass *Dactylon bermudagrass *Zoysiagrass
Moderate	7 – 8.5	Hard fescue Chewing fescue Creeping red fescue Bahiagrass Seashore paspalum St. Augustinegrass
High	8.5 – 10	Perennial ryegrass Kikuyugrass
Very high	>10	Tall fescue Creeping bentgrass Annual bluegrass *Kentucky bluegrass Rough bluegrass Annual ryegrass

[a] The ranges of ET are based on the most widely used cultivars of each species when grown in their respective climatic regions of adaptation and preferred culture regimes.
[b] Asterisk (*) indicates cultivars within these species may vary significantly.

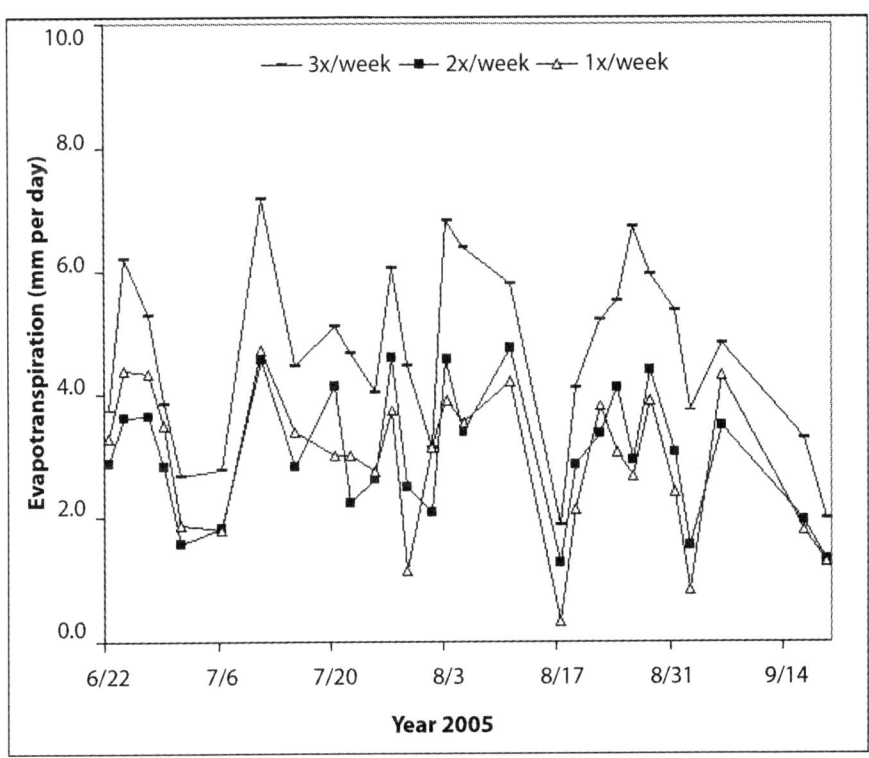

Figure 11.1. Evapotranspiration rate of creeping bentgrass (cv. Penncross) irrigated once per week (1x), twice per week (2x), and three times per week (3x). The test was performed in turfgrass grown in loamy soil and mowed at 3/8 inch height in field plots at the Rutgers University Turfgrass Research Farm, North Brunswick, New Jersey.

Cool-Season Turfgrasses

Commonly used cool-season turfgrasses in lawns, sports fields, parks, grounds, golf courses, and roadsides include *Festuca* L., *Poa* L., *Agrostis* L., and *Lolium* L. Within the cool-season turfgrasses, species vary in maximum ET rate, ranging from 7 to 8.5 mm per day to more than 10 mm per day (Table 11.1). Tall fescue (*Festuca arundinacea* Schreb.), creeping bentgrass (*Agrostis stolonifera* L.), and annual bluegrass (*Poa annua* L.) are considered to be the highest water users, whereas hard fescue (*Festuca longifolia* Thuill.), Chewing's fescue (*Festuca rubra* L. ssp. *commutata* Gaud.), and creeping red fescue (*Festuca rubra* L. ssp. *rubra*) are the lowest water users (Beard 1989, 1994). DaCosta and Huang (2006) in New Jersey compared water use among three bentgrass species and found that velvet bentgrass (*Agrostis canina* L.) used less water than creeping bentgrass and colonial bentgrass (*Agrostis capillaries* L.).

There can be as much variation in the water-use rate among cultivars within certain species as between turfgrass species (Beard 1989). The difference in ET between cultivars may range from 20 to 60% (Kjelgren, Rupp, and Kilgren 2000).

Warm-Season Turfgrasses

Warm-season turfgrasses include *Cynodon* L.C. Rich, *Buchloe* Engelm., *Zoysia* Willd., *Paspalum* L., *Eremochloa ophiuroides* [Munro] Hack., and *Stenotaphrum secundatum* [Walt.] Kuntze. Their maximum water use ranges from less than 6 mm per day to 8.5–10 mm per day (Table 11.1). Among warm-season grasses, bermudagrass (*Cynodon* spp.), zoysiagrass (*Zoysia* spp.), and buffalograss (*Buchloe dactyloides* [Nut.] Engelm.) have relatively low water-use rates, whereas seashore paspalum (*Paspalum vaginatum* Swartz.) and St. Augustinegrass (*Stenotaphrum secundatum* ([Walt.] Kuntze.) have relatively high water-use rates (Table 11.1). Within a warm-season turfgrass species, cultivars may vary in water-use rate.

Plant Growth Characteristics Affecting Water Use

Water use of the turfgrass canopy is affected by water loss through shoot transpiration and soil evaporation, and by water uptake from the soil via the root system. Therefore, turfgrass species variations in water-use rates are associated with differences in shoot and root characteristics, such as canopy configuration or leaf orientation, tiller or shoot density, growth habit, rooting depth, and root density (Beard 1973; Huang and Fry 1999).

Canopy/Shoot Characteristics

Species that have a prostrate shoot growth habit typically have a lower water-use rate than grasses with an upright growth habit (Kim and Beard 1988). The former group often has higher resistance to evapotranspiration (Johns, Van Bavel, and Beard 1981, 1983). Generally, turfgrasses with a rapid vertical shoot extension rate tend to have higher water-use rates than slower-growing or dwarf-type grasses because of increasing leaf area from which transpiration occurs (Kim and Beard 1988; Shearman and Beard 1973). Shearman (1986) reported that shoot vertical extension rate was positively correlated with water-use rate for 20 Kentucky bluegrass cultivars with upright growth pattern. No correlation was found, however, between water use and leaf extension rate within several warm-season turfgrass species that have a prostrate growth habit, such as bermudagrass (Beard, Green, and Sifers 1992) and zoysiagrass (Green, Sifers, and Beard 1991). The more horizontal growth characteristic may be dominant because of higher canopy resistance. Dense and compact type turfgrass canopies have lower water loss from soil evaporation than thin, open canopies. It is apparent that low-water-use grass species may possess at least one of the combined characteristics of slow vertical growth, prostrate growth pattern, and dense canopy. Under plant water stress, stomatal closure and cuticle formation can be key factors controlling water use (Beard 1973; Kim 1983).

Rooting Characteristics

An extensive, well-branched, deep root system is important for efficient water uptake from the soil. Plants with deep and dense root systems typically have a high capacity for water extraction from soil. Deep rooting enables plants to avoid water stress by taking up water from deeper in the soil profile when the surface soil is dry. Tall fescue, which develops a deep, extensive root system, has been shown to use more water than most other cool-season turfgrasses. But, not all grasses with extensive root systems are necessarily high water users. For example, bermudagrass has a prostrate growth habit, slow shoot growth rate, and a deep root system, but exhibits a low water-use rate (Kim and Beard 1988; Youngner 1985).

Root distribution strongly responds to spatial variations in water availability. Generally, roots tend to proliferate or extend in localized wet zones in a soil profile. For example, when soil surface is maintained wet constantly because of frequent irrigation or rainfall, plants develop extensive, shallow root systems. When a soil surface is allowed to dry periodically, production of roots increases considerably in the lower layer where water is available. This root response has been observed in various turfgrass species (Beard 1973; Huang 1999; Huang, Duncan, and Carrow 1997). The ability of roots to follow moisture into deeper layers of the soil profile conditions the ability of a plant to tolerate or avoid short and long periods of drought. Development of a deep root system could be related to a faster elongation rate of roots under drying conditions. When limited soil water is stored in deeper soil profiles, however, faster root extension into deeper soil profiles may be detrimental for plants because of rapid depletion of water.

As soil dries, root hairs increase in length and number (Huang and Fry 1998). Increases in root hairs in dry soil have a pronounced effect on total root surface area. This response may be an adaptive mechanism to maintain liquid continuity around the growing roots and to provide greater root surface for nutrient absorption because the rate of nutrient diffusion to the root decreases in drier soil. Root hairs can be sites for extensive mucilage production. Mucilage can enhance the ability of the hair to attach to soil particles and thereby prevent air gaps from developing between the soil and root surface when the soil dries; decrease water efflux from plants into drying soils; and ultimately delay root desiccation. Extensive development of root hairs enhances water uptake and facilitates water retention under soil-drying conditions.

Dormancy and Water Use

Another important plant characteristic of low water use is dormancy, a phenomenon in which turfgrass leaves may turn brown in response to a water deficit, but the meristematic crowns and stem nodes are not dead. Dormant turfgrass plants have limited or no transpirational water loss, and thus, have low water usage. Limited water in dormant plants may be concentrated in the crown and rhizomes. Dormancy is a mechanism of turfgrass escape from drought stress such that dormant plants survive (without growth) for extended periods of drought stress and resume growth when soil moisture becomes available (Beard 1973).

In general, dormant turfgrasses, especially those with rhizomes (underground stems), can survive without water for several weeks or months with limited damage, depending on the air temperature. After rainfall or irrigation, the grass will quickly recover. Allowing certain turfgrasses to go dormant in low maintenance areas can result in significant water savings without loss of turfgrass. The lengths of time turfgrasses can survive in a dormant condition vary with temperatures and turfgrass species. In general, turfgrasses can be expected to survive in a dormant condition for several weeks at temperatures at or below normal, but may survive in dormant conditions for a shorter period of time during the summer when temperature is elevated. Kentucky bluegrass is a good example of a species that has the capacity for survival during extended water stress because it has extensive rhizomes that generate new roots and shoots once soil moisture is replenished. But bunch-type turfgrasses such as perennial ryegrass and tall fescue are slow to recover to their full canopy upon rewatering, once the turf canopy becomes desiccated and thinned under nonirrigated conditions.

Environmental Factors Influencing Turfgrass Water Use

In addition to species and cultivar variations, water use of turfgrasses is influenced by many external factors in their growing environment, including temperature, wind, solar radiation, relative humidity, and edaphic factors such as soil moisture (Beard 1973). These factors affect both plant transpiration and soil evaporation. Understanding major environmental factors influencing water use is important for developing efficient cultural strategies for turfgrasses, especially in areas with a limited water supply.

Water loss through transpiration from turfgrass plants is controlled by three major processes: external boundary layer resistance, vapor pressure gradient between the leaf and air, and internal leaf diffusion resistance (Fu, Fry, and Huang 2004; Johns, Van Bavel, and Beard 1981, 1983). High transpirational water loss may be because of lower boundary layer resistance, higher vapor pressure gradient, and/or lower internal diffusion resistance. Internal leaf diffusion resistance is associated with stomatal density and conductance, intercellular space, cell size and density, and leaf cuticle thickness, which are all basically controlled by genetics. The boundary layer is a layer of stagnant air over the leaf surface, which creates resistance to water vapor escape from the leaf. Atmospheric environmental factors affect transpiration mainly though alteration of the boundary layer resistance and vapor pressure gradient between the leaf and air.

Solar Radiation

Water use of turfgrasses typically is much higher in areas exposed to full sun than in shaded or dark nocturnal conditions. Evapotranspiration is an energy-dependent process. If more energy is available, there is potentially a greater rate of evapotranspiration. A linear relationship between irradiance and water use rate has been reported in turfgrasses (Aurasteh 1983; Beard, Green, and Sifers 1992; Feldhake 1981; Kim and Beard 1988; Shearman and Beard 1973). Our study in New Jersey with creeping bentgrass, velvet bentgrass, and colonial bentgrass found a strong correlation between water-use rate and solar radiation level with a correlation coefficient of 0.81.

Temperature

Plants transpire more water at higher temperatures because water evaporates more rapidly with increasing temperatures under nonlimiting water availability. This is reflected by increasing water demand of turfgrass during midday and summer months. Increasing water use rate was highly correlated with increasing temperatures, with a correlation coefficient of 0.81 (1.0 = perfect correlation) in our New Jersey study. A leaf exposed to 30°C may transpire three times as fast as it does at 20°C when leaves are fully hydrated.

Temperature influences transpiration through its effect on vapor pressure, which in turn affects the vapor pressure gradient between the leaf and air (Beard 1973). Leaves of a well-watered plant maintain a near 100% relative humidity or high vapor pressure within the leaf. High temperature dries the air or lowers the relative humidity of air, creating a larger gradient in vapor pressure between the air and the leaf, thereby resulting in high transpirational water loss (Shearman and Beard 1973).

Relative Humidity

Relative humidity of the atmosphere affects water use mainly by influencing the vapor pressure gradient between the leaf and air. Low relative humidity, or dry air, surrounding the leaf causes rapid water loss from the leaf because of the increased vapor pressure gradient (Beard 1973). Therefore, water use typically increases with decreases in relative humidity. Carrow (1995) reported turfgrass ET was 40 to 60% less in a humid environment compared with the same cultivar in an arid environment.

Wind

Water-use rate of turfgrasses typically is higher on windy days than on calm days and in open areas compared with areas enclosed with trees, shrubs, and other structures. The effects of wind on water use are associated with changes in the water vapor pressure gradient between the leaf and air and the external boundary layer resistance, particularly the latter factor (Beard 1973). Wind above the leaf dries the air adjacent to the leaf and therefore causes increases in the vapor pressure gradient. Also, the thickness of the boundary layer is primarily a function of leaf size and shape, the presence of leaf hair, and wind speed. The thickness of the boundary layer decreases with increasing wind speed, which may lead to increases in the transpiration rate at a lower wind speed; however, high wind speed may cause stomatal closure and low transpiration rate before the leaves become desiccated (Hopkins 1999). Grace (1974) reported that the transpiration rate of a tall fescue cultivar increased as wind speed increased from

1 m s^{-1} to 3.5 m s^{-1}. Beard, Green, and Sifers (1992) also observed a positive relationship between water-use rate and wind speed.

Soil Moisture

Water-use rate can be restricted by water supply from the roots and in turn depends on the availability of soil moisture. When soil moisture supply is not enough to replace what is lost from the shoots, stomata will be closed and a plant cannot continue to transpire. Therefore, plants in general tend to use less water when the soil water content is low. Water-use rates of both cool-season and warm-season turfgrasses have been found to decline with decreases in soil water content (Biran et al. 1981; DaCosta and Huang 2006; Kim 1983). Not all water in the soil can be used by plants. Only water retained in the soil by capillary forces can be extracted by the plant. Available soil water is the water held by the soil between field capacity, defined as the amount of water remaining in the soil when drainage ceases, and permanent wilting point, defined as the soil water content below which there is no available water and plants wilt and do not recover. Soil moisture availability is affected by water supply through rainfall or irrigation. Efficient irrigation should maintain soil moisture above the permanent wilting point, but below field capacity. Irrigation above field capacity results in waste of water.

Frequently irrigated turfgrasses (soils that are kept wet constantly) use more water than turfgrasses that receive less frequent irrigation (allowing soil to dry between irrigation events) (Gibeault et al. 1985; Kneebone, Kopec, and Mancino 1992). For example, creeping bentgrass irrigated three times per week had a higher ET rate than that irrigated once per week or twice per week on many days from June to September 2005 (Fig. 11.1). Vertical shoot growth may be promoted with increasing irrigation frequency or quantity, resulting in increased demand for water. Maintaining wet soil constantly also promotes shallow root systems, which decreases water utilization in deeper soil profiles.

Turfgrass maintained under water deficit conditions typically uses less water than well-irrigated plants. Deficit irrigation can decrease water uses in various turfgrass species (see Section 12). The level of deficit irrigation varies with plant species, soil types, and climatic conditions. Many turfgrass species—such as Kentucky bluegrass, perennial grass, tall fescue, creeping bentgrass, velvet bentgrass, colonial bentgrass, zoysiagrass, and bermudagrass—are able to tolerate certain levels of deficit irrigation with little or no loss of aesthetic turfgrass quality.

Soil Texture and Physical Properties

The amount of water that turfgrass can actually use also is affected by soil type and texture. Both soil texture and type affect water retention and infiltration, and thereby influence water use and irrigation quantity or frequency. Generally, larger-particle-size soils, such as sandy soil, have better drainage and hold on to less water than fine-particle soils, such as clay and silt, and have about 50% of the soil water available to plants. Therefore, sands and sandy soil require more frequent irrigation to meet plant needs, but with smaller amounts of water per irrigation event. Conversely, fine-textured soils, such as clay loams and clay, hold larger amounts of water, but only about 30 to 35% of the total water in the soil is available to plants. Deep, infrequent irrigation of fine-textured soils may be needed to meet plant needs. Compared with plants grown in clay soils, Kentucky bluegrass grown in a sand-peat mix had a 6% higher ET in a study conducted during the summer in Colorado (Feldhake, Danielson, and Butler 1983).

Traffic and compaction are major problems in recreational turfgrass areas. Both stresses may adversely affect soil infiltration, water holding capacity, and plant growth, and thus, indirectly affect water use. Soil compaction increases bulk density, water retention, and soil strength, and decreases aeration porosity and oxygen needed for root growth. O'Neil and Carrow (1983) reported a 21 and 49 % decrease in water use for perennial ryegrass grown in moderate and severely compacted soils, respectively, than that under noncompacted soils in Kansas. Similarly, Kentucky bluegrass grown in compacted soils also exhibited a lower water-use rate (Agnew and Carrow 1985).

Summary

As discussed, turfgrass water use is a function of plant growth characteristics and environmental conditions. Therefore, an effective conservation program should be developed based on plant needs and environmental conditions. Use of less water and/or drought-resistant turfgrass species and cultivars is a primary means of decreasing water needs. Selection of turfgrass species and cultivars adapted to local climatic conditions can result in significant water savings. For example, in arid and semi-arid regions, warm-season turfgrasses provide a better turfgrass and use less water than cool-season turfgrasses. Quantification of actual water use by measuring evapotranspiration rate under local environmental conditions at different times of the year helps to decide how much to irrigate under different conditions, as has been described for San Antonio, Texas, in Section 15. Knowledge of critical plant physiological status and soil moisture content of different types of soils also is important for scheduling when to irrigate, how much water to apply by irrigation to replenish water lost through evapotranspiration, and how deep to irrigate in the soil (see also Section 15).

Literature Cited

Agnew, M. L. and R. N. Carrow. 1985. Soil compaction and moisture stress preconditioning in Kentucky bluegrass. I. Soil aeration, water use, and root responses. *Agron J* 77:872–877.

Aurasteh, R. M. 1983. A model for estimating lawn grass water requirement considering deficit irrigation, shading and application efficiency. PhD dissertation, Utah State University, Logan, Utah.

Beard, J. B. 1973. *Turfgrass: Science and Culture.* Prentice-Hall, Inc., Englewood Cliffs, New Jersey.

Beard, J. B. 1989. Turfgrass water stress: Drought resistance components, physiological mechanisms, and species-genotype diversity. *Proc Intl Turfs Res Conf*, Tokyo, Japan. 6:23–28.

Beard, J. B. 1994. The water-use rate of turfgrasses. *TurfCraft Australia* 39:79–81.

Beard, J. B. and H. J. Beard. 2004. *Beard's Turfgrass Encyclopedia for Golf Courses—Ground—Lawns— Sports Fields.* Michigan State University Press, East Lansing, Michigan.

Beard, J. B., R. L. Green, and S. I. Sifers. 1992. Evapotranspiration and leaf extension rates of 24 well-watered turf-type *Cynodon* genotypes. *HortSci* 27:986–998.

Biran, I., B. Bravdo, I. Bushkin-Harav, and E. Rawitz. 1981. Water consumption and growth rate of 11 turfgrasses as affected by mowing height, irrigation frequency, and soil moisture. *Agron J* 73:85–90.

Carrow, R. N. 1995. Drought resistance aspects of turfgrasses in the Southeast: Evapotranspiration and crop coefficients. *Crop Sci* 35:1685–1690.

DaCosta, M. and B. Huang. 2006. Minimum water requirements for creeping, colonial, and velvet bentgrass under fairway conditions. *Crop Sci* 46:81–89.

Feldhake, C. M. 1981. Turfgrass evapotranspiration and microenvironment interaction. PhD dissertation. Colorado State University, Fort Collins, Colorado.

Feldhake, C. M., R. E. Danielson, and J. D. Butler. 1983. Turfgrass evapotranspiration. I. Factors influencing rate in urban environments. *Agron J* 75:824–830.

Fu, J., J. Fry, and B. Huang. 2004. Minimum water requirements of four turfgrasses in the transition zone. *HortSci* 39:1740–1744.

Gibeault, V. A., J. L. Meyer, V. B. Youngner, and S. T. Cockerham. 1985. Irrigation of turfgrass below replacement of evapotranspiration as a means of water conservation: Performance of commonly used turfgrasses. *Proc Int Turfgrass Res Conf*, Avignon, France. 5:347–356.

Grace, J. 1974. The effect of wind on grasses. I. Cuticular and stomatal transpiration. *J Exp Bot* 25:542–551.

Green, R. L., S. I. Sifers, and J. B. Beard. 1991. Evapotranspiration rates of eleven zoysiagrass genotypes. *HortSci* 26:264–266.

Hopkins, W. G. 1999. *Introduction to Plant Physiology*, 2nd ed. John Wiley & Sons, Inc., New York.

Huang, B. 1999. Water relations and root activities of *Buchloe dactyloides* and *Zoysia japonica* in response to localized soil drying. *Plant Soil* 208:179–186.

Huang, B. and J. D. Fry. 1998. Root anatomical, morphological, and physiological responses of two tall fescue cultivars to drought stress. *Crop Sci* 38:1017–1022.

Huang, B. and J. Fry. 1999. Turfgrass evapotranspiration. Pp. 317–334. In M. B. Kirkham (ed.). *Water Use in Crop Production*. Food Products Press, New York.

Huang, B., R. Duncan, and R. Carrow. 1997. Root spatial distribution and activity for seven turfgrasses in response to localized drought stress. *Int Turf Res J* 7:681–690.

Johns, D., C. H. M. Van Bavel, and J. B. Beard. 1981. Determination of the resistance to sensible heat flux density from turfgrass for estimation of its evapotranspiration rate. *Ag Meteor* 25:15–25.

Johns, D., C. H. M. Van Bavel, and J. B. Beard. 1983. Resistance of evapotranspiration from a St. Augustinegrass turf canopy. *Agron J* 75:419–422.

Kim, K. S. 1983. Comparative evapotranspiration rates of thirteen turfgrasses grown under both non-limiting soil moisture and progressive water stress conditions. MS thesis, Texas A&M University, College Station, Texas.

Kim, K. S. and J. B. Beard. 1988. Comparative turfgrass evapotranspiration rates and associated plant morphological characteristics. *Crop Sci* 28:328–331.

Kjelgren R., L. Rupp, and D. Kilgren. 2000. Water conservation in urban landscapes. *HortSci* 35:1037–1040.

Kneebone, W. R., D. M. Kopec, and C. F. Mancino. 1992. Water requirement and irrigation. Pp. 441–472. In D. V. Waddington, R. N. Carrow, and R. C. Shearman (eds.). *Turfgrass Agronomy Mono* 32. ASA, CSSA, SSSA, Madison, Wisconsin.

O'Neil, K. J. and R. N. Carrow. 1983. Perennial ryegrass growth, water use, and soil aeration status under soil compaction. *Agron J* 75:177–180.

Shearman, R. C. 1986. Kentucky bluegrass cultivar evapotranspiration rates. *HortSci* 24:767–769.

Shearman, R. C. and J. B. Beard. 1973. Environmental and cultural preconditioning effects on the water use rate of *Agrostis palustris* Huds., cultivar Penncross. *Crop Sci* 13:424–427.

Youngner, V. B. 1985. Physiology of water use and water stress. Pp. 37–43. In V. A. Gibeault and S. T. Cockerham (eds.). *Turfgrass Water Conservation*. University of California Pub 21405, Riverside.

12

Turfgrass Cultural Practices for Water Conservation

Robert C. Shearman

Abstract

Cultural practices can be used, alone or in systems, to decrease water use, conserve water, and enhance drought resistance. Cultural practices that influence turfgrass canopy resistance and rooting depth, extent, and plasticity are of primary interest for integrated turfgrass management systems designed to conserve water. Water use declines as turfgrass canopy resistance increases and leaf elongation rate and leaf area decrease. Turfgrasses with deep, extensive root systems and demonstrated root plasticity, coupled with decreased water use, are more drought avoidant and have greater water conservation potential.

Mowing height and frequency, nutrition, and irrigation are primary cultural practices that directly impact vertical elongation rate, leaf surface area, canopy resistance, rooting characteristics, and water use. Secondary cultural practices, such as soil cultivation, topdressing, wetting agents, plant growth regulators, and pest management, also influence turfgrass top and root growth and subsequently influence potential water conservation. These primary and secondary cultural practices alone, or in combination, can be manipulated to decrease water loss and enhance water conservation. Information in the turfgrass research literature strongly supports the opportunity to develop integrated turfgrass management systems that conserve water without losing turfgrass quality or function.

Introduction

It is important to keep in mind that turfgrass water use varies among turfgrass species and within cultivars of species as discussed in Section 11. Even though there are inherent differences among turfgrasses in water-use rates, cultural practices can be manipulated to decrease a species' water use and enhance its drought resistance.

Johns, Beard, and Van Bavel (1983) introduced the concept of canopy resistance to turfgrass literature. Turfgrasses with dense, tight, uniform stands exhibit greater canopy resistance than those with open, uneven stands. Shearman (1986, 1989) and Kopec, Shearman, and Riordan (1988) demonstrated that verdure and shoot density were negatively correlated to evapotranspiration (ET) rates of Kentucky bluegrass and tall fescue cultivars, whereas leaf area and vertical elongation rate were positively correlated. Kim (1983) and Kim and Beard (1988) reported similar morphological effects on ET for seven warm-season turfgrasses, especially at the interspecies level. Water use increases with turfgrass leaf area and leaf vertical elongation rate, but decreases as verdure or shoot density increase. These results are of interest because characteristics such as leaf area, leaf elongation rate, shoot density, and verdure are influenced readily by turfgrass cultural practices, whether the turfgrasses are grown in arid, semi-arid, or humid environments.

Mowing Effects

Of all the cultural practices used in turfgrass maintenance, mowing has the greatest impact on turfgrass growth, physiology, and stress tolerance. Mowing height, frequency, and equipment influence turfgrass top and root growth, water use, and drought resistance (Beard 1973; Carrow, Shearman, and Watson 1990; Shearman 1985a).

Mowing Height

Increased water use is associated with higher mowing heights for cool- and warm-season grasses (Doss et al. 1962; Feldhake et al. 1983; Mitchell and Kerr 1966; Shearman and Beard 1973). Shearman and Beard (1973) reported a 56% increase in water use of Penncross creeping bentgrass (*Agrostis stolonifera* L.) when mowing height increased from 6 millimeters (mm) to 25 mm. Similarly, Mitchell and Kerr (1966) reported a 37% increase in water use for perennial ryegrass (*Lolium perenne* L.) when mowing height was increased from 25 mm to 50 mm. Common bermudagrass (*Cynodon dactylon* [L.] Pers.) used 3.0 mm of water per day at 25 mm mowing height and 4.8 mm at 150 mm (Doss et al. 1962). Shearman and Beard (1973) related the increased water use to increased leaf area exposed to desiccating atmospheric conditions.

Turfgrasses mowed at higher heights of cut have an open canopy with wide leaf blades and low shoot density, whereas low-cut turfgrasses form a tight canopy with narrow leaf blades and high shoot density (Beard 1973; Biran et al. 1981; Madison 1962b; Shearman and Beard 1973). Spak, DiPaola, and Anderson (1993) reported that tall fescue stand density declined from April to September in North Carolina by 31% for turfgrasses mowed at 95 mm and 61% for unmowed stands. Turfgrasses with an open stand lack canopy resistance and have more leaf area exposed to desiccation under high ET-demand conditions than turfgrasses with high stand density (Johns, Beard, and Van Bavel 1983). Shearman (1986, 1989) demonstrated that turfgrass ET was negatively correlated to shoot density and verdure. Therefore, higher mowing heights alone should not be considered as a water conservation practice.

It often is speculated that the increased water use associated with increased mowing height would be offset by a deeper root system and the plant's ability to draw on a greater soil volume for its water needs. Madison and Hagan (1962) related increased mowing height with increased depth and extent of rooting of Merion Kentucky bluegrass, and they demonstrated soil-moisture extraction from greater soil depth with increased mowing height. Certainly, deep root systems have been associated with drought-avoidance mechanism in turfgrasses (Carrow 1985; Carrow, Shearman, and Watson 1990; Kim 1987; Kopec, Shearman, and Riordan 1988; Salaiz et al. 1991). Grasses with deeper root systems can draw on a larger volume of soil for moisture and can maintain growth and development longer between irrigation, or rainfall, events than those with less extensive root systems (Erusha and Shearman 2002; Huang 1999; Huang and Fry 1998; Huang and Gao 2000; Huang, Duncan, and Carrow 1997; Kyoung-Nam, Shearman, and Riordan 1999). Turfgrasses mowed at higher heights of cut should respond in a similar way, because they would have deeper root systems than ones mowed at lower cutting heights. It must be emphasized, however, that this response may not relate to water conservation, because the higher water-use rate and deeper root system associated with the higher mowing height may simply deplete soil moisture over a larger portion of the soil profile. Root health, vigor, and plasticity certainly would influence this response, and these factors, in turn, would be impacted by soil physical and chemical conditions, as well as biotic and abiotic stresses.

Liu and Huang (2002) demonstrated that even small decreases in mowing height on a creeping bentgrass turf caused declines in root production, length, and new root growth. There is a need to consider carefully the effects of raising, or lowering, mowing heights within the accepted range for a particular species as it relates to water conservation and root growth, especially in light of potential abiotic stresses such as high temperature and soil compaction stress. More research is needed to relate increased mowing height and rooting responses with potential water conservation. This relationship would be confounded by interactions with mowing frequency, plant nutrition, and irrigation rate and scheduling, as well as with soil conditions and environmental stresses.

Mowing Frequency

Studies have indicated that leaf area, vertical elongation rate, and shoot size decrease, while shoot density and verdure increase, with increased mowing frequency (Beard 1973; Madison 1962a, b; Madison and Hagan 1962). Johns (1980) reported that, in Texas, turfgrass water use increased as the interval between mowing events increased. Shearman and Beard (1973) reported water use increased by 41% for creeping bentgrass when mowing frequency was changed from six times weekly to once every 2 weeks. Johns, Beard, and Van Bavel (1983), Shearman and Beard (1973), and Shearman (1986) found that more frequently mowed turfgrasses used less water. These authors hypothesized that frequent mowing results in less vertical elongation of turfgrass leaves, thus enhancing canopy resistance, which decreases the overall turfgrass water use.

It is apparent that mowing height and frequency can be manipulated to enhance water conservation through decreased ET. Mowing height and frequency also can be manipulated to enhance drought avoidance mechanisms through improved depth and extent of rooting. Johns (1980) demonstrated that excessive leaf area was eliminated and transpiration was decreased by manipulating mowing frequency. Krans and Beard (1985) reported greater rooting and improved root-to-shoot ratio when they mowed Kentucky bluegrass twice weekly as opposed to bi-weekly mowing. It is common practice to decrease the number of mowings when experiencing drought stress conditions. Research suggests, however, that mowing frequency should be maintained in an effort to conserve water and extend turfgrass performance and quality during periods of drought stress. Turfgrasses should not be mowed when there is no vertical elongation or when they are under visual moisture stress symptoms. More research is needed on the interaction between mowing height and frequency and their effects on turfgrass water conservation.

Plant Growth Regulators

St. Augustinegrass and bermudagrass were used in Texas to test plant growth regulators (PGRs) in different treatment combinations (Johns and Beard 1982). They observed 11 to 28% decrease in ET with flurprimidol treatments. Green, Beard, and Kim (1988) found ET decreases with treatments of flurprimidol and mefluidide. Flurprimidol treatments had an average ET decrease of 18%, decreased leaf elongation by 83%, but also decreased turfgrass quality. Shearman (1982b) assessed the ET rate of Kentucky bluegrass turfs treated with paclobutrazol, flurprimidol, and mefluidide under field conditions. The PGRs decreased ET by as much as 44% compared with the untreated control treatment. Paclobutrazol and flurprimidol were more effective than mefluidide. Ervin and Koski (2001) studied the effects of trinexapac-ethyl (TE) on Kentucky bluegrass ET and clipping yield from 1995 to 1997 in Colorado. The TE treatment decreased ET and clipping yield, and they concluded TE could be used to decrease clipping yield and ET rates. These results support the potential to use PGRs in a similar manner as mowing height and frequency to decrease water use and conserve water.

Mower Blade

Beard (1973) indicated that turfgrass quality, growth, and performance are impacted by mowing with reel versus rotary mowers, and dull versus sharp mower blades. He speculated that water loss was greater from turfgrasses mowed with a dull, or improperly adjusted, mower than those mowed with a properly adjusted, sharp mower. He felt that the potential increase in water loss was associated with increased tissue mutilation caused by the dull mower blade and the increased exposure of vascular tissue to desiccation. Steinegger and colleagues (1983) found that Park Kentucky bluegrass mowed with a dull mower blade had decreased turfgrass ET, vertical elongation rate, verdure, and turfgrass quality compared with turfgrasses mowed with a sharp blade. In their study, turfgrasses mowed with the dull mower had a slower leaf elongation rate and a decreased ET rate compared with those mowed with the sharp mower. Even though water use was decreased with the dull mower treatment, the loss in turfgrass quality made it unacceptable as a water conservation practice. No comparisons of rotary versus reel mower effects on turfgrass water use are available in the literature. It is commonly speculated that rotary mowers cause more leaf tip injury than reel mowers. This injury could lead to added water loss on a short-term basis, but no research results were found to support this speculation.

Turfgrass Nutrition

Nutrition affects turfgrass growth, leaf area, vertical elongation rate, depth and extent of rooting, water-use rates, and drought resistance. Shearman and Beard (1973) demonstrated that decreased water use was feasible through manipulation of nitrogen (N) nutrition. Feldhake, Danielson, and Butler (1983) reported Merion Kentucky bluegrass used 13% more water when fertilized with 4 kilograms N, 1,000 meter (m)$^{-2}$ month^{-1} versus the same application rate applied once per season. Kneebone and Pepper (1982) reported higher water use in bermudagrass and zoysiagrass (*Zoysia japonica* Steudel) receiving 5 grams (g) N m^{-2} monthly versus 5 g N m^{-2} every 2 months.

Schmidt and Breuninger (1981) studied the effects of N, phosphorus (P), and potassium (K) nutrition on Kentucky bluegrass drought recovery. They found that recovery from drought stress was greater for turfgrasses receiving autumn-applied N than for those receiving spring treatment. They also reported that P enhanced recovery but was dependent on N-rate and application timing. Potassium, on the other hand, benefited recovery regardless of N and P treatments. Powell, Blaser, and Schmidt (1967) reported that bentgrass produced more root weight with autumn N applications than with spring applications. Subhrajit and Trenholm (2005) found slow-release N carriers decreased water use in St. Augustinegrass turf. Petrovic and Baikan (1998) concluded from studies conducted in New York that banning fertilization as a means of decreasing the need for irrigation is ineffective, unless programs are applying N levels in excess of turfgrass nutritional needs.

As discussed earlier, faster-growing turfgrasses use more water than those with decreased leaf growth rates. Krogman (1967) demonstrated that increased N nutrition increased water use, and P applications caused a slight increase in water use for forage grasses. He indicated that N nutrition caused an increase in water-use efficiency relative to dry matter production. In turfgrasses, dry matter production is not of primary interest. Water-use efficiency relative to verdure production, turfgrass quality retention, and abiotic stress tolerance is an issue. More research is needed to discern the importance of water-use efficiency to turfgrass quality and water conservation issues.

Beard (1973) reported that K fertilization influenced turfgrass growth and water use. Markland and Roberts (1967) and Monroe, Coorts, and Skogley (1969) found increased rooting with increased K nutrition. Increased root growth, associated with K applications, would contribute to improved soil water use, drought resistance, and drought recovery. Shearman (1982a) evaluated N and K nutritional level effects on drought tolerance of Kentucky bluegrass and reported an interaction between N and K nutrition. Nitrogen nutrition increased visual wilting tendency under drought stress conditions, but K applications countered that response and decreased visual wilt symptoms. Miller and Dickens (1997) found K nutrition decreased leaf firing and enhanced drought stress recovery in bermudagrass. Ebdon and Petrovic (2000) concluded that minimal N and liberal K in turfgrass nutrition were key ingredients for decreasing water use. Schmidt and Breuninger (1981) indicated higher P and K tissue content levels were significantly correlated with improved Kentucky bluegrass drought stress recovery. Shearman, Erusha, and Wit (2005) concluded from their research, involving K nutrition and deficit irrigation, that K nutrition could be used to maintain turfgrass quality and enhance water conservation in Kentucky bluegrass maintained under fairway conditions even under drought stress. These research results support the potential to decrease water use and enhance drought resistance by manipulating turfgrass nutrition levels.

Certainly, interactions between turfgrass nutrition and other cultural practices are of interest where integrated management approaches for water conservation can be practiced feasibly. Nutritional programs should be adjusted to produce the least amount of excess top growth and the greatest amount of root growth and depth possible to improve water conservation and drought avoidance. Systems approaches to water conservation are needed.

Turfgrass Irrigation

Irrigation practices can play an important role in turfgrass water conservation. As a cultural practice, irrigation interacts with other cultural practices to influence water use (Danielson, Feldhake, and Butler 1981; Shearman, Erusha, and Wit 2005). Readers interested in water application and management can obtain more in-depth information from Section 13 of this publication. A common recommendation for turfgrass irrigation is to water deeply and infrequently. This recommendation is a bit perplexing because it does not provide a specific amount or frequency for irrigation application. Shearman and Beard (1973) reported a 33% decline in the water use of creeping bentgrass receiving water only when visual wilt symptoms were evident compared with those being watered three and seven times per week in controlled environment conditions. In Nevada, Tovey, Spencer, and Muckel (1969) applied water every 3, 7, or 10 days to a Kentucky bluegrass and fine fescue (*Festuca spp.*) polystand. They found this turfgrass stand could be maintained with high quality using the 3-day irrigation interval, and the 7-day interval was adequate on loam soils. The 10-day interval produced unsatisfactory turfgrass quality. In Israel, Biran and colleagues (1981) found that delaying irrigation on two cool-season and nine warm-season turfgrass species until wilt symptoms occurred resulted in a decrease in water use up to 35% without a decline in turfgrass quality compared with those that were not allowed to wilt.

Deficit Irrigation

Deficit irrigation of turfgrasses can be defined as irrigation applications below the maximum water demand of the turfgrass plant. To be most effective, this irrigation practice should not result in a significant loss of turfgrass quality and function. Deficit irrigation results in decreased soil-moisture content and decreased turfgrass water use. The practice is most effective in areas where rainfall is sufficient to recharge the soil profile periodically or where adequate irrigation is available to replenish soil moisture before moisture deficits impair turfgrass growth and development.

Research with cool- and warm-season turfgrasses has demonstrated the effectiveness of this strategy on turfgrass water conservation (DaCosta and Huang 2006; Fu, Fry, and Huang 2004; Kneebone and Pepper 1984; Qian and Engelke 1999; Shearman, Erusha, and Wit 2005). Kneebone and Pepper (1984), working in Arizona, found that ET increased by 75% when irrigation applications were increased from 16 mm to 52 mm day^{-1}. Their results suggest that higher irrigation application rates may result in higher water loss because of increased evaporation, runoff, and percolation below the rootzone of the grasses. In Kansas, tall fescue and bermudagrass were maintained under deficit irrigation conditions during the 2001 and 2002 growing conditions (Fu, Fry, and Huang 2004). These researchers found that tall fescue produced acceptable turfgrass quality when irrigated at 60 and 80% actual evapotranspiration (ET_a) in 2001 and 2002, respectively, whereas bermudagrass had acceptable quality at 60% ET_a in both growing seasons. Qian and Engelke (1999) reported that bermudagrass, St. Augustinegrass, and buffalograss produced acceptable turfgrass quality when irrigated at 55% ET_a under growing conditions in Dallas, Texas. DaCosta and Huang (2006) studied creeping bentgrass, velvet bentgrass, and colonial bentgrass under deficit irrigation practices in New Jersey. Their results indicated that irrigating at 60 or 80% of ET_a had no effect on water-use efficiency for any of the bentgrass species studied. They concluded that under New Jersey growing conditions bentgrass species could be irrigated at 60 to 80% ET_a, and such irrigation practices could result in considerable water conservation.

Shearman, Erusha, and Wit (2005), working in Nebraska, reported water savings of 21 and 40% when Kentucky bluegrass received deficit irrigation of 60 and 80% of potential evapotranspiration (ET_p) compared with 100% ET_p. Water conservation for 60 and 80% ET_p ranged from 2,387 and 1,225 m^3 hectare (ha)$^{-1}$ over the growing season, when compared with 100% ET_p. Under the conditions of their studies, all irrigation and K nutrition treatments provided acceptable turfgrass color and quality when compared with similar irrigation rates receiving no supplemental K treatments. In Colorado, Feldhake, Danielson, and Butler (1983) reported a 10% loss in turfgrass quality for Kentucky bluegrass receiving 73% of ET_p. This decline in turfgrass quality was relatively minimal based on the high ET demand conditions of Colorado. The acceptable turfgrass quality observed by Shearman, Erusha, and Wit (2005) with deficit irrigation supports the possibility of irrigating turfgrasses at lowered rates while maintaining an acceptable turfgrasses quality, particularly when K nutrition was incorporated as a management treatment option. Danielson, Feldhake, and Butler (1981) studied Kentucky bluegrass under limited water and N nutrition in Colorado. In their study, turfgrass quality declined linearly with decreased water application over a range from 100 to 40% of maximum ET. With adequate N nutrition, turfgrass quality was maintained at 70% of ET_p. These water conservation findings are encouraging, but need to be corroborated by additional research involving more diverse climatic conditions.

Deficit irrigation has been used to contribute to efficient use of water resources (Carrow, Shearman, and Watson 1990). In California, Henry and colleagues (2005) irrigated tall fescue, bermudagrass, zoysiagrass, and buffalograss, using crop coefficients (Kc) of 70 and 100%. Tall fescue did not produce acceptable turfgrass quality under deficit irrigation, whereas buffalograss (*Buchloe dactyloides* [Nutt.] Engelm.) and zoysiagrass performed best under these conditions. Their results demonstrate the importance of species selection and adaptation to climate and management when implementing deficit irrigation as a water conservation practice.

Irrigation Scheduling

A key component in water conservation strategies is efficient irrigation, preferably while maintaining an acceptable and functional turfgrass quality (Meyer and Camenga 1985). In southern California, Youngner and colleagues (1981) used tensiometers and evaporation pans to determine irrigation frequency needs. They concluded bermudagrass and St. Augustinegrass (*Stenotaphrum secundatum* [Walt.] Kuntze) maintained turfgrass quality with half the amount of water used for the control treatment, which reflected common irrigation practices used in the area. Research in Florida using microswitch tensiometers and electronic soil-moisture sensing devices for irrigation scheduling decreased water application by 89% over conventional approaches without decreasing turfgrass quality (Augustine, Snyder, and Burt 1981).

In Arizona, Kneebone and Pepper (1982) reported that water use of subirrigated turfgrasses followed Class A pan evaporation closely during periods of high water demand and active growth. Water use expressed as percentage of Class A pan evaporation ranged from 42 to 80%, depending on turfgrass species and intensity of management. They concluded that irrigating warm-season turfgrasses in excess of 50 to 80% of pan evaporative water loss and 60 to 85% for cool-season turfgrasses resulted in unnecessary water use. O'Neil and Carrow (1983) compared tensiometer-controlled irrigation to a set schedule of irrigation for Kentucky bluegrass grown in Kansas. The tensiometer-controlled irrigation technique decreased water use by 48% without impacting turfgrass quality. They reported the greatest savings of 66% in the autumn when ET demand was decreased, and the least savings (i.e., 11%) in August when ET demand was extremely high.

Reference evapotranspiration (ET_o) rate can be used to schedule irrigation (Brown 1995). Brown (1995) compared several equations commonly used to determine ET_o for irrigation scheduling in Arizona. He demonstrated that the equations commonly used differed by as much as 20% in their determination of ET_o, indicating a need to adjust irrigation based on the equation used for referencing the Kc.

Throssell, Carrow, and Milliken (1987) were the first to apply turfgrass canopy temperature as a predictive tool for irrigation scheduling on turfgrasses. Shearman, Erusha, and Wit (2005) demonstrated increased canopy temperatures during water stress in Kentucky bluegrass. Throssell, Carrow, and Milliken (1987) compared turfgrass canopy temperature measured with infrared thermometer with scheduled irrigation of Kentucky bluegrass and used stress degree day, crop water stress index (CWSI), and critical point model (CPM) indices to schedule irrigation.

Their irrigation treatments included well-watered irrigation with a soil water potential of -0.04 megapascal (MPa); slightly stressed irrigation at a soil water potential of -0.07 MPa; and moderately stressed irrigation with a soil water potential of -0.40 MPa. They observed water conservation of 50% between the tensiometer-scheduled irrigation and CPM, but turfgrass quality ratings were highest for CWSI and CPM scheduling. These quality differences reflected the greater amount of water applied with CWSI and CPM. The CWSI scheduling used 33% less water than the CPM technique. Park and colleagues (2005) used remote sensing to detect changes in turfgrass quality and water-stress expression in a hybrid bermudagrass (*C. dactylon* x *C. transvaalensis*), and demonstrated water conservation of 7% using these approaches. More research is needed comparing remote sensing techniques to detect soil moisture stress and irrigation scheduling needs.

There is an excellent potential to manipulate irrigation using Kc predictions along with other cultural practices, such as mowing frequency and nutrition, to decrease turfgrass water use. Systems approaches easily could be recommended for species based on the information available in the turfgrass literature. Sensors for computer control and scheduling of turfgrass irrigation are a reality. More information on soil-moisture sensors is available in Section 13 of this publication. Researchers are investigating remote sensing approaches to detect moisture stress in its earliest stages and enhance irrigation scheduling for water conservation purposes. Additional research information is needed to refine these systems approaches to maximize water conservation and minimize negative impacts on turfgrass quality and use.

Soil Cultivation, Topdressing, and Wetting Agents

Turfgrasses often are exposed to intense traffic and soil compaction stress. Soil compaction and its implications on turfgrass soils is discussed in detail in Section 6 of this publication. Soil compaction influences soil bulk density, water infiltration and retention, and macropore space, which in turn impact turfgrass top and root growth, water use, and drought resistance. Root and shoot growths are decreased as a result of soil compaction stress (Agnew and Carrow 1985; Beard 1973; Carrow 1980, 1986; O'Neil and Carrow 1983). O'Neil and Carrow (1983) reported soil compaction decreased oxygen diffusion rates (ODR), and as the ODR level declined, shoot growth was decreased more than those for root growth, indicating shoot growth was more sensitive than root growth to the declining soil oxygen. They reported that turfgrass water use declined with increased soil compaction and that this decline was reflected in decreased shoot growth activity. Rieke and Murphy (1989) reviewed the benefits of soil cultivation on alleviating soil compaction stress, enhancing depth and extent of rooting, and improving soil infiltration rate.

Soil Cultivation

Use of traffic-tolerant turfgrasses and soil cultivation techniques are important factors in water conservation management on intensively used turfgrass sites. Wiecko, Carrow, and Karnok

(1993) found that cultivation increased water use of Tifway bermudagrass grown on compacted soil by 22%. They concluded that soil cultivation practices enhanced rooting depth and improved shoot growth, which resulted in the increased water-use rate. Even though water use increased, they speculated that the improved turfgrass quality and performance would offset the water used by improving the plant's stress tolerance and drought resistance capabilities. These speculations open interesting opportunities for additional research involving management of turfgrass soil compaction and its interactions with turfgrass water use, water conservation, and drought resistance.

Topdressing

Topdressing can have positive and negative effects on turfgrass water use and drought avoidance. Carrow and Petrovic (1992) reviewed the impacts of soil layering on turfgrass water relations, rooting, and shoot growth. Soil layering can be created by a number of means, such as infrequent topdressing, sod with organic matter or soil textural differences, and soil deposition from wind and flooding. The implications of turfgrass soil layering in urban landscapes are discussed in detail in Section 6 of this publication. Regardless of the source, soil layering impedes water infiltration and percolation, impacts soil moisture retention and ODR, and impedes rooting depth. These adverse relationships to soil layering could impact turfgrass root growth and health negatively, and decrease turfgrass water-use efficiency and drought resistance. Proper topdressing approaches minimize thatch accumulation, decrease potential layering (Rieke et al. 1989), enhance turfgrass root growth and development, and enhance turfgrass drought resistance.

Decreasing thatch accumulation enhances water infiltration and percolation and improves soil moisture distribution (Carrow, Shearman, and Watson 1990). Johnson, Davis, and Qian (2005) topdressed Kentucky bluegrass with composted manure. They reported improved turfgrass quality and drought avoidance as a result of topdressing with composted manure. They concluded that the composted topdressing allowed more water to penetrate the soil after irrigations and enhanced the available soil moisture that turfgrasses need during water stress conditions. The potential benefits of topdressing to water conservation are not as well defined as those with some other cultural practices. There is ample evidence of topdressing effects on soil physical characteristics and soil water movement, but more research is needed to clarify the real benefits of topdressing to water conservation programs.

Wetting Agents

Water repellency has been observed in turfgrass soils and modified rootzones and has been associated with high-sand-content soils, plant residues, and microorganism activity (Dekker, Ritsema, and Oostindie 2004; Park et al. 2004; Waddington 1992). Hydrophobic conditions increase as soils dry and soil water repellency increases. These conditions can impact soil water infiltration and unsaturated flow, which in turn can impact turfgrass water use. Kostka and colleagues (1997) and Dekker, Ritsema, and Oostindie (2004) reported that wetting agents provided a means for enhancing soil wetting in hydrophobic soils and increased soil moisture content in the rootzone. Much of the turfgrass wetting agent research information relating to improving water penetration in hydrophobic turfgrass soils has not been published in peer-reviewed scientific journals. Enhanced soil wetting would be beneficial to improved soil water content, especially field studies, and soil-plant water relations. There is a need for more research with soil wetting agents and their effects on turfgrass soil water content, and for researchers to publish existing data in peer-reviewed journals.

Peterson, Shearman, and Kinbacher (1984) and Shearman (1985b) reported improved turfgrass quality and decreased ET for Kentucky bluegrass treated with wetting agents on a monthly basis. They reported decreased ET by as much as 28%, and they observed that treatment effects were transitory and essentially gone by four weeks after treatment. Park and colleagues (2005) reported surfactant treatments decreased irrigation requirements of bermudagrass by 71% without decreasing turfgrass quality. It is apparent that wetting agents could be used to manipulate water conservation and warrant further research to verify their role in systems approaches to turfgrass water management.

Antitranspirants, Biostimulants, and Pesticides

Chemicals such as pesticides, antitranspirants, and biostimulants may directly, or indirectly, influence turfgrass top and root growth, water use, and drought resistance. Antitranspirants have been used to control transpiration by controlling stomatal closure, covering the mesophyll surface with a thin layer, and coating the leaf surface with a water-impervious film. Phenylmercuric acetate (PMA) has been used to induce stomatal closure in species other than grasses (Shimishi 1963). Stahnke (1981) conducted controlled-environment research on creeping bentgrass and bermudagrass using absisic acid (ABA), B-napthozyacetic acid, and a mixture of PMA and wetting agent. She found that ABA decreased ET by 59% in creeping bentgrass and 11% in bermudagrass without causing visual injury to the plant or increasing leaf temperatures. The ABA treatment was effective, but its use on a large-scale basis was questioned because of costs. Monomolecular films have been used on water reservoirs to minimize water loss from evaporation and also have been used to decrease plant transpiration (Aubertine and Grosline 1964). Coatings that cover leaf surfaces have been used to decrease transpiration and serve as winter desiccation protection on ornamental plants (Wooley 1967). It is speculated that leaf surface coatings would not work effectively on turfgrasses because of leaf orientation and removal by mowing and traffic. Johns (1980) expressed skepticism about the potential use of antitranspirants on turfgrasses based on his research with St. Augustinegrass. He found ET was influenced more by environmental factors than stomatal aperture, and that turfgrass canopy resistance offered more potential to influence ET than antitranspirants.

In Virginia, applications of biostimulants, such as fortified seaweed extract, have been reported to improve turfgrass drought resistance (Sun, Schmidt, and Eisenbach 1997; Yan, Schmidt, and Orcutt 1997). They demonstrated improved turfgrass rooting as a result of the treatments. Yan, Schmidt, and Orcutt (1997) reported higher leaf-water potential in perennial ryegrass treated with fortified seaweed extract compared with the untreated control. Only limited information is available in the turfgrass literature regarding biostimulant effects on drought resistance, and no literature was found on their impact on water use.

Pesticides may influence turfgrass water use. For example, PMA has been discussed as an antitranspirant, but it also was used as a fungicide. Treatments of PMA as a fungicide could have an indirect effect on water loss from the treated plant material. Shearman, Kinbacher, and Reierson (1980) reported a decline in water use and water-use efficiency for tall fescue treated with the preemergence herbicide siduron. Siduron treatments decreased shoot and root growth and increased wilting and drought-stress injury in the treated turf. Judicious pesticide applications are an integral part of certain turfgrass management systems, especially on intensively used sites. Those concerned with water conservation should carefully select pesticides based on efficacy, environmental sustainability, and potential impacts on plant factors that might increase water loss.

Summary

Water availability and conservation is a priority for the turfgrass industry. There is adequate information in the turfgrass literature to substantiate the use of cultural practices alone, or in systems approaches, to decrease turfgrass water use, conserve water, and enhance drought resistance. These practices could be used immediately to conserve water and maintain turfgrass quality and use.

Literature Cited

Agnew, M. L. and R. N. Carrow. 1985. Soil compaction and moisture stress preconditioning in Kentucky bluegrass. I. Soil aeration, water use, and root responses. *Agron J* 77(6):872–878.

Aubertine, G. M. and G. W. Gorsline. 1964. Effect of fatty alcohol on evaporation and transpiration. *Agron J* 56:50–52.

Augustine, B. J., G. H. Snyder, and E. O. Burt. 1981. Turfgrass irrigation water conservation using soil moisture sensing devices. P. 123. In *Agronomy Abstracts*. American Society of Agronomy, Madison, Wisconsin.

Beard, J. B. 1973. *Turfgrass: Science and Culture*. Prentice-Hall, Englewood Cliffs, New Jersey. 672 pp.

Biran, I., B. Brando, I. Bushkin-Harav, and E. Rawitz. 1981. Water consumption and growth rate of 11 turfgrasses as affected by mowing height, irrigation frequency, and soil moisture. *Agron J* 73:85-90.

Brown, P. 1995. Turfgrass irrigation with municipal effluent: Nitrogen fate, turf Kc and water requirements. Pp 38-40. In *1995 Turfgrass and Environmental Research Summary*. USGA, Far Hills, New Jersey.

Carrow, R. N. 1980. Effects of soil compaction on the growth of three cool-season grasses. *Agron J* 72:1038-1042.

Carrow, R. N. 1985. Soil water relationships in turfgrass. Pp 85-102. In V. B. Gibeault and S. T. Cockerham (eds.). *Turfgrass Water Conservation*. Publ. 21405, University of California, Riverside, Division of Agriculture and Natural Resources.

Carrow, R. N. 1986. Water management on compacted soils. *Weeds, Trees and Turf.* 25(3):34, 38, 40.

Carrow, R. N. and A. M. Petrovic. 1992. Effects traffic on turfgrass. Pp. 286-325. In D. V. Waddington, R. N. Carrow, and R. C. Shearman (eds.). *Turfgrass Monograph 32*. American Society of Agronomy, Madison, Wisconsin.

Carrow, R. N., R. C. Shearman, and J. R. Watson. 1990. Turfgrass. Pp. 891-916. In B. A. Stewart and D. R. Nielsen (eds.). *Irrigation of Agricultural Crops*. ASA Monograph 30. American Society of Agronomy, Madison, Wisconsin.

DaCosta, M. and B. Huang. 2006. Minimum water requirements for creeping, colonial and velvet bentgrass under fairway conditions. *Crop Sci* 46:81-89.

Danielson, R. E., C. M. Feldhake, and J. D. Butler. 1981. Urban lawn irrigation and management practice for water saving with minimum effect on lawn quality. Completion Report to OWRT Project No. H-043-Colo. Colorado State University, Ft. Collins, Colorado.

Dekker, L. W., C. J. Ritsema, and K. Oostindie. 2004. Dry spots in golf courses: Occurrence, amelioration and prevention. *Acta Hort* 661:99-104.

Doss, B. D., O. L. Bennette, D. A. Ashley, and H. A. Weaver. 1962. Soil moisture regime effect on yield and evapotranspiration from warm-season perennial forage species. *Agron J* 54:239-242.

Ebdon, J. S. and A. M. Petrovic. 2000. Fertilizing for water conservation. *Golf Course Management* December:61-65.

Erusha, K. S. and R. C. Shearman. 2002. Kentucky bluegrass cultivar rooting responses measured in hydroponics. *Crop Sci* 42:848-852.

Ervin, E. H. and A. J. Koski. 2001. Trinexapac-ethyl effects of Kentucky bluegrass evapotranspiration. *Crop Sci* 41(1):247–250.

Feldhake, C. M., R. E. Danielson, and J. D. Butler. 1983. Turfgrass evapotranspiration. II. Factors influencing rate in urban environments. *Agron J* 75:824–830.

Fu, J., J. Fry, and B. Huang. 2004. Minimum water requirements for four Turfgrasses in the transition zone. *HortSci* 39(7):1740–1744.

Green, R. L., J. B. Beard, and K. S. Kim. 1988. The effects of Flurprimidol and mefluidide on ET, leaf growth and quality of St. Augustinegrass grown under two soil moisture levels. P. 151. In *Agronomy Abstracts*. American Society of Agronomy, Madison, Wisconsin.

Henry, J. M., S. N. Wegulo, V. A. Gibeault, and R. Autio. 2005. Turfgrass performance with reduced irrigation and nitrogen fertilization. *Intl Turf Soc Res J* 10:93–101.

Huang, B. 1999. Water relations and root activities of *Buchloe dactyloides* and *Zoysia japonica* in response to localized soil drying. *Plant and Soil* 208:179–186.

Huang, B. and J. Fry. 1998. Root anatomical, physiological, and morphological responses to drought stress for tall fescue cultivars. *Crop Sci* 38:1017–1022.

Huang, B. and H. Gao. 2000. Root physiological characteristics associated with drought resistance in tall fescue cultivars. *Crop Sci* 40:196–203.

Huang, B., R. R. Duncan, and R. N. Carrow. 1997. Drought resistance mechanisms of seven warm season turfgrasses under surface soil drying: II. Root aspects. *Crop Sci* 37:1863–1869.

Johns, D. 1980. Resistance to evapotranspiration from St. Augustinegrass (*Stenotaphrum secundatum* [Walt.] Kuntze). Ph.D. thesis. Texas A & M University, College Station, Texas. 88 pp.

Johns, D. and J. B. Beard. 1982. Water conservation—A potentially new dimension in the use of growth regulators. Texas Turfgrass Research Progress Report 4040.

Johns, D., J. B. Beard, and C. H. M. Van Bavel. 1983. Resistances to evapotranspiration from a St. Augustinegrass turf canopy. *Agron J* 75:419–422.

Johnson, G. A., J. G. Davis, and Y. L. Qian. 2005. Topdressing Kentucky bluegrass with composted manure: Soil and water quality impacts. In *Agronomy Abstracts*. American Society of Agronomy, Madison, Wisconsin.

Kim, K. S. 1983. Comparative evapotranspiration rates of thirteen turfgrasses grown under both non-limiting soil moisture and progressive water stress conditions. M.S. thesis. Texas A&M University., College Station, Texas.

Kim, K. S. 1987. Comparative drought resistance mechanisms of eleven major warm-season turfgrasses. Ph.D. Diss. Texas A&M University, College Station, Texas (Dissertation Abstracts 87-20914).

Kim, K. S. and J. B. Beard. 1988. Comparative turfgrass evapotranspiration rates and associated plant morphological characteristics. *Crop Sci* 28:328–331.

Kneebone, W. R. and I. L. Pepper. 1982. Consumptive water use by subirrigated turfgrasses under desert conditions. *Agron J* 74:419–423.

Kneebone, W. R. and I. L. Pepper. 1984. Luxury water use by bermudagrass turf. *Agron J* 76:999–1002.

Kopec, D. M., R. C. Shearman, and T. P. Riordan. 1988. Evapotranspiration of tall fescue turf. *HortSci* 23(2)300–301.

Kostka, S., J. Cisar, J. R. Short, and S. Mane. 1997. Evaluation of soil surfactants for the management of soil water repellency in turfgrass. *Intl Turf Soc Res J* 8:485–494.

Krans, J. V. and J. B. Beard. 1985. Effects of clipping on growth and physiology of 'Merion' Kentucky bluegrass. *Crop Sci* 25(1):17–20.

Krogman, K. K. 1967. Evapotranspiration by irrigated grass as related to fertilizer. *Can J Plant Sci* 47:281–287.

Kyoung-Nam K., R. C. Shearman, and T. P. Riordan. 1999. Top growth and rooting responses of tall fescue cultivars grown in hydroponics. *Crop Sci* 39:1431–1434.

Liu, X. and B. Huang. 2002. Mowing effects on root production, growth, and mortality of creeping bentgrass. *Crop Sci* 42(4):1241–1250.

Madison, J. H. 1962a. Mowing of turfgrass. II. Responses of three species of grass. *Agron J* 54:250–253.

Madison, J. H. 1962b. Turfgrass ecology. Effects of mowing, irrigation, and nitrogen treatments of *Agrostis palustris* Huds.; 'Seaside' and *A. tenuis* Sibth.; 'Highland' on population, yield, rooting and cover. *Agron J* 54:407–412.

Madison, J. H. and R. M. Hagan. 1962. Extraction of soil moisture by Merion bluegrass (*Poa pratensis* L. 'Merion') turf, as affected by irrigation frequency, mowing height and other cultural operations. *Agron J* 54:157–160.

Markland, F. E. and E. C. Roberts. 1967. Influence of varying nitrogen and potassium levels on growth and mineral composition of *Agrostis palustris* Huds. P. 53. In *Agronomy Abstracts*. American Society of Agronomy, Madison, Wisconsin.

Meyer, J. L. and B. C. Camenga. 1985. Irrigation systems for water conservation. In V. A. Gibeault and S. T. Cockerham (eds.). *Turfgrass Water Conservation*. Publication No. 21405. University of California, Riverside.

Miller, G. L. and R. Dickens. 1997. Water relations of two *Cynodon* turf cultivars as influenced by potassium. *Intl Turf Soc Res J* 8:1298–1306.

Mitchell, K. J. and J. R. Kerr. 1966. Differences in rate and use of soil moisture by stands of perennial ryegrass and white clover. *Agron J* 58:5–8.

Monroe, C. A., G. D. Coorts, and C. R. Skogley. 1969. Effects of nitrogen-potassium levels on the growth and chemical composition of Kentucky bluegrass. *Agron J* 61:294–296.

O'Neil, K. J. and R. N. Carrow. 1983. Perennial ryegrass growth, water use, and soil aeration status under soil compaction. *Agron J* 75(2):177–180.

Park, D. M., J. L. Cisar, K. E. Williams, and G. H. Snyder. 2004. Alleviation of soil water repellency in sand based bermudagrass in south Florida. *Acta Horticulturae* 661:111–115.

Park, D. M., J. L. Cisar, D. K. McDermitt, K. E. Williams, J. J. Haydu, and W. P. Miller. 2005. Using red and infrared reflectance and visual observation to monitor turf quality and water stress in surfactant-treated bermudagrass under reduced irrigation. *Intl Turf Soc Res J* 10:115–120.

Peterson, M. P., R. C. Shearman, and E. J. Kinbacher. 1984. Surfactant effects on Kentucky bluegrass evapotranspiration rates. In *Agronomy Abstracts*. American Society of Agronomy, Wisconsin.

Petrovic, A. M. and B. B. Baikan. 1998. Watering, fertilization can lower turfgrass water use and waste. *Landscape and Irrigation* 22(1):18–20.

Powell, A. J., R. E. Blaser, and R. E. Schmidt. 1967. Effect of nitrogen on winter root growth of bentgrass. *Agron J* 59:529–530.

Qian, Y. and M. C. Engelke. 1999. Performance of five Turfgrasses under linear gradient irrigation. *HortSci* 34:893–896.

Rieke, P. E. and J. A. Murphy. 1989. Advances in turf cultivation. *Int Turfgrass Soc Res J* 6:49–54.

Rieke, E., J. A. Murphy, J. N. Rogers, and M. T. McElroy. 1989. Topdressing program effects of soil properties of a putting green. *Ann Meet Abstr* 81:164.

Salaiz, T. A., R.C. Shearman, T.P. Riordan, and E.J. Kinbacher. 1991. Creeping bentgrass cultivar water use and rooting responses. *Crop Sci* 31(5):1331–1334.

Schmidt, R. D. and J. M. Breuninger. 1981. The effects of fertilization on recovery of Kentucky bluegrass turf from summer drought. Pp. 333–341. In R. W. Sheard (ed.). *Proceedings of the 4th International Turf Research Conference*. University of Guelph, Ontario, Canada.

Shearman, R. C. 1982a. Nitrogen and potassium nutrition influence on Kentucky bluegrass turfs. *Proceedings of the 7th Nebraska Turfgrass Field Day and Equipment Show*. Department of Horticulture Publication No. 82-2:67–70. University of Nebraska, Lincoln.

Shearman, R. C. 1982b. Turfgrass responses to plant growth regulators. *Proceedings of the 7th Nebraska Turfgrass Field Day and Equipment Show*. Department of Horticulture Publication No. 82-2:34-44. University of Nebraska, Lincoln.

Shearman, R. C. 1985a. Turfgrass culture and water use. Pp. 61–70. In V. B. Gibeault and S. T. Cockerham (eds.). *Turfgrass Water Conservation*. Division of Agriculture and Natural Resources Publication No. 21405. University of California, Riverside.

Shearman, R.C. 1985b. Wetting agents. *Lawn Landscape* 61(11):70–72.

Shearman, R. C. 1986. Kentucky bluegrass cultivar evapotranspiration rates. *HortSci* 21(3):455–457.

Shearman, R. C. 1989. Perennial ryegrass cultivar evapotranspiration rates. *HortSci* 24(5):767–769.

Shearman, R. C. and J. B. Beard. 1973. Environmental and cultural preconditioning effects on the water use rate of *Agrostis palustris* Huds., cultivar Penncross. *Crop Sci* 13:424–427.

Shearman, R. C., K. S. Erusha, and L. A. Wit. 2005. Irrigation and potassium effects on *Poa pratensis* L. fairway turf. *Int Turfgrass Soc Res J* 10:998–1004.

Shearman, R. C., E. J. Kinbacher, and K. A. Reierson. 1980. Siduron effects on tall fescue (*Festuca arundinacea*) emergence, growth and high temperature injury. *Weed Sci* 28:194–196.

Shimishi, D. 1963. Effect of chemical closure of stomata on transpiration in varied soil and atmospheric environments. *Plant Physiol* 38:709–712.

Spak, D., J. DiPaola, and C. Anderson. 1993. Tall fescue sward dynamics: I. Seasonal patterns of turf shoot development. *Crop Sci* 33:300–304.

Stahnke, G. K. 1981. Evaluation of antitranspirants of creeping bentgrass (*Agrostis palustris* Huds., 'Penncross') and bermudagrass [*Cynodon dactylon* (L.) Pers. x *C. transvaalensis* Burtt-Davy, 'Tifway']. M.S. thesis. Texas A&M University, College Station.

Steinegger, D. H., R. C. Shearman, T. P. Riordan, and E. J. Kinbacher. 1983. Mower blade sharpness effects on turf. *Agron J* 75:479–480.

Subhrajit, K. S. and L. E. Trenholm. 2005. Effect of fertilizer source on water use of St. Augustinegrass and ornamental plants. *HortSci* 40(7):2164–2166.

Sun, H., R. E. Schmidt, and J. D. Eisenbach. 1997. The effect of seaweed concentrate on the growth of nematode-infected bent grown under low soil moisture. *Intl Turf Soc Res J* 8:1336–1343.

Throssell, C. S., R. N. Carrow, and G. A. Milliken. 1987. Canopy temperature based irrigation scheduling indices for Kentucky bluegrass turf. *Crop Sci* 27(1):126–131.

Tovey, R., J. S. Spencer, and D. C. Muckel. 1969. Turfgrass evapotranspiration. *Agron J* 61:863–867.

Waddington, D. V. 1992. Soils, soil mixtures and soil amendments. Pp. 331–385. In D. V. Waddington, R. N. Carrow, and R. C. Shearman (eds.). *Turfgrass*. ASA Monograph 32. American Society of Agronomy, Madison, Wisconsin.

Wiecko, G., R. N. Carrow, and K. J. Karnok. 1993. Turfgrass cultivation methods: Influence on soil physical, root/shoot, and water relationships. *Intl Turfgrass Soc Res J* 7:451–457.

Wooley, J. T. 1967. Relative permeabilities of plastic films to water and carbon dioxide. *Plant Physiol* 42:641–643.

Yan, J., R. E. Schmidt, and D. M. Orcutt. 1997. Influence of fortified seaweed extract and drought stress on cell membrane lipids and sterols of ryegrass leaves. *Intl Turfgrass Soc Res J* 8:1356–1363.

Youngner, V. B., A. W. Marsh, R. A. Strohman, V. A. Gibeault, and S. Spaulding. 1981. Water use and turf quality of warm-season and cool season turfgrasses. Pp. 251–257. In R. W. Sheard (ed.). *Proceedings of the 4th International Turfgrass Research Conference*. University of Guelph, Ontario, Canada.

13

Achieving High Efficiency in Water Application via Overhead Sprinkler Irrigation

Michael T. Huck and David F. Zoldoske

Introduction

Water-use efficiency is the combination of water application uniformity and water management. To decrease or eliminate over-irrigation, both parameters must be optimized. This section addresses water uniformity as applied through sprinkler systems and efforts to improve water management by identifying controllers and moisture sensors that provide the best timing and amount of water required by the plant.

Water-Efficient Sprinkler Irrigation Systems

The key components in landscape irrigation include design (engineering); water management (when and how much water to apply); equipment (pipes, valves, emission devices, controllers, etc.); installation; and maintenance. If not done correctly, any one of these items or activities will have a negative impact on water-use efficiency. It is the sum of all the parts in an irrigation system that ensures water application uniformity. When purchasing a new irrigation system, one should be able to specify both the level of uniformity and irrigation control. The level of irrigation uniformity should be defined at the design stage, and then tested at system start-up after it has been installed per the plan.

Irrigation Design Guidelines for Water-Efficient Systems

Designing an efficient irrigation system is no accident. It requires planning, thought, and knowledge of both engineering (sprinkler performance, control systems, and hydraulics) and plant culture (soil science and horticultural/agronomic plant water requirements). Entire texts have been written on turfgrass and landscape irrigation design (Barrett et al. 2003; Choate 1994; Jarrett 1985; Pair et al. 1983). The following points summarize the main considerations in designing a water-efficient irrigation system.

Develop a Plan

A scaled plan of the actual site is needed for the design of an efficient irrigation system. The plan should identify all existing or proposed structures, fences, hardscapes, plant materials, soil textures, and distinct slopes. These items affect irrigation design and water requirements because of their effect on sun, shade, and wind exposure. Recognizing the locations of these site features is very important when designating hydrozones (individual irrigation control areas). Some residential irrigation systems are designed in the field by contractors with no drawn plans. This approach may offer initial monetary savings, but may not always deliver the same long-term efficiency and water savings as a well-thought-out and professionally designed system.

Hydrozoning

The system must be "hydrozoned" into areas of similar water requirements based on planting schemes, soil textures, slopes, and microclimates. This zoning allows for the design of individual control zones for management of water applications based on plant requirements. For example, a turfgrass area on the north side of a structure will have different watering requirements than a turfgrass area on the south side of that same structure, especially during spring and autumn when the sun is positioned lower on the southern horizon and casts a longer shadow on the northern exposure. Therefore, areas with different exposures and/or plant materials having different water requirements (lawn grasses versus ornamental grasses, trees, groundcovers, and shrubs) need to be zoned independently for efficient water conservation.

Efficient Sprinkler Head Layouts

The most efficient sprinkler head model, nozzle, head configuration (triangular over square spacing where possible), and head spacing distance combination must be considered for each hydrozone. Care should be taken in design and hardware selection to avoid overspray onto impervious surfaces (driveways, sidewalks, etc.). Ideally, the sprinkler and nozzle combination selected will have a precipitation rate that is less than or equal to the soil infiltration rate. If the soil infiltration rate is less than the sprinkler precipitation rate, the controller must be capable of programming for multiple cycles that allow "soak-in" time between each cycle. This is especially a concern where sloped or mounded terrain may contribute to inefficiency through runoff. Depending on the size of each hydrozone, different-sized (diameter) sprinklers may be required, but within each individual hydrozone the same sprinkler and nozzle combinations must be used so that precipitation rates are uniform throughout the area.

Hydraulics (Flow, Pressure, and Pipe Sizing)

Hydraulic performance is based on a combination of factors affecting operating pressure losses resulting from the friction created as water passes through each component in the irrigation system before exiting the sprinkler nozzle. Friction losses are calculated based on the flow rates measured in gallons per minute and flow velocity measured in feet per second (fps). As a general rule of thumb, velocity of flow should not exceed 5 fps so the speed of the water flowing through the pipe is less likely to cause system damage because of surge pressures. Also, friction losses increase dramatically as water is forced through pipe at higher speeds.

Pressure losses between components will vary depending on the individual component's internal size (diameter and/or length) and shape (straight through or containing various changes of direction), as well as on the material of construction (PVC, brass, copper, etc.) of the particular component's water passageway. Each component of the system must be considered, beginning at the water meter and continuing on through any backflow devices, remote control valves, pipes, change of direction fittings, and finally, the sprinklers. Manufacturers' catalogs supply the necessary specifications regarding friction losses, flow rates, and operating pressures so that irrigation designers can calculate pressure losses between the point of connection and the final sprinkler in the system.

Other factors also must be considered such as the impact of elevation, which will increase pressure when water flows downhill or decrease pressure as it flows uphill. These loss factors are applied to each component and then tallied for the worst-case sprinkler circuit, which typically is the circuit with the sprinkler located the greatest distance from the point of connection to the water source. The sum of the total loss allows the designer to calculate the approximate pressure difference between the first and last sprinkler in the system where:

$$Pr = Ps - (Po + Pls)$$

Ps = static pressure; Po = operating pressure for "worst-case" sprinkler; Pls = pressure loss throughout the system, mainline, worst-case lateral circuit and all outer valves; and Pr = pressure remaining after satisfying the total system requirement.

The importance of calculating flow rates and determining pressure losses as it relates to pipe size and other components cannot be overstated. Pressure loss concerns from undersized pipe and/or extremely long runs of pipe can be very problematic, especially in larger irrigation systems such as those found in parks or golf courses. It is necessary to calculate pressure losses so the operating pressure difference between sprinklers is not so excessive that sprinkler distribution profiles and the resulting precipitation rates between overlapping sprinklers are not significantly affected (Choate 1994; Pair et al. 1983) (Figure 13.1). To minimize uneven distribution, designers typically set a maximum pressure variation between the first and last sprinkler between ± 10 to $\pm 20\%$ of the manufacturer's desired sprinkler operating pressure. Most small residential systems can be designed using a rule of thumb that if velocity of flow in any section of pipe does not exceed 5 fps, then the total friction loss of a worst-case circuit will remain within the $\pm 20\%$ desired pressure range. This scenario occurs because the typical small residential sprinkler system will not have exceptionally long pipe runs or elevation changes that will affect pressure losses significantly. It is always a good idea, however, to calculate pressure losses for the worst-case circuit, even in residential designs.

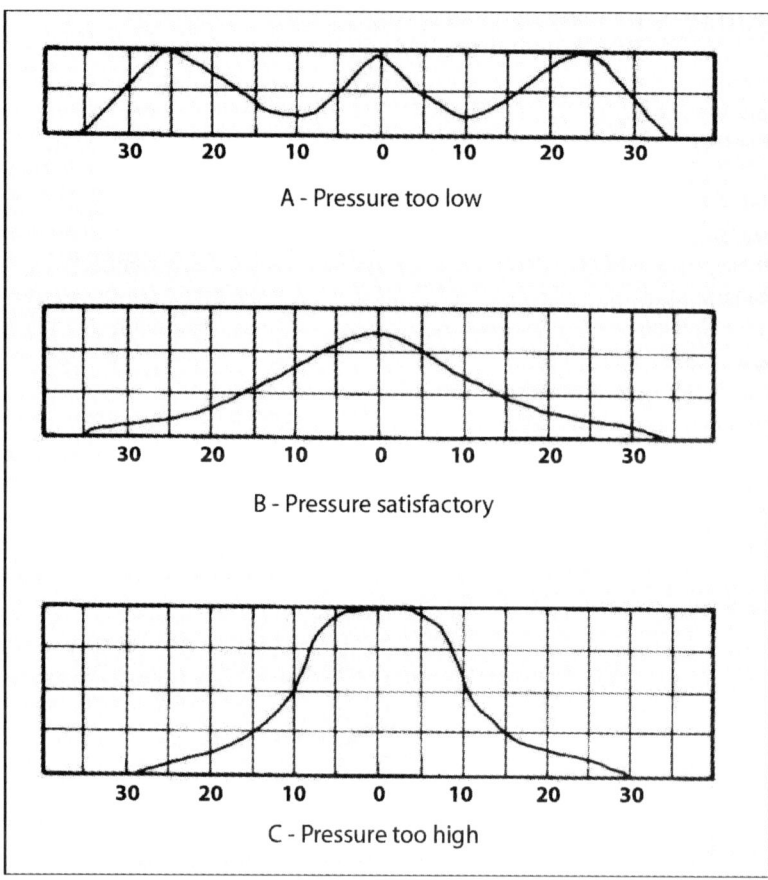

Figure 13.1. Effect of pressure on sprinkler distribution profiles (Choate 1994; Pair et al. 1983).

Other Requirements

Most municipalities always have required proper backflow prevention to protect potable water supplies from contamination. But other requirements are on the rise throughout the United States, particularly in arid regions where drought and water shortages are commonplace. Examples include the submission of landscape and irrigation plans for inspection before installation and the training or certifying of irrigation installation and maintenance contractors and their personnel.

Sprinkler and Nozzle Technologies

The sprinkler and associated nozzle usually are the most important components of the irrigation system because they distribute the water over the land. How uniformly that water is distributed has a large effect on the overall effectiveness and efficiency of the irrigation system.

There are three primary sprinkler technologies currently used in turfgrass and landscape irrigation systems: fixed-spray, impact rotor, and gear rotor heads. Fixed-spray sprinklers use a horizontally positioned flat fan pattern, slotted-spray nozzle to apply water. Impact and gear rotors can be found with a single- or multiple-nozzle arrangement. Depending on the specific design, multiple-stream designs may contain as few as two or perhaps as many as eight simultaneously operating nozzles and resultant streams. Multiple-stream/multiple-nozzle rotors usually are more efficient than single-stream rotors because they place water at different distances from the sprinkler throughout the radius of coverage.

Fixed-spray sprinklers commonly have been used in residential lawn, small landscape, and flowerbed irrigation systems. Nozzles are available for spacing distances between 3 and 18 feet apart. Fixed-spray sprinklers are notorious for both high precipitation rates (that contribute to unwanted runoff) and low-distribution uniformity when compared with rotors. Until recently, fixed-spray sprinklers were the only available option for irrigating small areas. Fortunately, various manufacturers have developed multiple-stream rotor nozzles for these small areas. The multiple-stream nozzles significantly improve the distribution uniformity and also apply water at lower precipitation rates than the fixed-spray nozzles. Both distribution uniformity and precipitation rates are brought into ranges similar to their larger rotor counterparts typically used in golf courses, parks, and athletic fields. Higher application uniformity and improved water-use efficiency now are available to homeowners for only a slight additional cost.

Single- and multiple-stream impact and gear rotors are found in medium- and large-radius sprinklers. Medium-radius types, generally spaced 18 to 60 feet apart, often are used for irrigated slopes, playgrounds, large landscape beds, and medium to large lawn areas. Larger commercial rotors will be spaced 50 to 100 feet apart depending on the application and system design. They most often are used in large turfgrass sites such as golf courses, parks, and athletic field complexes and typically produce high uniformity when properly matched to their application.

Current Sprinkler Performance Limitations

It has long been known that sprinkler performance and distribution uniformity vary depending on the type of sprinkler used. Between 1985 and 1987, the Irrigation Training and Research Center (ITRC) at the California State Polytechnic University in San Luis Obispo developed a Landscape Water Management, Scheduling, and Auditing Workshop under sponsorship of the

California Department of Water Resources and the Metropolitan Water District of Southern California. The workshop provides attendees a method to evaluate existing irrigation systems and to schedule water applications more efficiently. During development of that workshop the ITRC faculty and staff conducted hundreds of turfgrass and landscape irrigation uniformity audits. The collected data were summarized to identify performance by sprinkler type as reported in Table 13.1.

Table 13.1. Turfgrass and landscape sprinkler system field audit performance rankings by distribution uniformity and sprinkler type[a]

Sprinkler type (typical use)	Distribution uniformity (DU_{LQ}) and expected system performance				
	Excellent (achievable)	Very good	Good (expected)	Fair	Poor (needs improvement)
Multiple-stream gear and impact rotors (golf and large turfgrass areas)	85%	80%	75%	65%	55%
Single-stream gear and impact rotors (medium-sized landscape and turfgrass areas)	75%	70%	65%	60%	50%
Fixed-spray heads (small lawns and landscapes)	70%	65%	55%	50%	40%

[a] Developed by Cal Poly Irrigation Training and Research Center at California State Polytechnic University, San Luis Obispo. Funded by California Department of Water Resources and the Metropolitan Water District of Southern California. Adapted from Walker et al. 1988.

In reviewing Table 13.1, note that even the most efficient, state-of-the-art irrigation system rarely exceeds 85% distribution uniformity in the field. Also note that not even rain falls at 100% uniformity. This inefficiency is primarily because of the influence of wind, land slope, and pressure variations found between individual sprinklers across an irrigated property.

Future Improvements

Advancements in nozzle and sprinkler designs are ongoing. Single nozzles that perform similarly to multiple nozzles or that operate more effectively at a wider range of operating pressures are under development. These nozzles combine a modified orifice, as shown in Figure 13.2, that is slotted, notched, or of other design, combined with a controlled amount of internal turbulence generated within the nozzle bore. The turbulence draws a small amount of water from the outside perimeter of the stream through the notched orifice to place it at various locations within the radius of throw. Precision manufacturing is required to accomplish this effect without sacrificing distance of the central stream's radius. The placement of water throughout the pattern is controlled by the size and shape of the slots and the amount of turbulence generated while passing through the nozzle. These nozzle designs can be adapted to many styles of gear and impact drive rotors.

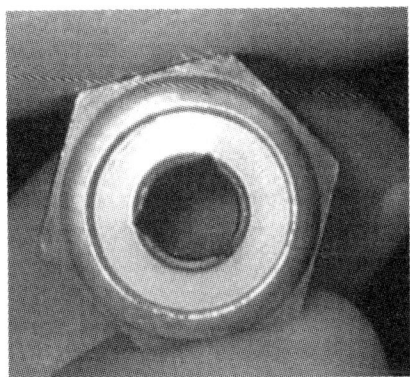

Figure 13.2. A nozzle with a modified orifice. Note three small notches in outside perimeter of the nozzle bore. This design was developed to improve distribution uniformity of a single-stream impact sprinkler used primarily in agriculture, but the basic design also could be adapted to turfgrass and landscape rotors.

Role of Software in Irrigation Design

Less than 20 years ago, the selection of an efficient sprinkler, nozzle, and spacing combination could be based only on the past experiences of an irrigation designer. More recently, however, the availability of personal computer software allows independent agencies (or self-conducted testing) to provide information for more informed decisions on sprinkler, nozzle, spacing, and configuration selections. Although the software cannot predict all possible variations of wind, pressure, and/or slope found in the field, it does provide greater insight to the designer than previously available. This innovation is one of the more important advancements of sprinkler irrigation design in many years.

Modeling Sprinkler Coverage

The basic concept behind irrigation uniformity is to apply the water as evenly as possible. Most irrigation scheduling is driven by the "dry spots," the areas that receive the least amount of water. Applying more water to the dry spots, however, over-irrigates the rest of the plant material and often wastes water in the process. Decreasing the difference between the minimum and maximum wetted area is the goal of highly uniform water application.

The uniformity of irrigation systems now can be easily modeled to determine the expected uniformity based on site and design considerations. The basis for modeling the irrigation uniformity of an irrigation system is derived from a single-leg sprinkler profile test. These tests can be performed either in an indoor laboratory or outside in the field.

In a single-leg profile test, catch-cans are placed at equal distances, starting at the nearest sprinkler head and extending beyond the wetted radius of the sprinkler. The distance between the catch-cans will vary depending on sprinkler radius. Only one sprinkler is operated during the test period. The test duration is established by the application rate of the sprinkler. The International Organization for Standardization has published an industry-accepted method for testing agricultural irrigation sprinklers (*Agricultural Irrigation Equipment* 2004).

The single-leg profile data, as shown in Figure 13.3, can be collected based on a variety of possible combinations including sprinkler make or model, nozzle size, and operating pressure(s). The software then can model coverage at any spacing distance and/or configuration (e.g., square versus triangle) as determined by the operator. The water application uniformity as measured in the overlap area can be calculated statistically in numerous ways. Note that statistical uniformity evaluations share a common weakness in that they do not take into account the exact location of the wetter and/or drier applications, and whether those drier or wetter values are concentrated in a localized area or dispersed throughout the pattern where a surrounding high value may benefit an adjacent low value.

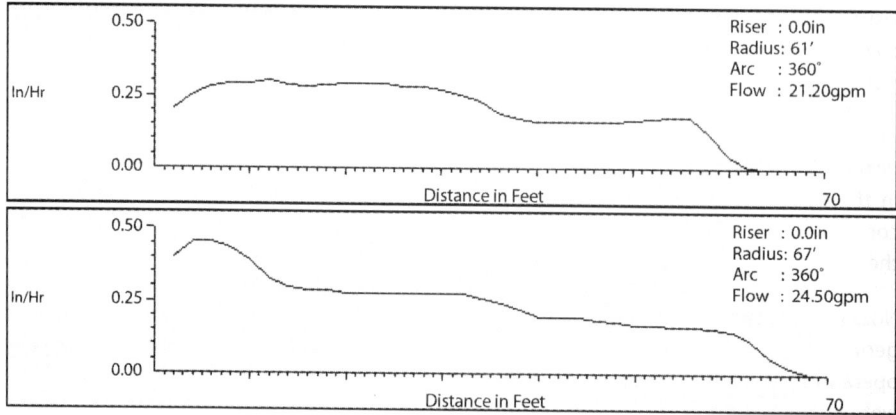

Figure 13.3a (top image) and b (bottom image). Examples of different-shaped single-leg profile data resulting from testing the same sprinkler equipped with different nozzle combinations and nozzle geometry, but operated at the same pressure.

The Coefficient of Uniformity (CU), as defined by J. E. Christiansen, is historically one of the most referenced measures of uniformity in agricultural irrigation. When used to analyze turfgrass irrigation, however, this standard of measurement has a weakness: it treats excessive and deficient irrigation the same, but most turfgrass irrigation management is driven by the dry spots.

Therefore, Distribution Uniformity (DU), based on the low quarter, is a method commonly used in the turfgrass irrigation field audit process, such as the previously mentioned Table 13.1. It is a measure of the low quarter or driest 25% of the coverage area compared as a ratio with the average. Although DU focuses on the under-irrigated area by providing the average precipitation, there is still no reference to the size, shape, or location(s) of the dry area(s).

A third measure of sprinkler irrigation uniformity is the Scheduling Coefficient (SC). The SC uses a ratio of the average application rate compared with the average found in the driest continual application area (usually specified as 1, 5, or 10% of the pattern area). This ratio, which must be 1 or greater, is used to estimate how long the irrigation system must run to apply the minimum needed water to the driest area. The larger the SC number, the longer the system must operate to keep the dry spots green. An irrigation system with an SC of 1.5 would have to run 50% longer than a perfectly uniform system with an SC of 1.0 to apply equal amounts of water to the driest part of the coverage area.

Computer software provides an alternative, or perhaps more accurately, a supplemental method of evaluating uniformity with visual graphic representations. In Figure 13.4 (see page I-14), two examples of a graphic called a densogram each show the overlapping coverage of three complementing sprinklers based on the single-leg profile data. The densogram allows visual assessment of both the location(s) and size(s) of the wet and/or dry spot(s) based on the color density of the graphic. Lighter areas represent drier areas, whereas the darker areas represent the wetter areas.

In Figures 13.3a and 13.4a, the profile and densogram of each graphic represent water distribution and uniformity of coverage from an existing sprinkler/spacing combination found in the field. Note that the densogram (Figure 13.4a) shows the three sprinkler heads contributing to the repeating coverage area. The red boxes indicate the driest continual 5% of the coverage area, and the green boxes indicate the wettest continual 5% area.

Nozzle configurations can vary from a standard, round taper bore to a wide collection of geometric shapes including triangle, square, hexagon, etc. The nozzle configuration and operating pressure determine how the water stream breaks up and is distributed along the wetted radius. Additionally, variables such as wind speed, wind direction, and sprinkler rotation speed influence how uniformly water is applied during the overlap process.

The sprinkler parameters used in Figures 13.3b and 13.4b are identical to Figures 13.3a and 13.4a except for the nozzle geometry used in the sprinkler. The standard, straight-bore nozzle used in Figures 13.3a and 13.4a is dependent on water pressure to provide the breakup necessary to distribute the water along the sprinkler radius. The nozzle used in Figures 13.3b and 13.4b uses both water pressure and nozzle geometry to achieve higher application uniformity based on an improved sprinkler profile shape.

Irrigation or water application uniformity is a function of the sprinkler profile, pressure at the nozzle, and spacing distance and configuration (triangular, square, etc.) of the sprinkler spacing in the field. The graph in Figure 13.5 depicts changes in distribution uniformity as a function of field spacing. The far left column characterizes various measures of uniformity (93% CU, 88% DU, and 1.4 SC) at an 18.3-meter (m) by 15.9-m (60 feet [ft] by 52 ft) triangular spacing. Located at the far right column is the same sprinkler spaced at a 21.3-m by 18.6-m (70 ft by 61 ft) triangular spacing. The uniformity is decreased to 79% CU, 64% DU, and 2.2 SC. The degradation of uniformity as impacted by an increased distance between the sprinkler heads represents a 24% decrease in DU. The other two uniformity measurements reflect similar changes. Other data points and spacings are represented in the graph between these two extremes.

The data presented are collected under "no wind" conditions. This circumstance allows for a direct comparison of proposed changes in the sprinklers' operating conditions. Although it is true that sprinklers operate under various wind conditions in the field, it is considered impossible to compare performance data if wind conditions are not identical. Wind most typically deteriorates uniformity and rarely, if ever, improves it.

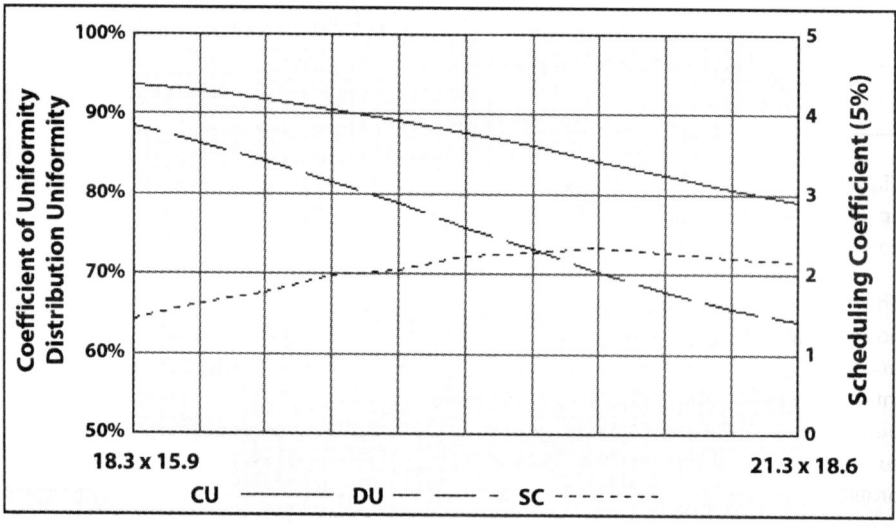

Figure 13.5. Application uniformity versus sprinkler spacing distance for the nozzle and sprinkler combination and resultant profile presented in Figure 13.3a. All spacing combinations represent equilateral triangular sprinklers by row spacing in meters.

A different sprinkler/nozzle combination, and thus a different profile shape, produces a substantially different result in Figure 13.6. The far left column characterizes various measures of uniformity (89% CU, 86% DU, and 1.2 SC) at an 18.3-m by 15.9-m (60 ft by 52 ft) triangular spacing. At the far right column is the same sprinkler spaced at a 21.3-m by 18.6-m (70 ft by 61 ft) triangular spacing. A high level of uniformity is maintained at this significantly greater spacing of 88% CU, 84% DU, and 1.3 SC. In this instance, the uniformity degradation is decreased by only 4% as measured as DU. Again, the other two measures of uniformity are only slightly affected by the change in sprinkler spacing.

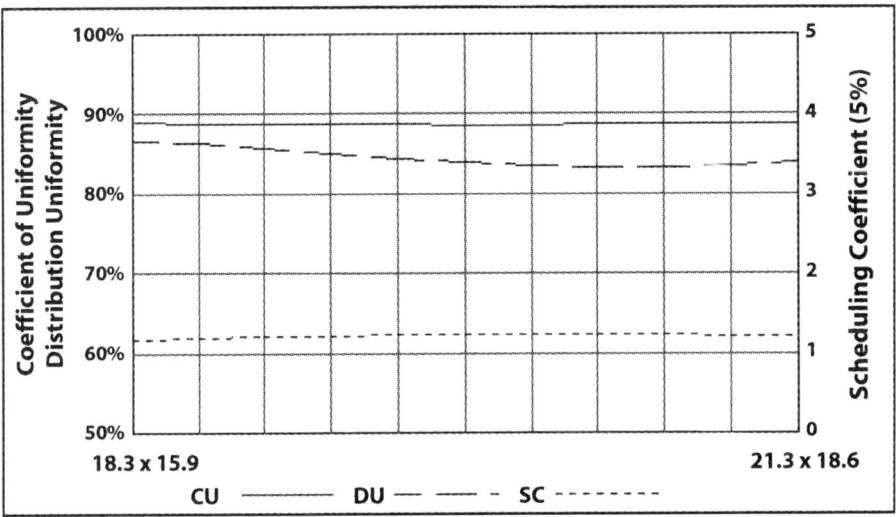

Figure 13.6. Application uniformity versus sprinkler spacing distance for the nozzle and sprinkler combination and resultant profile presented in Figure 13.3b. All spacing combinations represent equilateral triangular spacing in meters.

The message in comparing these two graphs is that not all sprinkler profiles are created equal. Some are much more forgiving over a greater spacing distance as found in the field. It is not uncommon for the field spacing of sprinklers to vary slightly when mature trees or hardscape interfere with sprinkler placement. So knowing how a sprinkler profile reacts to a range of potential field spacings can help with selecting products that will perform the best. We can summarize that the sprinkler profile shown in Figure 13.5 is not as forgiving as the sprinkler profile shown in Figure 13.6. Given that all other considerations are equal, selecting the sprinkler profile shown in Figure 13.6 would be the best choice of products and performance. Sprinkler performance data are available from most sprinkler manufacturers and from independent testing laboratories such as the Center for Irrigation Technology, or can be collected by individuals.

In general, a wedge-shaped sprinkler profile is considered one of the best shapes for providing high values of uniformity when properly spaced in the field. But variation in droplet size and its effect on wind drift can alter sprinkler coverage in the field. It is always best to model sprinkler performance first and then to audit the results in the field to identify acceptable water distribution uniformity. Modeling software and available sprinkler performance data make it feasible for irrigation designers to evaluate sprinkler options in a matter of minutes, whether for the smallest residential design or for a much larger park or golf course.

Development of Smart Water Application Technologies™ (SWAT™)

Most inground irrigation systems are operated by a timer/controller. The basic design of these controllers requires frequent input from the operator (homeowner) to adjust irrigation run times during the year. It has been noted that much over-irrigation occurs during the autumn season when plant water demand is dropping off and the corresponding irrigation run times are not decreased accordingly.

The development of Smart Water Application Technologies (SWAT) was initiated in November 2002 by water purveyors who wanted to improve residential irrigation water scheduling. Studies reported in the California Water Plan Update (California Department of Water Resources 2005) indicate a water savings of 17% and runoff decrease of 50% through the adoption of controllers automatically adjusted to reflect daily changes in ET. The hope is that the widespread adoption of smart controllers and soil moisture sensors would conserve a significant portion of the excess water applied.

SWAT is a national initiative coordinated through the Irrigation Association (IA) and designed to achieve landscape water-use efficiency through the use of irrigation technology. SWAT identifies, researches, and promotes technological innovations and related management practices that advance the principles of efficient water use.[1]

Irrigation Controllers

The evolution of the smart irrigation controller has ushered in a new era of technology that promises to remove the homeowner from the irrigation scheduling equation. The premise of the smart controller is to monitor changing plant water demand continually and apply water when it is required. Smart controllers also must recognize rainfall and its water contribution to the root zone in the irrigation schedule. Further, these controllers are programmed to minimize runoff and deep percolation below the root zone where water is lost.

Although the measurement of on-site weather conditions has been done for more than 20 years in the golf course industry, it had not been applied widely (or cost effectively) to homeowners. Additionally, many of the newer methods rely on "virtual" ET estimates rather than on-site weather stations. Because of increased water prices, homeowners can invest more money into controllers and still receive a financial benefit.

---

[1] For additional information regarding the SWAT™ program and technologies, visit http://www.irrigation.org/SWAT/default.asp.

Protocols were developed by the IA to evaluate the performance of smart controllers.[2] The Center for Irrigation Technology at California State University in Fresno is the unbiased "third party" currently evaluating controllers registered in the program. The test results allow water purveyors and end users to access unbiased results reporting the controller's ability to address key areas of water management inefficiencies.

Because controller manufacturers have taken different approaches to the exact technology used to monitor environmental conditions needed to adjust irrigation applications, the protocol defines a procedure for characterizing the overall efficacy of the irrigation controllers. Examples of the various types of controllers being evaluated include:

1. Controllers that store historical ET data;

2. Controllers that use an on-site sensor(s) as the basis for calculating real-time ET;

3. Controllers that use a central weather station as a basis for ET data calculations and transmit the data to individual residences from remote sites via satellite of telephone modems;

4. Controllers that use rainfall and temperature sensors; and

5. Control technology that is added onto existing time-based controllers.

The art and science of applying irrigation water to turfgrass and landscape areas is a practice developed over time. Whereas general procedures based in science give an appropriate framework for determining irrigation amount and frequency, the "fine tuning" of the irrigation schedule often is developed as a site-specific practice. This fine tuning allows for adjustments in the irrigation schedule for shading, hardscapes, or other site-specific features that impact plant water demand and/or irrigation uniformity by the contractor or end user.

The test protocol was developed to mimic typical and problematic irrigation landscapes found anywhere in North America. It is recognized that the virtual yard used in the IA's testing protocol cannot represent every conceivable irrigated landscape. But it also is known that certain conditions contribute more to wasted water than others. The protocol attempts to simulate many of those problematic conditions in its evaluation.

The IA protocol does not have a pass/fail rating. The procedure is designed to evaluate the controller's performance against an established "ideal" standard. The protocol uses the Environmental and Water Resource Institute of the American Society of Civil Engineers' study on the standardization of reference ET formulas for the baseline (Penman-Monteith reference ET equation). Other widely recognized standards also are cited in the protocol.

---

[2] The protocol is in its 6th draft, and all protocol versions are available at the IA website www.irrigation.org.

Efforts are under way to identify the appropriate agency (state or federal) that can set the expected performance bar for controllers. Once a pass/fail limit is set, incentives designed to accelerate the adoption of smart controllers can be implemented by water purveyors.

The promise of significant water savings offered by the widespread adoption of smart controllers has led to the recommendation that beginning in 2010 all new irrigation controllers sold in California will have to meet the requirements of the IA Controller Testing protocol. This recommendation comes from a State Task Force created by Assembly Bill 2717, which was charged with developing new landscape irrigation guidelines and policy recommendations for the legislature. It also was recognized that a minimum performance level for smart controllers must be set, and the expectation is that this level would be established by the California Department of Water Resources.

Soil Moisture Sensors

Soil moisture sensing is not a new concept. It has been available and used successfully for decades by large-scale irrigators such as agriculture. It has been slow to gain acceptance in turfgrass and landscape applications because of cost, increased maintenance, reliability, and other limitations related to microclimatic factors unique to turfgrass and landscape sites.

As an example, the number of sensors per unit area increases in turfgrass and landscape applications as compared with agriculture. In agriculture, a pair of sensors (one turning irrigation on and one turning it off) may adequately manage irrigation of many acres of a specific crop grown under uniform soil textural and climatic conditions. In a landscape, however, each hydrozone and soil textural condition will require a pair of these sensors. There are other limitations, advantages, and disadvantages to various types of soil moisture sensors as presented in Table 13.2. As technology advances, these sensors are improving rapidly and becoming less costly, including decreased need for calibration and maintenance. They soon may become common to turfgrass and landscape irrigation control systems.

New generation soil moisture sensors are a promising technology for irrigation scheduling. These sensors can provide closed-loop feedback to time-based system controllers, allowing controllers to recognize soil moisture levels and terminate irrigation events when soil moisture reaches predetermined levels. More sophisticated controllers can have the ability to interpret soil moisture readings to determine frequency and duration of irrigation events.

The IA currently is developing a Soil Moisture Sensor protocol to evaluate sensors under laboratory conditions. These evaluations will look at the sensor's responses under different levels of moisture, soil type, and salinity. The test will be designed to expose the sensor to a wide range of conditions that exist in the field. The test results will be bracketed by confidence limits, which indicate whether or not the sensitivity of the sensor will allow for informed decisions on maintaining an adequate soil moisture balance. A draft of test protocols is in a second phase of review, and final protocols should be published soon. The cost of these new-generation soil moisture sensors, although not directly identified in the testing protocol, also will influence their rate of adoption.

Table 13.2. Current and potential soil moisture sensor technologies for landscapes

Sensor technology	Operating principle	Advantages	Disadvantages
Tensiometer	Functions much like an artificial root. A porous ceramic cup in contact with the soil allows soil water to reach equilibrium inside the device by a reference pressure maintained inside the device. Changes in pressure trigger irrigation events when extremes are reached.	Not affected by temperature and/or conductivity (salinity) of the soil or soil water solution.	Frequent service/maintenance required (3- to 4-week intervals in dry climates).
Granular matrix sensor (gypsum block)	Measures resistance of electrodes embedded in a porous (gypsum) block that absorbs soil water. Changes are proportional to water content and calibrated to reflect the soil water content.	Low cost, easy to use.	Affected by changes in soil, water conductivity (such as fertilizer applications). Calibration to site's soil/soil water conductivity is required. Gypsum blocks dissolve over time requiring replacement.
Dielectric sensor	Measures how easily soil conducts electrical signals based on and calibrated to soil water content.	Provides nearly instantaneous (real time) readings, no need to wait for water to be absorbed. Low to no maintenance required.	Higher cost ($100 to $500), and susceptible to influences by temperature, salinity, and soil properties. Sensor affected by air gaps in the soil near sensing probes.
Heat dissipating sensor	Temperature in a porous block is measured before and after application of a small heat pulse. Heat flow is proportional to amount of moisture contained in the porous block. The resulting temperature change is calibrated to reflect soil moisture.	Small size. Not affected by soil/soil water conductivity.	Commercial availability unknown (was developed for NASA). Has a larger power requirement when compared with other sensors.

Future of SWAT

A committee on the future of SWAT has been established through the IA. Members of the committee represent water purveyors, industry, and government agencies. The committee's role is to expand the list of product categories that can demonstrate high water-use efficiency. Currently a list of five product categories has been identified. Beyond the initial scheduling technologies (e.g., controllers, sensors), four additional product categories have been identified, including:

- Overhead irrigation technologies (e.g., sprinklers, sprayers, nozzles)

- Low-volume irrigation technologies (e.g., emitters, distribution systems)

- Hydraulic management devices (e.g., pressure management, check valves)

- Malfunction abatement technologies (e.g., high-flow shutoffs, self-cleaning filters)

Nearly every irrigation product can be assigned to one of these five categories. It has been proposed that the categories will be ranked high to low, with the number one ranking having the greatest water-efficiency potential. Each succeeding lower ranking will signify a diminished potential in water-efficiency savings.

Initial funding has been identified for the review of the proposed product categories based on their potential water savings. After completion of the review, protocols will be developed for each category, starting with those products found in the number one category. The process will continue until protocols have been developed for all water-efficient products.

It is anticipated that eventually the SWAT process will include approved design, installation, and maintenance requirements. The long-term goal is that a "SWAT-designated irrigation system" will include approved design methodologies and products and will be installed and maintained according to published guidelines. The end result will be an irrigation system that achieves the highest possible water-use efficiency under prevailing economic conditions. This goal fully recognizes the importance of water as a finite resource.

Irrigation System Installation

The efficiency of even the best design and sprinkler selections can be compromised by a bad installation. Reasons for installation problems may include poor workmanship and a lack of staff training, supervision, or ignorance of the installer. Profit-driven economizing by a contractor to increase profits or cost-cutting measures by short-sighted property owners often result in the substitution of lower-quality equipment and/or in-field design changes and shortcuts taken during the installation process. All these issues can and will affect the overall efficiency of the design. Specific examples include:

- Expanding sprinkler spacing distances with intentions to decrease the total number of sprinklers, control stations, and resulting hydrozones, and remote control valves needed to construct the system.

- Installing sprinklers without uniform spacing distances and configurations within the same hydrozone. Ideally, either triangular or square spacing will be used as much as possible to maintain uniform distribution.

- Substituting undersized or other components (smaller-sized pipe, valves, and less-costly sprinklers) that are not "equal" in performance to those specified in the design plan to save on material costs. These substitutions affect the hydraulics of the design and result in less uniform precipitation rates between the first and last sprinkler within the same hydrozone.

- Mixing full- and part-circle sprinklers in the same hydrozone when using sprinkler models that are not designed by the manufacturer to deliver "matched precipitation rates."

- Combining smaller hydrozones into a single, larger control zone to economize on clock size (number of total stations) and total number of remote control valves required.

- Omitting specified pressure regulation equipment to save on expense.

- Not properly adjusting pressure regulators when they are installed. Operating sprinklers at higher pressure than suggested by the manufacturer results in more fine droplets and mist that are easily affected by wind. High pressure also affects sprinkler distribution uniformity.

- Improper adjustment of part-circle rotor sprinkler heads or improper selection of fixed-spray sprinkler nozzle arcs that allow overspray into other hydrozones and/or onto bordering hardscapes.

- Installing mismatched sprinkler heads or nozzles that do not deliver similar precipitation rates.

- Installing sprinklers in a tilted or sunken position where the spray pattern is interrupted by surrounding terrain or the turfgrass.

- Improper initial programming of the clock after installation. Even smart technology initially must be programmed properly or it, too, can waste water.

- Failing to flush soil and other debris from distribution lines before installing sprinklers to avoid plugging nozzles and affecting performance.

- Planting trees and/or shrubs randomly without consideration for sprinkler locations where they may interfere with sprinkler spray patterns and coverage uniformity.

Irrigation System Maintenance

Like any other mechanical device, an irrigation system needs periodic maintenance to continue delivering optimum performance. An annual walk-through inspection of the system and operating each station to observe its condition is suggested. Items to observe at each station while operating include the following:

- Missing or broken sprinkler heads
- Leaking seals, lateral lines, or fittings
- Leaking check valves allowing lateral line drainage at the lowest sprinkler
- Sprinkler arc misalignment causing overspray onto hard surfaces
- Plugged or worn nozzles
- Tilted/sunken sprinklers with spray deflected from nozzles
- High or low operating pressure
- Unusually high or low sprinkler rotation speed
- Malfunctioning remote control valves or control stations

Additionally, the annual sprinkler system inspection is an excellent opportunity to evaluate turfgrass and landscape cultural practices to determine any need for additional soil aeration or thatch removal to enhance irrigation efficiency further by decreasing water runoff.

Summary

There are many elements to high water-use efficiency in irrigation. These elements begin at design and continue on through the installation, management, and maintenance of the irrigation system. One critical element is application of the proper amount of water when the landscape needs the water to avoid both deep percolation and runoff. This practice may include cycling of the valves to avoid the surface movement of applied water.

A second critical element to high water-use efficiency is to apply water as uniformly as possible. Although current technology does not allow for 100% uniformity of applied water, improved sprinkler designs for turfgrass and drip/micro-irrigation for landscape plants have improved irrigation uniformity significantly in recent years, when properly designed and installed.

Tools now are available for designers to model sprinkler application uniformity before the system is purchased and installed in the ground. Given this fact, it is reasonable to specify irrigation application uniformity in a contract before purchasing an irrigation system. Auditing can be used to verify the system performance after installation.

The combination of SWAT (primarily smart controllers) with highly uniform sprinkler and/or drip irrigation systems will produce high water-use efficiency (leading to significant water savings over conventional practices). Optimizing *only one* of these options, however, potentially could lead to significant over-irrigation. High efficiency in water use can be best summarized by the two basic tenets of (1) Only apply water in the amounts and times the plants require (See Sections 11 and 12), and (2) Apply the water as uniformly as possible.

Literature Cited

Agricultural Irrigation Equipment – Sprinklers, Part 3: Characterization of Distribution and Test Methods. 2004. ISO 15886-3:2004. International Organization for Standardization, Geneva, Switzerland. 15 pp.

Barrett, J., B. Vinchesi, R. Dobson, P. Roche, and D. Zoldoske. 2003. *Golf Course Irrigation: Environmental Design and Management Practices.* John Wiley & Sons, Hoboken, New Jersey. 452 pp.

California Department of Water Resources. 2005. *California Water Plan Update, Vol.2.* Regional Management Strategies, Sacramento, California.

Choate, R. B. 1994. *Turf Irrigation Manual.* 5th ed. Telsco Industries, Dallas, Texas. 396 pp.

Jarrett, A. R. 1985. *Golf & Grounds Irrigation and Drainage.* Reston Publishing Company, Reston, Virginia. 246 pp.

Pair, C. H., W. H. Hinz, K. R. Frost, R. E. Sneed, and T. J. Schlitz. 1983. *Irrigation.* The Irrigation Association, Arlington, Virginia. 686 pp.

Walker, R., M. Lehmkuhl, G. Kah, and P. Corr. 1988. *Landscape Water Management Handbook Series: Principles, Auditing, & Budgeting.* Irrigation and Training Research Center, California State Polytechnic University, San Luis Obispo.

14

Recycled, Gray, and Saline Water Irrigation for Turfgrass

M. Ali Harivandi, Kenneth B. Marcum, and Yaling Qian

Introduction

In dry regions of the country, and in highly populated metropolitan areas where water is a limited natural resource, irrigation with municipal, recycled water, untreated household gray water, or other low-quality (saline) water is a viable means of coping with potable water shortages. In recent years, large quantities of municipal recycled water have been used in arid and semi-arid metropolitan areas to irrigate golf courses, parks, cemeteries, athletic fields, and other urban landscape sites. Some western U.S. states have public policies encouraging or requiring the use of recycled or other saline-source water for irrigation of turfgrass landscapes (Arizona Department of Water Resources 2000; California Water Resources Control Board 1993). As more salt-tolerant turfgrasses become available, and the demand for high-quality water rises, the need to use other less-than-premium quality water (e.g., household gray water, brackish water, saline groundwater) is likely to increase. The use of alternative water for irrigation to conserve potable water is dictated by concerns about water shortages/cost in both high-rainfall areas and regions of recurring drought.

Against this backdrop, the quantity and quality of water available to irrigate turfgrasses has become a significant focus of urban water management. Irrigation water quality, a function of the volume and type of dissolved salts present in the water, affects the chemical and physical properties of soil and therefore plant-soil-water relations. These interrelationships can be monitored, and in many situations managed, by frequent chemical analysis.

Alternative Sources of Water

All irrigation water, regardless of source, contains dissolved minerals. Most water used for domestic purposes, such as water from rivers, lakes, reservoirs, and aquifers, contains low levels of dissolved minerals. These waters are preferred for irrigation in terms of their chemical, physical, and biological composition. But water from alternative sources almost always has a significantly higher mineral content and is less optimal for turfgrass, landscape plants, and landscape soils. For example, residential and commercial uses of water, plus water treatment chemicals, typically add about 200 to 400 milligrams per liter (mgL^{-1}) or parts per million (ppm) of dissolved salts, and these salts remain after wastewater treatment (Harivandi 1994). Currently, the most commonly used alternative irrigation water nationwide is treated municipal sewage water, also known as wastewater, effluent water, reclaimed water, and recycled water. To a much lesser extent, untreated household gray water and brackish (sea) water also are used.

Recycled Water

"Recycled water" refers to any water that has undergone one cycle of (human) use and then received sufficient treatment at a sewage treatment plant to be made suitable for various reuse purposes, including irrigation. Recycled water may be primary, secondary, or advanced (tertiary) treated municipal or industrial wastewater (Asano, Smith, and Tchobanoglous 1984). Primary treatment usually involves a screening or settling process that removes organic and inorganic solids from the wastewater. Primary treated sewage effluent is not suitable for turfgrass irrigation.

Secondary treatment is a biological process in which complex organic matter is broken down to less complex organic material and then metabolized by simple organisms that later are removed from the wastewater. Secondary treatment can remove more than 90% of the organic matter in incoming sewage. Currently, reclaimed water used for turfgrass and landscape irrigation must be at least secondary effluent water.

Advanced wastewater treatment consists of processes similar to potable water treatment, such as chemical coagulation and flocculation, sedimentation, filtration, adsorption of compounds by a bed of activated charcoal, or reverse osmosis. Because advanced treatment usually follows secondary treatment, it is sometimes referred to as "tertiary treatment." These processes produce highly purified (and pathogen free) waters, especially if followed by chlorination or ultraviolet treatment for disinfection.

Gray Water

"Gray water" refers to water that has gone through one cycle of use, usually in homes, but not including water from toilets and dishwashers (or other kitchen water potentially contaminated by food residues), and which has not been treated. Although regulatory authorities in most urban (or drought-stricken) communities accept, and even encourage, use of recycled water for irrigation, use of gray water generally is prohibited. This restriction is in place because the human disease-causing organisms destroyed during recycled water treatment remain a potential component of gray water. Nevertheless, use of household gray water for irrigation around homes is gaining popularity as potable water becomes more scarce and expensive. Several states, including California, Arizona, New Mexico, and counties within other states now allow home landscape irrigation with gray water (Arizona Department of Water Resources 2000; California Water Resources Control Board 1993). Although recycled water is used most commonly in large landscape facilities such as golf courses, parks, etc., gray water irrigation offers a way for people to reuse wastewater generated in their own homes. Furthermore, reusing gray water can decrease the volume of wastewater to be treated in the wastewater treatment plant and save the cost associated with wastewater treatment and delivery. Although the actual number of households irrigating with gray water is unknown, a 1998 survey by The Soap and Detergent Agency reported that 7% of U.S. households used gray water for irrigation (NPD Group 1999).

Saline Water

In U.S. coastal areas where brackish, or sea, water is plentiful, its use for landscape irrigation is a major topic of conversation. Seawater comprises approximately 97% of all water on Earth, but high salinity ($\approx$80% sodium chloride [NaCl]) makes it uniformly useless for irrigation. Fresh water contamination by saline water also occurs through tidal flow into coastal freshwater sources (i.e., rivers), resulting in brackish water, and through intrusion along permeable subsurface layers that contaminates impounded and groundwaters. These salt-contaminated waters also are generally unsuitable for irrigation, although in some areas coastal turfgrass developments are required to use on-site brackish wells for irrigation. Development of highly salt-tolerant grasses, however, may make future irrigation with brackish, or sea, water possible (Duncan, Carrow, and Huck 2000). But serious concerns remain because long-term use of seawater irrigation may make soils permanently useless because of soil salt accumulation. Regardless, advances in economic water desalination, combined with more intensive turfgrass management, also will lead to increased irrigation with saline water.

In arid metropolitan areas of the southwestern United States (e.g., Arizona, Nevada) one available source of nonpotable water is shallow, perched saline aquifers. This water (urban agricultural drainage water) exists because of the overwatering of urban landscapes by homeowners and highly stratified sediments of low permeability present in the vadose zone (Schaan et al. 2003).

Agronomic Concerns

Much recycled water and all brackish water used for irrigation contains elevated concentrations of dissolved salts that potentially are toxic to turfgrasses. Consequently, periodic monitoring with chemical water analysis is a key component of sound irrigation management. In addition to dissolved salts, biological and nondissolved suspended solids also are of concern and should be evaluated when household gray water is used for irrigation.

The most important parameters determined by chemical analysis of irrigation water will be discussed in the next sections.

Salinity

Currently, all recycled waters contain dissolved mineral salts. Some soluble salts are nutrients and therefore beneficial to turfgrass growth. Other soluble salts, however, may be phytotoxic or may become so when present in undesirable soil concentrations. The rate at which salts accumulate to potentially toxic levels in a soil depends on their concentration in the irrigation water, the amount of water applied annually, the annual precipitation (rain plus snow), and the soil's physical/chemical profile characteristics.

Water salinity is reported as Total Dissolved Solids in units of either ppm or mgL^{-1}. It also may be reported as electrical conductivity (EC$_W$) in terms of milimhos per centimeter (mmhos cm^{-1}) or decisiemens per meter (dSm^{-1}). Most waters of acceptable quality for turfgrass irrigation contain from 0.3 to 1.2 dSm^{-1} (200 to 800 mgL^{-1}) soluble salts (Harivandi 2004a,b).

Only careful management can prevent deleterious salt accumulation in a soil irrigated with a high EC$_W$ water. Water with an EC$_W$ value higher than 0.7 dSm^{-1} (or 450 mgL^{-1} equal to 450 ppm) increases salinity problems. Water with an EC$_W$ higher than 3 dSm^{-1} (2,000 mgL^{-1} equal to 2,000 ppm) should be avoided entirely or diluted with less-saline water before use, although highly salt-tolerant turfgrasses may tolerate this water salinity level if adequate soil drainage and leaching are maintained (Carrow, Huck, and Duncan 2000; Harivandi 1994). Good soil permeability and drainage allow a turfgrass manager to leach excessive salt from the root zone by periodic, prolonged irrigations.

Soil physical characteristics and drainage—both important factors in determining root zone salinity—also must be considered when determining the suitability of a given recycled irrigation water. For example, water with an EC$_W$ of 1.5 dSm^{-1} (1,000 mgL^{-1} equal to 1,000 ppm) may be used successfully on turfgrass grown on sandy soil with adequate drainage (and thus high natural leaching), but proves injurious within a very short time if used to irrigate the same grass grown on a clay soil, or soil with limited drainage, because of salt buildup in the root zone.

Sodium

Sodium (Na) content is another important factor in evaluating recycled irrigation water quality. Plant roots absorb Na and transport it to leaves where it can accumulate and cause injury. Any irrigation water with high levels of Na salts can be particularly toxic if applied to plant leaves by overhead sprinklers because salts can be absorbed directly by leaves (Devitt 2005; Jordan et al. 2001). Field observations have shown that among grasses grown on golf courses, annual bluegrass and bentgrass are the most susceptible to Na phytotoxicity.

Although Na can be directly toxic to plants, its most frequent hazard is its effect on soil structure (Oster et al. 1992; Richards 1954; Westcot and Ayers 1984). The high Na content common to recycled water can cause deflocculation, or breakdown of soil clay particles, thereby decreasing soil aeration and water infiltration and percolation. In other words, soil permeability is decreased by recycled irrigation water high in Na. The likely effect of a particular irrigation water on soil permeability can be gauged best by the water's Sodium Adsorption Ratio (SAR), expressed as the proportion of Na to calcium (Ca) plus magnesium (Mg) ions in the water. The formula to calculate SAR is:

$$SAR = \frac{Na}{\sqrt{\frac{Ca + Mg}{2}}}$$

In this equation, values for all the ions are expressed in milliqualent per liter (meqL^{-1}).

SAR, the best indicator of a water's likely Na effect, should be provided in all laboratory water analyses. In general, water with an SAR lower than 3 is considered safe for turfgrass and other ornamental plants, whereas waters with an SAR higher than 9 can cause severe permeability problems when applied to fine-textured (i.e., clay) soils over a period of time. In coarse-textured (i.e., sandy) soils, permeability problems are less severe and an SAR of this magnitude may be tolerated. But SAR is an important enough factor in water evaluation to merit thorough understanding rather than reliance on general "cut off" values. For example, there are instances when a water with an SAR of more than 6 may be used for irrigation without a significant impact on soil permeability, as long as it contains soluble salts of more than 1.2 dSm$^-$ (770 mgL^{-1} equal to 770 ppm) (Harivandi 2004a). For recycled waters high in bicarbonate, some laboratories "adjust" the calculation of SAR (yielding a number called "adjusted SAR" or "Adj. SAR") because bicarbonate affects soil Ca and Mg concentrations. In simplest terms, Adj. SAR reflects the combined effects of water's Ca, Mg, Na, and bicarbonate, as well as the water's total salinity (Ayers and Westcot 1985; Westcot and Ayers 1984).

Bicarbonate and Carbonate

Recycled waters are prone to excessively high bicarbonate (HCO$_3$) levels. Elevated HCO$_3$ in irrigation water can increase soil pH and affect soil permeability. In addition, the HCO$_3$ ion may combine with Ca and/or Mg to precipitate as Ca and/or Mg carbonate in the root zone, causing SAR increases in the soil solution by lowering the dissolved Ca and Mg concentrations.

The HCO$_3$ hazard of recycled water is measured by Residual Sodium Carbonate (RSC), calculated as follows: RSC = (HCO$_3$ + CO$_3$) − (Ca + Mg). In this equation, ion concentrations are in meqL^{-1} (Eaton 1950). Generally, recycled water with an RSC value of 1.25 meqL^{-1} or lower is safe for irrigation, water with an RSC between 1.25 and 2.5 meqL^{-1} is marginal, and water with an RSC of 2.5 meqL^{-1} and higher probably is not suitable for turfgrass irrigation, unless it is treated with acids to neutralize excess bicarbonates (Carrow, Duncan, and Huck 1999; Harivandi 2004a).

Hydrogen Ion Activity (pH)

Water's acidity and alkalinity are measured in pH units. The scale ranges from 0 to 14, with pH 7 representing neutral (i.e., water with a pH of 7 is neither acidic nor basic). From pH 7 to pH 0, water is increasingly acidic. From pH 7 to pH 14, water is increasingly basic (or "alkaline"). pH is easily determined and provides useful information about a water's chemical properties. Although seldom a problem in itself, a very high or low pH warns users that water needs evaluation for other constituents. The desirable soil pH for most turfgrasses is 5.5 to 7.0. The pH of most irrigation water across the United States, however, ranges from 6.5 to 8.4 (Ayers and Westcot 1985; Richards 1954). Depending on the soil on which the grass is grown, an irrigation water pH range of 6.5 to 7 is desirable. Waters with a pH outside the desirable range must be carefully evaluated for other chemical constituents.

Chloride

In addition to contributing to the total soluble salt concentration of irrigation water, chloride (Cl) may be directly toxic to plants. Although Cl is not particularly toxic to turfgrasses, many trees, shrubs, and groundcovers are sensitive to it. In sensitive plants, Cl toxicity leads to necrosis—leaf margin scorch in minor instances or total leaf kill and abscission in severe instances. Irrigation waters with a Cl content above 355 mgL^{-1} (355 ppm) are toxic when absorbed by roots, whereas a Cl content higher than 100 mgL^{-1} (100 ppm) can damage sensitive ornamental plants if applied to foliage (Ayers and Westcot 1985; Devitt 2005; Farnham, Hasek, and Paul 1985). Chloride salts are soluble and therefore may be leached from well-drained soils with adequate subsurface drainage.

Chlorine

Municipal recycled water may contain excessive residual chlorine (Cl_2), a potential plant toxin. Chlorine toxicity almost always is associated with recycled waters that have been disinfected with Cl-containing compounds. Chlorine toxicity may occur only if high levels (more than 5 mgL^{-1} equal to 5 ppm) of Cl_2 are sprayed directly onto foliage of nonturf ornamental plants, a situation likely to occur only where recycled water goes straight from a treatment plant to an overhead irrigation system. Free Cl_2 is unstable in water. It will dissipate rapidly if stored in ponds between treatment and application to plants. Residual Cl_2 is of concern at levels above 5 mgL^{-1} (5 ppm) in irrigation water (Asano, Smith, and Tchobanoglous 1984). Although Cl_2 toxicity on nonturf plants may occur, no Cl_2 toxicity (injury) to turfgrasses has been reported.

Boron

Boron (B) is a micronutrient essential for plant growth, although it is required in small amounts. Even at concentrations as low as 1 to 2 mgL^{-1} (1 to 2 ppm) in irrigation water, it is phytotoxic to most ornamental plants and capable of causing leaf burn (Farnham, Hasek, and Paul 1985). Injury is most obvious as a necrosis on the margins of older leaves. Turfgrasses generally are more tolerant of B than other plants grown in urban landscapes and may grow at soil B levels as high as 10 $mg\ kg^{-1}$ (10ppm) (Oertli, Lunt, and Youngner 1961).

Nutrients

Recycled waters contain a range of micro- (trace) elements sufficient to satisfy the need of most turfgrasses for these substances. They also may contain enough macro- (major) nutrients (i.e., nitrogen [N], phosphorus [P], and potassium [K]) to figure significantly in the fertilization program of turfed areas (Broadbent and Reisenauer 1984; Harivandi 1994; Page and Chang 1984). As an example, 0.123 hectare–meter (1 acre–feet) of a recycled water containing 23 mgL^{-1} (23 ppm) of N delivers 28.5 kg (62.5 pounds) of N when applied to a hectare (acre) area. The economic value of these nutrients can be substantial. They can be used very efficiently by turfgrasses because they are applied on a regular basis.

Irrigation System Issues

The most important aspect of irrigation system design when alternative saline water is used is uniformity of water application. An irrigation system must be designed to apply water efficiently and uniformly, regardless of the type of water applied. Because both harmful (salts) and beneficial (nutrients) substances may be applied with recycled irrigation water, every effort should be made to distribute water uniformly. An irrigation system audit is necessary for urban landscape facilities contemplating recycled or other saline water use. In addition to allowing for correction of nonuniform application, such an evaluation can help determine actual water need (use) and the volume of water required to leach root zone salts.

Filtration is another crucial irrigation system component because of the suspended matter content in recycled and brackish waters. Filtration is a must, particularly if recycled water is stored in ponds where algal bloom is a constant problem (Devitt et al. 2005). Algae and other suspended matter can plug irrigation nozzles, thereby decreasing irrigation efficiency and uniformity and requiring additional labor for repeatedly unplugging heads.

If recycled water is stored in a lake/pond before use, lining of the reservoir often is required to prevent potential groundwater contamination. Similarly, if a site irrigated with recycled water is located above a drinking water aquifer, a comprehensive groundwater quality monitoring program may be required to make sure excessive nutrients (nitrogen, phosphorous), heavy metals, or pathogens are not reaching the groundwater.

Depending on local regulations, sites irrigated with recycled water also may be required to protect adjacent properties, streams, creeks, or any other body of water from runoff or overspray from their irrigation. Compliance with such regulations may mean redesigning an irrigation system to allow irrigation of the site perimeter with fresh water.

When recycled water or brackish water is used, and a site's irrigation system is connected to a potable water system or any dedicated fire line using potable water, protection of cross-connection systems is necessary. In general, all physical connections between the recycled/brackish water irrigation system and the potable water system should be disconnected. On-site lakes, wells, and creeks whose water is used for potable purposes also should be protected from overspray or runoff from recycled water irrigation. Local regulations may require modification or redesign of the irrigation system to ensure these requirements are met.

When irrigating with recycled water, careful irrigation scheduling can prevent ponding and runoff. Modern irrigation controllers, which allow multiple start times, provide effective control of these problems on clay or compacted soils. Because corrosion from salts may be an issue with recycled or brackish water irrigation, corrosion-proof, stainless steel cabinets mounted with stainless steel bolts are essential.

In most states, the color purple has become the unofficial color of recycled water application. Almost all landscape irrigation components are now manufactured and marketed in this color, and this distinction alerts users to the presence of recycled water. On sites retrofitted for recycled water, all buried components of the existing irrigation systems often are "grandfathered in." But all visible irrigation system components may require labeling with purple tape, tags, paints, etc. Signs warning of recycled water also may be required.

Irrigation systems designed for application of household gray water pose their own unique challenges. Residential sites interested in using gray water for landscape irrigation need to be retrofitted with a gray water collection and storage system. Early applications of residential gray water likely began with homeowners hand-bailing gray water—such as shower water and washer water—to help irrigate landscape plants during times of drought. Current gray water systems range from simple collection without treatment to more complex systems that simulate water treatment on a miniature scale. Typically, a gray water system consists of a plumbing system to bring gray water out of the house, a surge tank to temporarily hold drain flows from washing machines and bathtubs, a filter to remove particles that could clog the irrigation system, and a pump to move water from the surge tank to the irrigation field.

Pathogenic microorganisms from feces may enter gray water during showering, bathing, and laundering. Because the majority of gray water reusers do not treat the gray water before watering, the potential for human exposure to pathogenic microorganisms exists. Therefore, many states permitting gray water for landscape irrigation require the use of flooding, subsurface, and/or drip irrigation methods. In no circumstances should an overhead sprinkler irrigation system be used in applying gray water. This constraint significantly decreases its use for turfgrass irrigation.

Managing Alternative Sources of Water

Field observations and several recent surveys indicate that using saline or recycled water containing elevated dissolved salts can have negative impacts on both soils and plants, including turfgrasses (Devitt et al. 2004; Qian and Mecham 2005). In most instances, however, problems associated with saline water irrigation are not insurmountable. A full discussion of problems associated with using low-quality irrigation water is beyond the scope of this article. From a practical standpoint, however, four of the most important solutions to salinity/sodicity problems posed by alternative water sources are (1) growing salt-tolerant turfgrass species, (2) blending the alternative source with better-quality water, (3) drainage and salt leaching, and (4) chemical treatment (amending) of water/soil to correct specific salinity parameters.

Salt-Tolerant Turfgrasses

Turfgrass salinity tolerance is complex, being influenced by a number of plant, environmental, and soil factors. Tolerance often differs with the stage of plant development. For example, tolerance at seed germination often is not a good predictor of tolerance for the mature plant (Harivandi, Butler, and Wu 1992). Both temperature and relative humidity influence plant responses to salinity, with tolerance decreasing under hot, dry conditions because of increased evapotranspirational demand that causes increased salt uptake. Soil water content changes have a direct effect on turfgrass salinity responses. Soil salinity varies with time and depth, increasing as the soil dries between irrigations and as depth increases (Maas and Hoffman 1977).

Because of many interacting factors, the "absolute" salinity tolerance of a turfgrass species cannot be determined. Different turfgrasses can be compared, however, with relative salt tolerances given in terms of the acceptable salt content of the soil root zone, or electrical conductivity of soil water extract (EC_e). A general guide to the salt tolerance of turfgrass species (substantial differences in salt tolerance exist among cultivars within species) is presented in Table 14.1 (Harivandi, Butler, and Wu 1992; Marcum 1999). Soils with an EC_e lower than 3 dSm^{-1} are considered satisfactory for growing most turfgrasses. Kentucky bluegrass tolerates soil salinity at EC_e levels up to 3 dSm^{-1}. Soils with an EC_e higher than 10 dSm^{-1} successfully support only salt-tolerant turfgrass species.

Table 14.1. Relative tolerances of turfgrass species to soil salinity (EC_e)

Sensitive (<3 dSm^{-1})	Moderately sensitive (3 to 6 dSm^{-1})	Moderately tolerant (6 to 10 dSm^{-1})	Tolerant (>10 dSm^{-1})
Annual bluegrass	Annual ryegrass	Perennial ryegrass	Alkaligrass
Bahiagrass	Buffalograss	Creeping bentgrass (cultivars 'Mariner' and 'Seaside')	Bermudagrass
Carpetgrass	Creeping bentgrass		Fine-leaf (Matrella type) zoysiagrasses
Centipedegrass	Slender creeping, red, and Chewings fescues	Coarse-leaf (Japonica type) zoysiagrasses	Saltgrass
Colonial bentgrass			Seashore paspalum
Hard fescue		Tall fescue	St. Augustinegrass
Kentucky bluegrass			
Rough bluegrass			

Much work has been done in screening existing cultivars or ecotypes for salinity tolerance, including these turfgrass species: *Agrostis stolonifera* (Marcum 2001), *Buchloe dactyloides* (Wu and Lin 1994), *Cynodon* spp. (Dudeck et al. 1983; Marcum 1999), *Distichlis spicata* (Marcum, Pessarakli, and Kopec 2005), *Festuca* spp. (Horst and Beadle 1984), *Lolium perenne*, (Rose-Fricker and Wipff 2001), *Paspalum vaginatum* (Dudeck and Peacock 1985; Duncan 2003; Lee, Duncan, and Carrow 2004), *Poa pratensis* (Qian and Suplick 2001; Rose-Fricker and Wipff 2001), *Puccinellia* spp. (Harivandi, Butler, and Soltanpour 1982, 1983), *Stenotaphrum secundatum* (Dudeck, Peacock, and Wildmon 1993), and *Zoysia* spp. (Marcum, Anderson, and Engelke 1998; Qian, Engelke, and Foster 2000). Such work is important and needs to be updated at regular intervals as new cultivars become available.

Blending Saline Water with Less Saline Water

Frequently, saline water can be used for irrigation if better-quality water also is available. The two waters can be pumped into a reservoir and mixed before irrigation. Water quality should improve in proportion to the mixing ratio (e.g., when equal volumes of two waters are mixed—one with an EC_w of 5 dSm^{-1} and the other with an EC_w of 1 dSm^{-1}—the salinity of the blend should be approximately 3 dSm^{-1}) (Harivandi 1994).

Drainage and Leaching of Salts

When managing soil irrigated with saline waters, the goal is to move as much salt out of the root zone as is added through irrigation. This, of course, assumes the original salinity level in the soil is not high enough to affect turfgrass growth significantly.

The Leaching Requirement (LR) specifies the amount of extra water needed to leach salt below the turfgrass root zone (and thus maintain a suitable level for a specific turfgrass). LR is calculated as follows (Ayers and Westcot 1985):

$$LR = \frac{EC_w}{5(EC_e) - EC_w}.$$

LR is the fraction of the plant's normal water requirement that must be added to the amount required solely for leaching purposes. EC_w is the electrical conductivity of the irrigation water being applied (presumably a saline water), and EC_e is the electrical conductivity of soil extract tolerated by the plant grown. For example, if a turfgrass that can tolerate a salinity of 3 dSm^{-1} (EC_e) is irrigated with a water having a salinity of 2 dSm^{-1} (EC_w), the leaching requirement would equal

$$LR = \frac{2}{5(3) - 2} = 0.15.$$

To prevent irrigation water salt from accumulating to hazardous levels for this specific turfgrass species, the volume of each irrigation should be increased by approximately 15%. This extra water will continually leach the salt, assuming drainage is adequate. If a hard or clay pan soil is present, the soil profile must be modified to improve water percolation, and if shallow water tables are a problem, or the soil does not drain well for any other reason, artificial drainage can be installed. Obviously, leaching does not occur without drainage. In addition, any changes in system input, such as rainfall, also can affect the amount of water needed for successful leaching significantly.

Chemical Treatment (Amending) of Water and Soil

Soils irrigated with sodic (high Na//bicarbonate) waters, eventually may become impermeable to water and air. Turfgrass growth is significantly restricted on such soils. Chemical materials (amendments) often used to treat sodic soils include gypsum and sulfur. These amendments increase the soil supply of Ca either directly (gypsum), or indirectly (sulfur and sulfur-containing materials), making Ca more soluble. Once available in the soil solution, Ca replaces Na on clay and organic matter particles, preventing excess Na accumulation. Subsequent leaching will flush out Na salts accumulated in the root zone. The amount of sulfur amendment used depends on the SAR of the irrigation water, the quantity of water used, soil texture, and type of amendment. Materials such as calcium chloride, sulfuric acid, sulfates of iron, and aluminum also are available and effective on sodic soils (Ayers and Westcot 1985; Carrow and Duncan 2002; Harivandi 1990). Their use is limited, however, because they are either too costly or, in the case of acids, hazardous to people and plants. Regardless of the amendment chosen, the two major factors in successful sodic soil reclamation are the incorporation of amendments into the soil's top 12 to 24 centimeters (5 to 10 inches) and the presence of internal drainage to facilitate the leaching of sodium ions from the root zone.

Adding amendments to irrigation water is increasingly popular because it is effective and relatively inexpensive (including minimal labor cost). Various forms of acid are injected into irrigation water to neutralize bicarbonate content and lower pH. Water-soluble forms of gypsum also make gypsum injections common.

Salinity/sodicity problems associated with recycled water irrigation are not as severe as in gray water irrigation. Because household gray water is not treated for pathogens when used for irrigation, it poses its own unique management challenges related to human health. As discussed previously, gray water needs to be applied via flooding, subsurface, or drip irrigation systems in residential landscapes. Because gray water usually is not the exclusive irrigation water source for plants around homes, alternating potable and gray water irrigation could mitigate potential negative effects that gray water salts may have on landscape plants.

Conclusions

Many years of practice and field observation confirm that recycled or brackish water can be used successfully to irrigate turfgrasses. Water conservation resulting from this practice far outweighs the potential negative impacts. Nonetheless, recycled or brackish water quality must be evaluated thoroughly before developing appropriate cultural strategies for its use. Although water quality involves a complex set of factors, and each irrigation water must be analyzed on an individual basis, very few water sources are absolutely unsuitable for turfgrass irrigation. Determination of the precise nature and magnitude of a water quality problem may require more than just water analysis: knowledge of climate, soil chemistry and physics, use patterns, and turfgrass quality expectations all are essential to analyzing and correcting any problem.

Switching from fresh, potable water to recycled/saline water for irrigation will add substantially to operating costs at a site. Along with labor associated with new agronomic tasks, any, or all, of the following also will increase costs:

- Irrigation system redesign, upgrade, and maintenance;
- Construction and maintenance of water storage facilities;
- Irrigation water blending;
- Protection of adjacent properties/bodies of water;
- Ground/surface water monitoring;
- Equipment labeling and painting;
- Equipment deterioration because of salt;
- Consultant and laboratory fees;
- Purchase of additional soil amendments and equipment.

Field observations and informal surveys indicate that a 10 to 20% increase in operating costs is common, with the actual figure varying with the quality of water used. (Devitt et al. 2004; Harivandi, 2004a,b; Emerson, S., Desert Mountain Golf Club, Arizona. Personal communication; Huck, M., Irrigation and Turfgrass Sciences. Personal communication).

Currently, the use of household gray water for irrigating home landscapes is not widely practiced. More research is needed to determine the most effective, least expensive, and safest (vis-à-vis human health) methods for using such water.

Literature Cited

Arizona Department of Water Resources. 2000. *Third Management Plan: 2000–2010.* Arizona Department of Water Resources, Phoenix.

Asano, T., R. G. Smith, and G. Tchobanoglous. 1984. Municipal wastewater: Treatment and reclaimed water characteristics. Pp.2:1–2:26. In G. S. Pettygrove and T. Asano (eds.). *Irrigation with Reclaimed Municipal Wastewater—A Guidance Manual.* Lewis Publishing, Chelsea, Michigan.

Ayers, R. S. and D. W. Westcot. 1985. *Water Quality for Agriculture.* FAO Irrigation and Drainage Paper 29, Rev. 1. Food and Agriculture Organization of the United Nations, Rome, Italy.

Broadbent, F. E. and H. M. Reisenauer. 1984. Fate of wastewater constituents in soil and groundwater: Nitrogen and phosphorus. Pp.12:1–12:16. In G. S. Pettygrove and T. Asano (eds.). *Irrigation with Reclaimed Municipal Wastewater—A Guidance Manual*. Lewis Publishing, Chelsea, Michigan.

California Water Resources Control Board. 1993. Porter-Cologne Act provisions on reasonableness and reclamation promotion, Section 13552-13577, California Water Code. California State Water Resources Control Board, Sacramento.

Carrow, R. N. and R. R. Duncan. 2002. *Salt-Affected Turfgrass Sites: Assessment and Management*. John Wiley & Sons, Hoboken, New Jersey.

Carrow, R. N., R. R. Duncan, and M. Huck. 1999. Treating the cause, not the symptoms. *USGA Green Section Record* 37(6):11–15.

Carrow, R. N., M. Huck, and R. R. Duncan. 2000. Leaching for salinity management on turfgrass sites. *USGA Green Section Record* 38(6):15–24.

Devitt, D. A. 2005. Foliar damage, spectral reflectance, and tissue ion concentrations of trees sprinkle irrigated with waters of similar salinity but different composition. *HortSci* 40(3):819–826.

Devitt, D. A., R. L. Morris, D. Kopec, and M. Henry. 2004. Golf course superintendents' attitudes and perceptions toward using reuse water for irrigation in the southwestern United States. *HortTechnol* 14(4):577–583.

Devitt, D. A., R. L. Morris, M. Baghzouz, M. Lockett, and L. K. Fenstermaker. 2005. Water quality changes in golf course irrigation ponds transitioning to reuse water. *HortSci* 40(7):2151–2156.

Dudeck, A. E. and C. H. Peacock. 1985. Effects of salinity on seashore paspalum turfgrasses. *Agron J* 77:47–50.

Dudeck, A. E., C. H. Peacock, and J. C. Wildmon. 1993. Physiological and growth responses of St. Augustinegrass cultivars to salinity. *HortSci* 28:46–48.

Dudeck, A. E., S. Singh, C. E. Giordano, T. A. Nell, and D. B. McConnell. 1983. Effect of sodium chloride on Cynodon turfgrasses. *Agron J* 75:927–930.

Duncan, R. R. 2003. Seashore paspalum (Paspalum vaginatum Swartz). Pp. 295–307. In M. D. Casler and R. R. Duncan (eds.). *Turfgrass Biology, Genetics, and Breeding*. John Wiley & Sons, Hoboken, New Jersey.

Duncan, R. R., R. N. Carrow, and M. Huck. 2000. Effective use of seawater irrigation on turfgrass. *USGA Green Section Record* 38(1):11–17.

Eaton, F. M. 1950. Significance of carbonates in irrigation waters. *Soil Sci* 69:123–133.

Farnham, D. S., R. F. Hasek, and J. L. Paul. 1985. *Water Quality: Its Effects on Ornamental Plants*. University of California Cooperative Extension Leaflet 2995. Division of Agriculture and Natural Resources, Oakland, California.

Harivandi, M. A. 1990. Sulfur, soil pH and turfgrass management. *Calif Turfgrass Culture* 40(1–4): 9–11.

Harivandi, M. A. 1994. Wastewater quality and treatment plants. Pp. 106–129. In J. Snow, M. P. Kenna, K. S. Erusha, M. Henry, C. H. Peacock, and J. R. Watson. (eds.). *Wastewater Reuse for Golf Course Irrigation*. Lewis Publishing, Chelsea, Michigan.

Harivandi, M. A. 2004a. Evaluating recycled waters for golf course irrigation. *USGA Green Section Record* 42(6):25–29.

Harivandi, M. A. 2004b. Considerations in retrofitting a golf course for recycled water irrigation. *USGA Green Section Record* 42(6):30–33.

Harivandi, M. A., J. D. Butler, and P. N. Soltanpour. 1982. Effects of sea water concentrations on germination and ion accumulation in alkaligrass. *Commun Soil Sci Plant Anal* 13:507–517.

Harivandi, M. A., J. D. Butler, and P. N. Soltanpour. 1983. Effects of soluble salts on ion accumulation in *Puccinellia* spp. *J Plant Nutr* 6:255–266.

Harivandi, M. A., J. D. Butler, and L. Wu. 1992. Salinity and turfgrass culture. Pp. 208–230. In D. V. Waddington, R. N. Carrow, and R. C. Shearman (eds.). *Turfgrass*. Monograph 32. American Society of Agronomy, Madison, Wisconsin.

Horst, G. L. and N. B. Beadle. 1984. Salinity effects germination and growth of tall fescue cultivars. *J Amer Society Hort Sci* 109:419–422.

Jordan, L. A., D. A. Devitt, R. L. Morris, and D. S. Newman. 2001. Foliar damage to ornamental trees sprinkler-irrigated with reuse water. *Irrig Sci* 21:11–15.

Lee, G., R. R. Duncan, and R. N. Carrow. 2004. Salinity tolerance of seashore paspalum ecotypes: Shoot growth responses and criteria. *HortSci* 39:1138–1142.

Maas, E. V. and G. Hoffman. 1977. Crop salt tolerance–Current assessment. *Drainage Div ASCE* 103:115–134.

Marcum, K. B. 1999. Salinity tolerance in turfgrasses. Pp. 891–906. In M. Pessarakli (ed.). *Handbook of Plant and Crop Stress*, 2nd ed. Marcel Dekker, New York.

Marcum K. B. 2001. Salinity tolerance of 35 bentgrass cultivars. *HortSci* 36(2):374–376.

Marcum, K. B., S. J. Anderson, and M. C. Engelke. 1998. Salt gland and ion secretion: A salinity tolerance mechanism among five zoysiagrass species. *Crop Sci* 38:806–810.

Marcum, K. B., M. Pessarakli, and D. M. Kopec. 2005. Relative salinity tolerance of 21 turf-type desert saltgrasses compared to bermudagrass. *HortSci* 40(3):827–829.

NPD Group. 1999. *Graywater Awareness & Usage Study*. Soap and Detergent Association, Washington, D.C.

Oertli, J. J., O. R. Lunt, and V. B. Youngner. 1961. Boron toxicity in several turfgrass species. *Agron J* 53:262–265.

Oster, J. D., M. J. Singer, A. Fulton, W. Richardson, and T. Prichard. 1992. *Water Penetration Problems in California Soils*. Kearney Foundation of Soil Science, Division of Agriculture and Natural Resources, University of California, Davis.

Page, A. L. and A. C. Chang. 1984. Fate of wastewater constituents in soil and groundwater: Trace elements. Chapt. 13, Pp. 1–6. In G. S. Pettygrove and T. Asano (eds.). *Irrigation with Reclaimed Municipal Wastewater—A Guidance Manual*. Lewis Publishing, Chelsea, Michigan.

Qian, Y. L. and B. Mecham. 2005. Long-term effects of recycled wastewater irrigation on soil chemical properties on golf course fairways. *Agron J* 97:717–721.

Qian, Y. L. and M. R. Suplick. 2001. Interactive effects of salinity and temperature on Kentucky bluegrass and tall fescue seed germination. *Inter Turf Sci Res J* 9:334–339.

Qian, Y. L., M. C. Engelke, and M. L. V. Foster. 2000. Salinity effects on zoysiagrass cultivars and experimental lines. *Crop Sci* 40:488–492.

Richards, L. A. (ed.). 1954. *Diagnosis and Improvement of Saline and Alkali Soils*. USDA Handbook 60. U.S. Government Printing Office, Washington, D.C.

Rose-Fricker, C. and J. K. Wipff. 2001. Breeding for salt tolerance in cool season turfgrasses. *Inter Turf Sci Res J* 9:206–212.

Schaan, C. M., D. A. Devitt, R. L. Morris, and L. Clark. 2003. Cyclic irrigation of turfgrass using shallow saline water. *Agron J* 95:660–667.

Westcot, D. W. and R. S. Ayers. 1984. Irrigation water quality criteria. Pp.3–1, 3–37. In G. S. Pettygrove and T. Asano (eds.). *Irrigation with Reclaimed Municipal Wastewater—A Guidance Manual*. Lewis Publishing, Chelsea, Michigan.

Wu, L. and H. Lin. 1994. Salt tolerance and salt uptake of diploid and polyploid buffalograsses (Buchloe dactyloides). *J Plant Nutr* 17:1905–1928.

15

San Antonio Water Conservation Program Addresses Lawngrass/Landscapes

Calvin Finch

Introduction

A water conservation program can be very effective. It can be based on science, and it can be embraced by the citizens of the community. The water conservation program in San Antonio, Texas, fits that description.

Since 1982, the per capita water use in San Antonio has been decreased from a high of 851 liters/person/day (LPD) (225 gallons) to a level of 458 LPD (121 gallons) in 2004 (Figure 15.1). The San Antonio LPD is calculated by dividing all water pumped by the population of the San Antonio Water System (SAWS) service area. The LPD calculation includes all industrial water and even the small amount of wholesale water sold by SAWS. The San Antonio LPD is believed to be the lowest of any large city in the West, and the progress to the current level is believed to be the most successful voluntary reduction ever accomplished by a city. The success of the conservation program has meant that San Antonio has not used any more water per year in recent years than it did in the 1990s, despite a population growth of more than 300,000 people.

The program's success also means that in 2005, San Antonio was able to reassess its 50-year (yr) water plan; eliminate several large, expensive water resource projects; and delay other infrastructure expansion. In 2005, SAWS also revised its expectations for conservation. The bottom-line goal of 492 LPD (130 gallons) in 2025 established in the 1998 plan was changed to a goal of 439 LPD (116 gallons), which is expected to be achieved in 2016. At that time (2016), it is believed that SAWS water users will have decreased indoor and landscape water use to the lowest levels possible without adversely affecting quality of life and without diminishing San Antonio's growth (2% per yr) and economic development.

The characteristics of the San Antonio water conservation program that contributed to its success include the following factors:

- San Antonio is motivated to decrease water use because of the need to protect the Edwards Aquifer. The Federal Courts and the Texas Legislature intervened to give impetus to the conservation program.

- San Antonio has embraced the idea that conservation works best if everyone is on the conservation team. This means that programming is diverse and extensive, and leadership comes from groups that could just as easily be opponents of conservation programming.

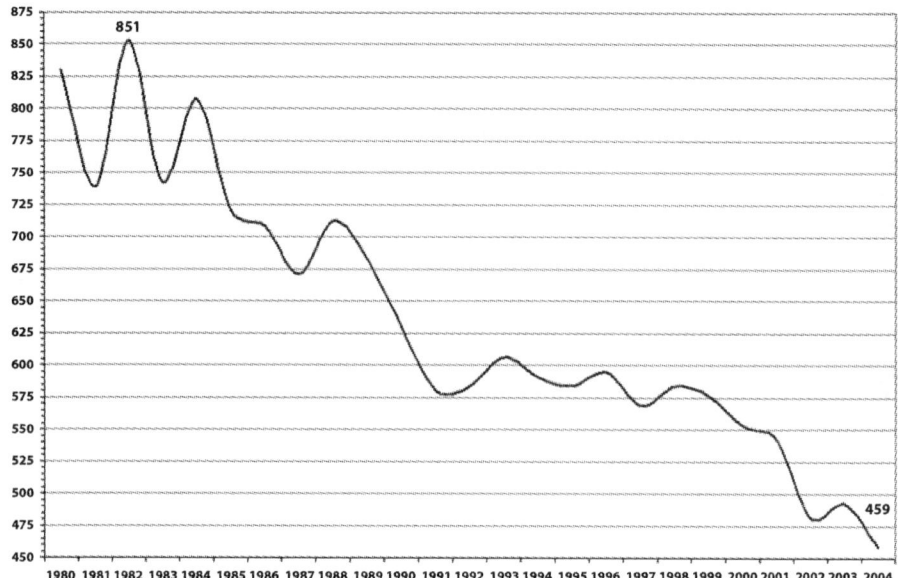

Figure 15.1. San Antonio Water System water use by year, liters per person per day (LPD).

- San Antonio treats water conservation as a water supply program. SAWS spends $5 per resident on conservation. Conservation funds are derived from dedicated sources to ensure that adequate funds are available to achieve the conservation goals.

- San Antonio addresses water conservation in a positive manner. The goal is to eliminate water waste, not to decrease water availability for uses that improve quality of life or economic activity. In San Antonio, low-water-use gardening is exciting gardening. If water use on gardens and landscapes is restricted, the rules are based on science and include local testing and public input.

Why Is Turfgrass a Target of Water Purveyors Seeking to Decrease Water Use?

Landscape watering is a large part of a community's water use in arid and semi-arid climates; in many cities, it is 50 to 60% of the total water use (Mayer et al. 1999). In a city such as San Antonio where per capita water use has been lowered by 40% during the last 20 yr, landscape water use was still 25% of total water use in 2005.

Most communities relegate landscape water use to the category of discretionary use. In order of priority, it is below health and safety, industrial, agricultural, and environmental flow uses. On some lists, it has less priority than recreational use.

Lawn and other landscape watering is very much related to weather. There is a high peak of use during the hottest part of summer. Prudent planning requires that water purveyors have the infrastructure necessary to deliver this peak demand, which often is double the normal demand (Mayer et al. 1999).

The idea of expecting a lush, green lawn is relatively new and was promoted for the general population in and after the 1920s. Based on the informal testimony of long-time city residents, irrigated lawns were not common in San Antonio before World War II. The city has an average of 81 centimeters (cm) (32 inches) of rain per year and 140 cm (55 inches) of evaporation. In areas such as San Antonio, a lush summer lawn has no identifiable relationship with the native landscape and is not sustainable in terms of our common definition of the concept.

The Beneficial Aspects of Lawns

- In many areas of the United States, lawns are relatively easy to maintain without excessive labor, pesticides, and/or irrigation.

- People have come to love the look and feel of lawns. Generally it is believed that when asked to describe the "ideal family home," most individuals will mention a lawn (Vickers 2001). A well-maintained landscape that includes a lawn is a major factor in determining property values (Bennett and Hazinski 1993).

- A large industry has developed around creating and maintaining lawns and other landscape features. A report to the National Urban and Community Advisory Committee stated that $147.8 billion of economic activity, which employs 1,964,339 people, is related to landscapes (Hall, Hodges, and Hayden 2005). It is an industry with some high-paying jobs and many entry-level positions that provide opportunities for unskilled individuals entering the job market.

- In certain situations, well-maintained lawns are an important fire-prevention factor (Youngner 1970).

- Lawngrass is cited as an important contributor to clean air and decreased pollutants in water runoff and water percolation into aquifers (Beard 1989; Mugass, Agnew, and Christains 2005).

- Lawns and other landscape features are more than just a luxury because of the opportunities they provide for exercise and the contribution they make to psychological health. Individuals who partake in lawn work and gardening are less likely to be involved in crimes or to have mental disease and are more involved in civic affairs (Sempik, Aldridge, and Becker 2005).

The City of San Antonio and SAWS have adopted the attitude that, in the short term at least, lawns are here to stay, and any radical approach to eliminating lawngrass will not be successful. It is believed that the majority of the citizens of San Antonio and other cities value lawns and will do battle to maintain their right to have and maintain one. The task of leaders of the water conservation effort in such cities becomes identifying ways to decrease water use without challenging the inclination of its citizens to maintain a lawn. The acceptance of this idea is a basic tenet of the San Antonio concept that water conservation is most effective when everyone is on the "Conservation Team."

The good news is that in a climatic area with 81cm of rain/yr (average) there are many ways to address the interest in lawns and landscapes that result in decreased water use. Science is an ally in the San Antonio experience—an experience that has left quality of life intact, supports a "growth is good" policy, and has lowered per capita water use by 40% in two decades.

San Antonio has had success, and in 2006 as water use continued to drop, support for the water conservation effort increased. It may look like the path to conservation success was well marked for its entire length, but that is not true. Many people inside and outside of the city described the community as being slow and reluctant to accept its responsibility to address its water challenge. The conservation effort that seems so defined and popular now evolved after years of missed opportunities, inaction, bitter arguments, and, finally, the intervention of the Federal Courts and the Texas Legislature. Leaders stepped forward from government, industry, and environmental groups to mold the program that now exists—a program that expects to reach its goal of 439 LPD in 2016 after a high-use level of 851 LPD in the early 1980s.

Other Sources of Savings

Landscape water use is a major source of discretionary water use in most communities in arid and semi-arid climates and must be addressed, but it is not the only source of water savings. Some other sources are much easier to address than changing landscape water use. Two of the most obvious targets in San Antonio were infrastructure improvements and removal of high-flow toilets in favor of the modern low-flow versions.

Infrastructure Improvements

One-third of the LPD reduction in San Antonio is because of improvements in infrastructure, including leak detection and repair. San Antonio's "lost water" figure was decreased from approximately 25% in 1982 to 8% in 2004. The rate varies depending on weather conditions (more main breaks in hot, dry weather), but never increases above 12%, the American Water Works Association's target for U.S. cities.

The lost water figure reflects how much water is lost between pumping from the groundwater or surface water source to selling water to customers. It is difficult to imagine asking a water system's ratepayers to cut back on water use when a major source of loss is in the system's pipes, connections, and meters. The characteristics of a water purveyor's lost water can be determined by a water audit. The audit identifies when losses occur, and then it is relatively easy to calculate the economics of decreasing the loss. In San Antonio, the replacement of polybutylene pipe and aggressive leak detection repairs have improved the situation.

Low-Flow Toilets

It has been U.S. law since 1994 (1992 in Texas), that all toilets manufactured must work with 6.0 liters ([L];1.6 gallons) of water per flush, instead of the 13.2 to 26.4 L per flush of old toilets. That means that a new toilet in residential homes uses approximately 41,640 L less water per year than the old toilet did. In commercial settings, the total savings often is 90,850 L per year. Once the change-out is made, the savings are permanent (except for a few brands that can become high-flow again if the wrong flapper is used). The low-flow toilets save 55 to 78% of the water used by high-flow toilets.

In San Antonio, any customer may receive up to two free low-flow toilets. The toilets cost SAWS approximately $50 each, wholesale, and the cost/m^3 of water saved is $.16 ($200/acre ft) during a 10-yr period. The 20,000 low-flow toilets distributed each year decrease water use by 942,973 m^3 every year without any change in lifestyle.

Turfgrass-Related Programming

Conservation Pricing of Water

One of the basic rules of the marketplace is if the price of a commodity is increased, users of that commodity examine their use and usually become more efficient. Translated, that means if you increase the price of water, consumers may eliminate waste, change discretionary use, and decrease overall consumption. The problem of charging a high price is that water is an essential survival item. Most communities believe that all families deserve to have a certain amount of clean, safe water for drinking, cleaning, and other household tasks. The resultant compromise between the right to a limited supply of high-quality water at a reasonable price and the effectiveness of limiting extravagant use of water by charging high prices is "conservation pricing."

In San Antonio, there are four tiers of water prices. The first tier is priced at $.0878 per 379 L (100 gallons) for levels of use below 32,173 L (8,500 gallons)/month. The amount of water identified in the first tier is the estimated water needed for a family of four for health and safety. The next 19,844 L (5,236 gallons)/month (2nd tier) is priced at $.1379/379 L, and the next 17,010 L (4,488 gallons)/month (3rd tier) is priced at $.2148/379 L. In the summer, the 4th tier—water use over 65,132 L (17,206 gallons)/month—is priced at approximately $.41 per 379 L. Pricing extravagant water use at a higher rate results in decreased water use.

In a study conducted in Austin, Corpus Christi, and San Antonio, increasing prices and water use showed a .20 relationship—double the cost of water, and it will decrease water use by 20% (TWDB 1999). In San Antonio and many other cities, individuals and businesses that water their lawns must be willing to pay a high premium for that privilege.

General Education

Turfgrass advocates often have cited the concept that having a lawn does not automatically result in high water use. They identify the problem of high water use as one of insufficient education and resultant poor irrigation habits by the homeowner or business caring for the lawn (Beard and Green 1994). It is true that many lawns receive more water than they require. The conservation program in San Antonio recognizes this fact. In addition to the Seasonal Irrigation Program (SIP), which makes it easy for citizens to use Potential Evapotranspiration (PET) information to determine lawn water needs, there is a general education effort directed at lawn water use.

The Media

The education program is delivered in many ways. Paid media includes public service announcements, billboards, TV, radio, websites, bill inserts, and newspaper commercials. SAWS spends nearly $1 million on paid conservation media. Probably just as effective is the SAWS free media effort. SAWS conservation staffers write four weekly newspaper columns, five monthly newsletter columns, and a monthly water bill insert. The staff does 4 hours of weekend radio call-in shows and is always available for TV, radio, and newspaper interviews. SAWS conservation staff also makes more than 100 public presentations per year on conservation topics. Water use and resources are newsworthy in San Antonio, and there are many opportunities to discuss issues and offer technical information. The messages revolve around SAWS conservation programming, effective watering, turfgrass care, and the theme "Low Water-Use Gardening is Exciting Gardening."

In terms of turfgrass care, one of the messages is that if you select bermudagrass, buffalograss, zoysiagrass, or certain cultivars of St. Augustinegrass, your lawn can go dormant in a drought. The dormancy option is not promoted heavily in any organized manner, but more than 50% of SAWS customers do no lawn watering.

SAWS has developed a 3-week (wk) plan to establish new sod and seeded lawns, which serves as the basis of a variance for new lawns established in a critical period (drought restrictions). At the end of 3 wk, the expectation is that the new lawn will be able to survive with watering on the once-per-wk schedule required in times of drought.

The goal of SAWS conservation programming is to ensure that San Antonio will decrease its per capita water use by 3.79 L/day (d) (1 gallon) each year until the ultimate goal of 439 LPD is achieved. A secondary goal is to have a program diverse enough that everyone can participate on the conservation team and contribute to achievement of the conservation goals. In the area of landscape programming, the following opportunities are offered: irrigation audits, irrigation system design rebates, SIP, water saver landscape rebates, and rain sensor rebates.

Low-Water-Use Landscape

Turfgrass is an important part of landscape water use, but trees, shrubs, and perennials also are important and should be addressed in any comprehensive water conservation effort. The message about the low-water-use option is positive. It is believed in San Antonio that telling people what they "should not do" is less effective than offering exciting opportunities in low-water-use gardening. The theme of "Low-Water-Use Gardening Is Exciting Gardening" is the guiding principle. It is backed by financial support for plant research and outreach. Every year in San Antonio new plant cultivars or cultural practices for old plant species are introduced that make the plants more fun to grow. In 2005, "Texas Lilac Vitex" was an example of such a promotion. Individuals were encouraged to use improved Vitex selections and prune and deadhead in a manner to increase blooms to attract butterflies and provide cut flowers. Other recent offerings have included "Belinda's Dream" roses, firebush, bluebonnet transplants, "Gold Star" esperanza, and "Texas Gold" columbine.

The palette of plants that prosper in an 81-cm rain zone, even with a high evaporation rate, is large enough to support such a concept. Coupled with the seven Xeriscape landscape principles, the large choice of native ornamentals and well-adapted plants available lends itself to the development of educational messages that offer the choices, variation, challenges, complexity, and aesthetics gardeners seek.

An important part of the success of "Low-Water-Use Gardening Is Exciting Gardening" in San Antonio is the absence of an organized opposing message. The gardening media, landscape professionals, and groups such as Master Gardeners accepted early that the city faced a water challenge and—to their credit—recognized the value of contributing to the solution. They quickly grasped the concept that if the landscape options offered contributed to meeting the water use (LPD) goals, it was an acceptable option to conservation advocates. It was even better if the option contributed to people being excited about gardening and landscaping and resulted in a large profitable landscape industry.

The "Exciting Gardening" idea is supported through SAWS-sponsored events such as Festival of Flowers, Spring Bloom Giveaway, and Garden Jazz. Brochures on Wildscape, Butterfly Gardening, Native Plants, Xeriscape Conversion, and Landscape Maintenance (Figure 15.2, see page I-15) also attract interest and provide information.

Irrigation Programming

Too often the effort to decrease water use on landscapes relies entirely on choosing plants. Irrigation practices, technology, and maintenance may be more important than plant choice in decreasing water use on landscapes. In fact, the presence of an irrigation system translates to more water use if all other things are equal. A study funded by the U.S. Bureau of Reclamation found that irrigation systems are associated with a 38% (500 LPD) increase of water use (136 LPD standard error) over landscapes without irrigation (Gregg et al. 1999).

Among the programs that address this reality in San Antonio are regulations that all irrigation systems must have rain sensors, all systems on properties of 5,000 square feet or more must complete a system review (simple audit) every year, all irrigation systems on newly constructed homes must be zoned by plant type, all irrigation designers must be locally licensed, and all irrigation designs must be reviewed and the work inspected before a home can receive a permit of occupancy. An Irrigation Design Rebate is available to businesses and homeowners who use the free SAWS irrigation audit and then take the auditor's advice concerning design improvements. And a Water Saver Rebate pays a bonus on properties with irrigation systems after they prove decreased water use for 1 yr.

The city of Seattle has a more extensive program of rebates for irrigation design improvements and water conservation technology additions than San Antonio. They list 12 specific improvements that merit rebates and offer rebates for other changes that result in "dependable, consistent water savings." The Seattle program is described on the website of the Saving Water Partnership (SWP 2007).

Seasonal Irrigation Program

The seasonal irrigation program provides homeowners and businesses with lawn irrigation information based on PET. The SIP uses information from a weather station in San Antonio and lawn characteristics provided by the homeowner to make a weekly watering recommendation for individuals who sign up to receive the information via e-mail or phone message. Individuals also may receive the irrigation recommendation from the *San Antonio Express-News*, the SAWS website, the KLUP Radio Gardening Show, or from a dedicated phone line.

The program uses a single weather station—the "modified" Penman equation—and an experimentally developed crop coefficient for the hot-weather grasses regularly used in San Antonio. The crop coefficient of St. Augustinegrass and zoysiagrass in the sun is 0.6. St. Augustinegrass and zoysiagrass in the shade, and bermudagrass in the sun, use a 0.4 coefficient. Buffalograss in the sun uses a 0.3 coefficient. Watering recommendations are rounded off to the nearest 0.64 cm (0.25 inch), and any recommendations less than 1.3 cm (0.5 inches) are postponed until the next week.

SAWS partners with the Texas Cooperative Extension, Bexar County Master Gardeners, and Garden Volunteers of South Texas to deliver the program. The volunteer organizations are charged with the responsibility to make community presentations to recruit new SIP enrollees. Individuals who commit to try SIP receive a SIP kit that includes the equipment necessary to measure irrigation system output.

The specifics of the program, including the adequacy of using one weather station, the accuracy of the crop coefficients, and the sufficiency of watering once per wk, all were developed through 4 yr of field tests in San Antonio.

Five thousand households take advantage of the personalized SIP service. Evaluations by the Extension Service and SAWS over the last 8 yr indicate that SIP enrollees use approximately 20% less water than they did before they enrolled. The program lowers the long-accepted recommendation on St. Augustinegrass from 2.54 cm of irrigation water per wk to 1.9 cm per wk, without reducing appearance to an unacceptable level.

The city of San Diego has a related program. Their Landscape Watering Calculator uses a historical evapotranspiration calculation to produce an irrigation schedule based on input to the San Diego Water Department website (City of San Diego 2007), on geographic location within the area, plant type in the landscape, soil characteristics, and irrigation system output.

The SIP is the first level of a SAWS water conservation program directed at lawn water use. It targets homeowners and businesses that want to maintain a green lawn without wasting any water. The SIP research program also verified that St. Augustinegrass could survive a San Antonio summer with an irrigation application every 2 wk. The lawn would not be attractive, but it would be alive.

The 2-wk survival capability became the basis for the third level of water restrictions in severe droughts. Irrigation restrictions based on street address, which decreased watering to every 2 wk, not only would protect a SAWS ratepayers' investment in a St. Augustinegrass lawn, but also would decrease water use in an emergency situation significantly.

Water Saver Rebate

SAWS spends $5 per individual on conservation programming. A portion of those funds is dedicated to the encouragement of SAWS ratepayers who are willing to convert their St. Augustinegrass-dominated landscape to a Water Saver landscape.

SAWS defines a Water Saver landscape as one that is less than 50% turfgrass, less than 10% high-water-use plants, and, if there is an irrigation system, the system is zoned by plant type and is in a well-maintained state. The rebate amount is based on lot size.

The bonus is awarded to homeowners who finish the year using less water on the landscape than PET would predict. In 2004, 53% of Water Saver rebate recipients received the bonus. The record of water savings as a result of landscape retrofits was not always as good. In 2003, SAWS evaluated the water-use results of 1,000 homeowners who received the rebate. Overall, there was not a net water-use savings. Approximately 50% used the same water as before the conversion, and 25% used more water than they had before. One-quarter of the rebate recipients decreased water use to the expected levels. As a result of this dismal performance, the rebate requirements were changed to match the characteristics of the landscapes in the successful 25%. These characteristics included zoned irrigation, well-maintained irrigation, less than 50% turfgrass, and less than 10% high-water-use plants (St. Augustinegrass, annual beds, acid-loving plants, etc.). The new requirements are reflected in the Water Saver landscape rebate table (Table 15.1).

Table 15.1. SAWS water saver rebate awards

Lot size in hectares (acres)	Lot size in m² (ft²)	Landscape with no irrigation system and/or >30% preserved native[a]	Landscape with irrigation systems and/or no preserved native[a]
		Premium rebate	Standard rebate
1.21 (0.091) or less	1,220 (4,000) or less	150	100
1.22–1.73 (0.092–0.13)	1,221–1,830 (4,001–6,000)	225	150
1.74–2.39 (0.14–0.18)	1,831–2,440 (6,001–8,000)	300	200
2.40–3.06 (0.19–0.23)	2,441–3,051 (8,001–10,000)	375	250
3.07–6.23 (0.24–0.46)	3,051–6,101 (10,001–20,000)	450	300
6.24 (0.47) or more	6,101 (20,000) or more	525	350

[a] Preserved native does not mean newly planted native plants. It means existing, established native trees, understory, and associated soils left undisturbed.

Meanwhile, El Paso is in the 20-cm/yr rain zone. That community decided that turfgrass is not an appropriate groundcover for most situations. The water utility offers a $1 per 366 cm² (1 square foot) rebate to citizens who convert their turfgrass to a landscape that is more sustainable in a desert environment. The El Paso Public Utilities Board Water Service website has more information (El Paso Public Utilities Board Water Service 2007).

Water Conservation Ordinance

Most of SAWS conservation programming is voluntary, often encouraged by financial incentives. For a number of years, however, it has been against the law to waste water on the landscape, generally defined as water leaving the yard or running down the street. It also is against the law to water between 10:00 a.m. and 8:00 p.m. This time prohibition relates to evaporation rates and wind, which would be highest during that period. It also relates to peak water use. Midday is when most industrial and commercial use occurs. By moving landscape watering to the evening or early mornings, the water-use peak is leveled and water system infrastructure needs are decreased. In January 2005, another variation from the incentive idea occurred when the Water Conservation Ordinance was passed unanimously by the San Antonio

City Council after 4 yr of work by the SAWS staff and conservation advocates. A consensus was reached in the community that some conservation practices are implemented most efficiently by ordinance rather than education or incentives. The ordinance has 16 provisions, a number of which affect landscape watering.

The "Turf News," published by Turf Producers International, documented the development of the San Antonio Water Conservation Ordinance in its January–February 2004, March–April 2004, May–June 2004, and January–February 2005 issues. The ordinance passed with a unanimous vote of the City Council in January 2005. The 16 provisions of the ordinance are expected to contribute 1.2 billion gallons of savings/ yr by 2016.

With the support of the new homebuilding industry—the Greater San Antonio Builders Association—several of the provisions in the ordinance address turfgrass issues. The provisions include:

- All landscapes with lawngrass must have at least 4 inches of soil under the entire extent of the lawn.

- All landscapes that include irrigation systems on new homes constructed after January 1, 2006 must be zoned by plant type.

- New home developments that have model homes must include a model home that offers a Xeriscape landscape model with less than 50% turfgrass by area.

- Beginning in January 2007, lawngrasses used on new homes must be capable of surviving 60 d of drought.

- Since 1992, all new irrigation systems are required to include a rain sensor. The ordinance requires that all systems without sensors, no matter when they were built, must be retrofitted for a sensor.

The ordinance provisions are effective because they address requirements that will save water in the long term. Some people would go so far as to say that some of the provisions are radical. The provisions will save future water use, they are consistent with the alternating drought and high-level rain cycle in San Antonio, and they have been developed by partnerships between SAWS and the stakeholders involved. The "60-day survival" provision is particularly interesting. The provision was supported by the Turfgrass Producers of Texas (TPT). In their testimony at the San Antonio City Council meeting, the TPT and other supporters cited the following reasons for their support:

- San Antonio is faced with a water challenge, and all parts of the community must do their fair share to decrease water use.

- There are many species and cultivars of turfgrass that can survive 60 d of drought if they are well established.

- New drought-tolerant cultivars of turfgrass are being introduced every year, and the ordinance allows for the list to be expanded if these new cultivars pass the 60-d test (Figure 15.3, see page I-16).

- San Antonio has a relatively high average rainfall at 32 inches/yr, but it also has a very high evaporation rate (55 inches/yr). The rainfall varies considerably from year to year, so it is important that plants, including grasses, are capable of surviving periods of drought.

- SAWS has agreed to partner with TPT and Texas A&M University turfgrass researchers to organize and conduct the tests to confirm which grasses can survive 60 d of drought.

Drought Restrictions and Enforcement

San Antonio has opted to follow a policy of encouraging property owners to decrease water use on landscapes without threatening property values or reducing San Antonio to a "brown city." Landscapes, gardening, and lawns are important in San Antonio, but the overriding philosophy is that low-water-use gardening is exciting gardening. That being said, the city also recognized that a major reduction in water use can be achieved almost immediately by making it more difficult to water lawns. If the restrictions are imposed carefully, the emergency reduction in water use can be achieved without threatening the long-term health of the landscape.

When the Edwards Aquifer falls to the level of 650 ft above sea level at the J-17 test well in San Antonio, Critical Period Rules (Drought Restrictions) are imposed. The primary rule is that lawn watering is limited to 1 day of the week based on address. When the Aquifer level falls to 640 ft, the legal times to irrigate during the designated day are limited to early morning and late evening. In Stage Three, 630 ft, lawn watering is limited to once every 2 wk. In Stage Four, the city manager may ban lawn watering. In 2000, during the last long-term drought, Critical Period Rules (Stage One and Stage Two) were in place for 6 months. Water use was decreased by 14.9% over the period based on the use predicted by PET.

The Critical Period Rules and Water Conservation Ordinances are enforced by five part-time San Antonio Police Officers supervised by the SAWS Conservation Director. The ordinances give the officers the power to ticket violators without a warning, but SAWS policy is only to ticket violators who ignore an educational contact by the officers, or other SAWS personnel, and are reported for a second violation. These violators are placed on the Water Waster list. The SAWS Conservation Enforcement Officers concentrate their efforts on violators on the Water Waster list. Water violations are a misdemeanor in San Antonio. The penalty for a first violation is a fine of $50 to $100. The second violation merits a fine of $250 to $500, and the third or more has a fine of $1,000 to $2,000.

Conclusion

The number of communities in the United States faced with water supply challenges increases every year. Water conservation often is an important way to help meet those challenges. Every water conservation effort attempts to identity where discretionary use exists that can be decreased. In some arid and semi-arid climates, landscape water use, and particularly lawn watering, is identified as a discretionary use that can and should be decreased. Because lawns figure highly in the concept of an ideal landscape for most individuals, there is an opportunity for a clash between citizens who value lawns and water purveyors who deem them as discretionary water use.

San Antonio is a community in a semi-arid climate. The city has decreased per capita water use by more than 40% since the early 1980s and has avoided conflict over landscape watering. Success has been achieved because SAWS recognized the value of lawns to its citizens and worked with them to develop a comprehensive water conservation program that addressed infrastructure improvements, inefficient plumbing, industrial technology, and other water-saving opportunities along with savings in landscape watering. The landscape watering savings are based on opportunities identified in outside research and local studies that resulted in changes in turfgrass management, variety or cultivar selection, and irrigation technology without attempting to eliminate lawns. Every community's situation is different, so the formula for decreasing water use also can differ. The example provided by San Antonio shows that water use can be decreased in a manner that takes advantage of the benefits of turfgrass and yet is consistent with positive attitudes about turfgrass use that exist in the city.

Literature Cited

Beard, J. B. 1989. Turfgrass water stress; Drought restrictions components, physiological mechanisms, and species genotype diversity. Pp 23–28. In H. Takotah (ed.). *Proceedings of the Sixth International Turf Research Conference, 31 July–5 August, 1989*. Japan Society of Turf Science, Tokyo.

Beard, J. B. and R. L. Green. 1994. The role of turfgrasses in environmental protection and their benefits to humans. *J Environ Qual* 23:452–460.

Bennett, R. E. and M. S. Hazinski. 1993. *Water Efficient Landscape Guidelines*. American Water Works Association, Denver, Colorado.

City of San Diego. 2007. Water Department: Landscape Watering Calculator page, http://sandiego.gov/landcalc/start.do (11 January 2007)

El Paso Public Utilities Board Water Service. 2007. Conservation: Turf Rebate page, http://epwu.org/conservation/turf_rebate.html. (11 January 2007)

Gregg, T., D. Strub, C. Grisby, and W. deHerrera. 1999. *Xeriscaping: Sowing the Seeds for Reducing Water Consumption*. Report for the United States Bureau of Reclamation. City of Austin, Texas. 7 pp.

Hall, C. A., W. Hodges, and J. W. Hayden. 2005. *Economic Impacts of the Green Industry in the United States*. Report to the National Urban and Community Advisory Committee. U.S. Department of Agriculture–Forest Service, Washington, D.C.

Mayer, P. W., W. B. DeOreo, E. M. Opitz, J. C. Kiefer, W. Y. Davis, B. Dziegielewski, and J. O. Nelson. 1999. *Residential End Uses of Water*. AWWA Research Foundation, Denver, Colorado. 86 pp.

Mugaas, R. J., M. L. Agnew, and N. E. Christains. 2005. *Benefits of Turfgrass*. Extension Service Publication BU 5726. University of Minnesota, Minneapolis–St. Paul.

Saving Water Partnership (SWP). 2007. *Seattle Participating Area Water Utilities: Sprinkler Rebates Overview,* http://savingwater.org/outside_sprinklers.htm (11 January 2007)

Sempik, J., J. Aldridge, and S. Becker. 2005. *Health, Well-being and Social Inclusion Therapeutic Horticulture in the UK*. The Policy Press, Bristol. 138 pp.

Vickers, A. 2001. *Handbook of Water Use and Conservation: Homes, Landscapes, Businesses, Industries, Farms*. Water Plow Press, Amherst, Massachusetts. 446 pp.

Texas Water Development Board (TWDB). 1999. *Water Price Elasticities for Single-Family Homes in Texas*. Stratus Consulting Co., Boulder, Colorado.

Youngner, V. B. 1970. Landscaping to protect homes from wildfire. *California Turfgrass Culture* 20(4):28–32.

16

Best Management Practices for Turfgrass Water Resources: Holistic-Systems Approach

Robert N. Carrow and Ronny R. Duncan

The Real Issue

According to the Environmental Literacy Council (ELC 2005), environmental knowledge and practice necessary to address environmental problems positively "requires a fundamental understanding of the systems of the natural world, the relationships and interactions between living and the nonliving environments, and the ability to deal sensibly with problems that involve scientific evidence, uncertainty, and economic, aesthetic, and ethical considerations." Or, to put this in the context of addressing turfgrass water use and quality concerns positively, perceived environmental problems must not be addressed in isolation, but in terms of all interrelationships and stakeholders associated with these landscapes.

Green spaces can have detrimental effects on the environment, just as an agricultural enterprise or a factory or urban hardscapes can, but green spaces also contribute to society and the local community through environmental, recreational, aesthetic, and economic benefits (Beard and Green 1994; Carrow 2006; CAST 2002, 2008). Perennial grasses are significant contributors to overall environmental stewardship, which encompasses conserving, maintaining, and improving our natural resources to ensure sustainability of air, soil, water quality/quantity, climate, natural ecosystems, energy, and endangered species.

Based on a holistic mindset as defined by the ELC (2005), the real question or concern becomes, "What is the best approach for achieving water quantity and quality stewardship" within the nation, states, watershed/basin, community, specific sites, and regulatory realms that impact green landscapes? Or, in more concise terms, "What approach will maximize turfgrass benefits while minimizing potential environmental problems?" The answer to these questions has profound implications for all direct and indirect stakeholders influenced by green spaces.

For protection of surface and subsurface water quality from pesticides, nutrients, and sediments, an excellent model has evolved over the past 35 years in the form of "best management practices" (BMPs) fostered by the U.S. Environmental Protection Agency (EPA) Clean Water Act (Rawson 1995; USEPA 2005a). For landscape water-use efficiency/conservation and for protection of water resources from irrigation water constituents, however, a widely adopted or consensus approach has not evolved that is integrated into the regulatory realms and site-specific landscape levels. A critical first step in addressing societal concerns related to these issues is to develop a successful, accepted approach. Certain characteristics have made the BMPs approach for protection of water quality from pesticides, nutrients, and sediments the premier means of dealing with this complex environmental problem. Understanding these characteristics is crucial to understanding how this science-based approach can be adopted as a model for other environmental issues, including water-use efficiency/conservation and water quality concerns related to irrigation water constituents.

Best Management Practices

History

The first federal initiative stating the term "best management practices" came from the 1977 amendment to the Clean Water Act (CWA), which established BMPs as soil conservation practices to protect water quality (Gold 1999). The BMPs focused on a holistic, systems approach that addressed concerns for pesticides, nutrients, and sediments as related to water quality protection and has culminated in comprehensive regulations and supporting BMPs within agriculture (USEPA 2003) and urban landscapes (USEPA 2005a).

In addition to BMPs for protection of water quality, other systems approaches to alleviate environmental problems have proved to be effective, such as:

- **Integrated Pest Management (IPM)** approach was developed in the late 1960s and early 1970s in response to how best to develop science-based pest control strategies that could include judicious use of pesticides, but within a system of pest control through other means—cultural, pest-resistant plants, or pest predators. In 1972, the U.S. Department of Agriculture funded the first major IPM research effort.

- **Sustainable Agriculture** was formalized in 1985 with the Food Security Act. It was another milestone in the whole-systems approach to addressing multiple environmental problems (Gold 1999). This was enhanced in 1988 by funding of the Low-Input Sustainable Agriculture (LISA) program. The LISA concept was expanded in 1990 to become the Sustainable Agriculture Research and Education Program (SARE).

- **Precision Agriculture**, although not a whole-systems approach, does highlight critical components, including the concepts that inputs should be applied only on the site where they are needed, at the rate required, and only when needed, and that site-specific information is the basis for site-specific management. It recognizes the great spatial variability that must be dealt with when managing a site—and illustrates why cultural practices must be based on educated, site-specific decisions.

Characteristics

Although IPM (pesticides), sustainable agriculture (soil quality, water issues, air quality, etc.), and precision agriculture (efficient use of inputs) concentrate on somewhat different environmental aspects than the BMP focus on water quality protection, all these approaches have certain inherent, common characteristics that are essential to achieve successful environmental stewardship (ELC 2005). These characteristics are as follows:

- **Science-based.** All are science-based and have inherent, foundational principles involving application of inputs only on the site where needed, when necessary, and only at the quantity required. The very definition of BMPs illustrates why this approach is effective: (1) "best" is used to imply the best combination of strategies that can be adopted on a site or for a particular situation with current technology and resources; (2) "management" denotes that environmental problems must be managed, and that management decisions by trained personnel can maximize success; and (3) "practices" implies that multiple strategies are necessary to make a positive difference. Thus, whether called a BMP, IPM, or SARE, all emphasize wise and efficient use of resources using a science-based and flexible philosophy. These approaches can be documented, and accountability can be monitored.

- **Holistic or whole-systems-based.** These approaches recognize that no "silver-bullet," or single practice, can achieve successful stewardship with regard to a specific environmental problem because we work within whole dynamic ecosystems. In contrast, rigid regulations (or command-and-control approach) are based on limited strategies and a "one-size-fits-all" concept, ignoring the principle that successful environmental stewardship must consider interactions among ecosystem components (ELC 2005). The ecosystem includes soil, plant/landscape, atmosphere/climate, turfgrass manager's expertise, irrigation system, irrigation water, precipitation/stormwater, surface/subsurface waters, hydrology, the positive/negative impacts that any practices have on all stakeholders, and any other related aspects.

- **Holistic in considering all stakeholders and implications relative to potential environmental and economic effects.** The holistic and multiple-stakeholder dimensions as components of the CWA are noted by: "Evolution of CWA programs over the last decade has also included something of a shift from a program-by-program, source-by-source, pollutant-by-pollutant approach to more holistic watershed-based strategies. Under the watershed approach equal emphasis is placed on protecting good quality waters and restoring impaired ones. A full array of issues are addressed, not just those subject to CWA regulatory authority. Involvement of stakeholder groups in the development and implementation of strategies for achieving and maintaining state water quality and other environmental goals is another hallmark of this approach" (USEPA 2006). The latter aspect notes that practices focused on a single environmental goal may result in unintended, adverse environmental consequences (CAST 2008).

- **Educated site-specific choices and management.** Because there is no single factor that will achieve maximum environmental benefits on a site, adjustments within the whole ecosystem are the basis of the BMPs model, and educated decision-making is important. BMPs encourage professionalism and education of the turfgrass manager, including continuing education. Each site is different, and adjustments, therefore, must be site-specific and account for system changes over time. Also, regional differences in climate and soil will modify site-specific BMPs.

- **Fosters entrepreneurial development and implementation of new technology and concepts.** BMPs encourage ongoing integration of new technology, plants, concepts, and products to achieve the "best" practices. Guideline templates can be developed and updated over time.

BMPs for protection of water quality are at multiple levels, starting at the federal level with the CWA, but also at state, regional, watershed, urban, and site-specific levels (FDEP 2002; EIFG 2006; USEPA 2005a). For perennial grass landscapes, the site-specific levels may be home lawns, general grounds, seed or sod production farms, parks, golf courses, or other areas using grasses. At the site-specific level, the BMPs model is exhibited in the diversity of state IPM and BMPs programs for different landscapes (UFL 2006; UGA 2006). It is important that the site-specific BMPs or IPM approaches maintain their multiple-strategy, science-based nature, rather than reverting to a mentality of banning (rigid regulations, command and control) pesticides or nutrients. Instances at the state or local levels involving political pressures for a command and control approach have occurred and will likely continue to occur. But the vast majority of EPA and state regulatory agencies have recognized that long-term, successful ecosystem management for protection of water quality must be based on incorporation of all stakeholders in a positive, interactive, and participatory (true partnership) manner.

Interestingly, the EPA recently has attempted to avoid the BMPs notation in favor of a rather neutered term "management practices or measures" (USEPA 2005a), citing as its reason that "The word 'best' has been dropped…because the adjective is too subjective. The 'best' practices in one area or situation might be entirely inappropriate in another area or situation." The initial and long-term meaning of BMPs, however, traditionally has been to denote the best combination of practices to resolve water quality issues on an area or specific situation. It never was meant to identify a single "best" practice. Therefore, we strongly prefer the continued use of BMPs rather than "management practices," which could imply good or bad practices relative to the particular issue.

BMPs Applied to Other Water Issues

Three interrelated water issues arise in urban perennial landscapes, and all can be addressed by a BMPs approach with the characteristics as defined in the previous section:

- BMPs for Protection of Water Quality

- BMPs for Irrigation Water Quality Management

- BMPs for Landscape Water-Use Efficiency/Conservation

As noted, the "BMPs for Protection of Water Quality" concept arose out of water quality concerns, and the traditional focus of BMPs has been on protection of surface and groundwater quality from applied nutrients, pesticides, and sediments. Considerable progress has been made toward landscape BMPs in this area at the national level (USEPA 2003, 2005a) and within the turfgrass industry (Cohen et al. 1999; Dodson 2005; EIFG 2006; FDEP 2002). Because BMPs in this arena are more developed than in other areas, the current paper will focus on the remaining two areas.

A significant development in recent years has been the BMPs terminology and concept being adopted and expanded into the water conservation area (Carrow, Duncan, and Waltz 2005; Carrow, Duncan, and Wienecke 2005a,b; Cathy 2003; CUWCC 2005; EIFG 2006; GreenCO and Wright Water Engineers 2004; IA 2005). The increasing inclusion of the BMPs concept/terminology into ordinances, regulations, and management manuals to deal with all water issues is a significant step toward defining a unified, science-based approach. The BMPs terminology/concept likely will be used for an expanded array of environmental concerns beyond water issues, such as for soil quality/health and wildlife protection within the turfgrass industry, as well as within the regulatory arena.

BMPs for Irrigation Water Quality Management

Decreased quantities of available potable water combined with increasing domestic demand emphasize the need to irrigate with recycled or other nonpotable alternative water resources of lesser quality relative to potable sources. Use of alternative irrigation water sources, rather than potable water supplied by a municipal water treatment system, is not a new practice to many large turfgrass areas and now is becoming the normal practice in many areas as competition for potable water increases (because of population increases and demand) and as grasses are developed that can tolerate much poorer water quality (Carrow and Duncan 1998; Duncan and Carrow 2000; Harivandi, 1991; Huck, Carrow, and Duncan 2000; Marcum 2005; Pettygrove and Asano 1985; Snow 1994; Thomas et al. 1997).

The umbrella terms of nonpotable and alternative irrigation water sources include a diversity of sources—e.g., brackish or saline surface or groundwater, reclaimed, recycled, stormwater, gray water, harvested water, or any other water source that is nonpotable. Specific water quality concerns often are associated with particular irrigation water sources (AWA 2000; Ayers and Westcot 1994; Pettygrove and Asano 1985; Snow 1994). Each source may exhibit chemical, biological, or physical constituents that can challenge landscape plant performance in the short term and require specific cultural practices for long-term environmentally safe use. The most prevalent constituents in many alternative water sources, which often are higher in concentration than found within potable sources, are soluble salts and nutrients, but many biological, physical, or chemical contaminants are possible depending on the source, such as the following:

- **Biologicals**—human pathogens, plant pathogens, algae, cyanobacteria, iron and sulfur bacteria, nematodes, weed seed
- **Physical contaminants**—total suspended mineral or organic solids, turbidity, color, temperature, odor
- **Chemical constituents**—total soluble salts, specific salt ions, nutrient ions, potential root or foliage toxic/problem ions, metal and trace ions, total dissolved solids, alkalinity, oxygen status, biodegradable organics, nonbiodegradable (refractory or resistant) organics, free chlorine residual, hydrogen sulfide gas

Irrigation water constituents as potential pollutants logically would seem to come under the "BMPs for Protection of Water Quality" area. In much of the literature, however, the emphasis is on irrigation practices as they may affect runoff or drainage water, and not on irrigation water constituents as a potential contributor to pollutants (Barton and Colmer 2005; USEPA 2003, 2005a). Irrigation water constituents can be very diverse, and quality guidelines have evolved that incorporate environmental, health, and agronomic considerations (AWA 2000; Ayers and Westcot 1994; Carrow and Duncan 1998; Yiasoumi, Evans, and Rogers 2003). Additionally, development of halophytic (salt-tolerant) grasses has allowed the use of poorer-quality water than previously used for agronomic or turfgrass situations, and maintenance strategies for managing salinity in the ecosystem and in adjusting management to these new grasses have become a priority (Duncan and Carrow 2000).

Depending on the chemical, physical, and biological characteristics of the irrigation water, the problem that confronts the landscape manager may occur at different points within the spectrum of water movement: from the initial source location, on-site storage, delivery system, turfgrass plant, soil profile, runoff areas, and underlying geo-hydrology. There may be multiple water quality challenges that can occur within the hydrological cycle on a particular site, not only from the irrigation water source, but also from other hydrological aspects such as tidal influences, water table depth and fluctuations, and stormwater flooding or surges. BMPs must be developed that encompass all possible problems and are sustainable for water, soils, and aquatic/wetland systems across the spectrum of water movement. An in-depth treatment of irrigation water quality issues across the whole water delivery spectrum using a BMPs approach is under way and slated for completion by spring 2008 (Duncan, R. R., R. N. Carrow, and M. Huck. Personal communication).

General management protocols are reasonably well developed in terms of overall concepts for saline irrigation water uses in agriculture and for turfgrass landscapes (Carrow and Duncan 1998; Hanson, Grattan, and Fulton 1999; Marcum 2005; Oster 1994; Rhoades, Kandiah, and

Mashali 1992). But more detailed BMPs need to be developed and presented in a BMPs format for perennial grass landscapes in urban areas. With more saline irrigation water being used on turfgrass sites, it is essential that potentially detrimental effects of salinity loading, accumulation, or movement in the environment be mediated by sound, integrated BMPs (Carrow and Duncan 1998; Duncan and Carrow 2000; FAO 2005).

Irrigation on landscape sites with reclaimed waters has received increasing attention as pressure for water conservation and water-use efficiency increases. Problems associated with reclaimed irrigation water have received extensive discussion (Bond 1998; Carrow and Duncan 1998; Duncan, Carrow, and Huck 2000; Harivandi 1991; Pettygrove and Asano 1985; Scott, Faruqui, and Raschid-Sally 2004; Snow 1994; Stevens et al. 2004; Thomas et al. 1997; USEPA 2004; WHO 2005). As with saline irrigation water sources, more in-depth BMPs protocols to deal with specific problems need to evolve and be targeted to urban landscape sites using perennial warm- and cool-season grasses.

BMPs for Water-Use Efficiency/Conservation on Specific Sites

As previously noted, the BMPs approach recently has been applied to water-use efficiency/conservation (Carrow, Duncan and Wienecke 2005a, b; CUWCC 2005; EFIG 2006; GreenCO and Wright Water Engineers 2004; IA 2005; Vickers 2001). In this section, the focus is on BMPs for water-use efficiency and conservation on a site-specific basis, especially for larger turfgrass landscapes such as parks, seed- and sod-production farms, golf courses, and large business grounds. In the next section, additional components of BMPs programs for community, regional, or watershed level water-use efficiency/conservation will be addressed.

At this point, the urban landscape industry cannot assume that environmental and water regulatory personnel understand the full scope of BMPs for water conservation, because BMPs terminology only recently has been applied to turfgrass water conservation in the regulatory realm. For example, it is not unusual for individuals or groups to view "turfgrass water conservation" as involving only one or two strategies—i.e., change the grass species, use only native grasses, decrease the area of irrigated turfgrass, improve irrigation design, Xeriscape™, or use weather-based means (evapotranspiration) for irrigation scheduling. BMPs for turfgrass water conservation, however, must be defined to include the widest set of potential strategies and not be limited to only one or two. Therefore, it is important to develop a consistent understanding of BMPs related to turfgrass water-use efficiency/conservation so that confusion does not arise.

One important BMPs aspect is to maintain the emphasis on inclusion of all stakeholders. Water conservation programs should include consideration of practices on water-use efficiency, the economy, environment (other environmental influences or unintended adverse environmental effects), jobs, and specific long-term site use. The customer, or user/manager/owner of a site, is not the only stakeholder potentially affected by water conservation measures. Others include the supply side (water authorities, suppliers); demand side (site user, site manager, agriculture industry, etc.); and those affected by environmental and economic water conservation measures (society in general, local economy, health related aspects, impact on soil quality, sustainability) (Beard and Green 1994; Carrow 2006; CAST 2002, 2008; Gibeault 2002). The importance of avoiding the use of water conservation as the sole determination when considering a BMPs plan is illustrated by the EPA water conservation plan guidelines for water system planners presented in Table 16.1 in which multiple considerations are noted (USEPA 1998). Similar considerations afforded to the public utilities realm should be included in a site-specific BMPs plan. Proponents of rigid regulations (command and control) for water conservation often give little attention to those factors that can affect all stakeholders.

Table 16.1. Criteria that can be used by water systems planners in selecting conservation measures for implementation on a community-wide or watershed basis (USEPA 1998) (Illustrates multiple considerations are required and not just a water conservation target.)

Program costs	Environmental and social justice
Cost-effectiveness	Water rights and permits
Ease of implementation	Legal issues and constraints
Budgetary consideration	Regulatory approvals
Staff resources and capability	Public acceptance
Environmental impacts	Timeliness of savings
Ratepayer impacts	Consistency with other programs

Carrow, Duncan, and Waltz (2005) in their BMPs workbook have defined "BMPs for turfgrass water conservation" when applied to a specific site as involving three primary activities: (1) Site Assessment and Planning—information gathering and planning aspects for the entire ecosystem; (2) Identify, Evaluate, and Select Water Conservation Options—options are all within the ten core water conservation strategies; and (3) Assess Benefits and Costs—of water conservation measures on all stakeholders (Table 16.2). These activities are presented in the following sections as an initial template to develop more detailed BMPs documents for water-use efficiency/conservation that are holistic and science-based.

Table 16.2. Outline of the planning process and components of a golf course BMPs for water-use efficiency/conservation

A. **Initial planning and site assessment.**
   1. Identify water conservation measures that already have been implemented by a golf course, including costs of implementation—this initial step aids in clarifying for the golf course management team and club members exactly what is entailed in BMPs water conservation measures. Also, when the final document/program is shared with regulatory agencies, this information is very valuable in pointing out that golf courses are not starting from "zero" in this arena but have been implementing BMPs for many years.
   2. Determine the purposes and scope of the site assessment. Site assessment is necessary to determine the best options for the specific golf course.
   3. Site assessment and information collection.
      - Current water-use profile
      - Irrigation/water system distribution audit
      - Additional site infrastructure assessment information—evaluation of alternative irrigation water sources; golf course design modifications; irrigation system design changes; microclimate soil/atmospheric/plant conditions affecting irrigation system design/zoning/scheduling; drainage needs for leaching of salts or any hydrological considerations that may arise from use of any particular irrigation water source
   4. Determine future water needs and identify an initial water conservation goal.

B. **Identify, evaluate, and select "water conservation strategies" and options.**
   1. Selection of turfgrasses and other landscape plants.
   2. Use of nonpotable water sources for irrigation—alternative water sources; water harvesting/reuse; water treatment if necessary.
   3. Efficient irrigation system design and devices for water conservation.
   4. Efficient irrigation system scheduling/operation. Both irrigation system design and irrigation scheduling in the future will require much more site-specific information. Sensor technology integrated into a Geographic Information Systems/Global Positioning Systems approach will assist in development and interpretation of information for improved irrigation system distribution efficiency and scheduling.
   5. Golf course design for water conservation.
   6. Altering management practices to enhance water-use efficiency—soil amendments, cultivation, mowing, fertilization, etc.

Continued on next page

Table 16.2. *(continued)* Outline of the planning process and components of a golf course BMPs for water-use efficiency/conservation

> 7. Indoor water conservation measures in facility buildings. Conservation strategies for landscape areas other than the golf course and immediate facilities.
> 8. Education. Plan for initial and continuing education on water conservation/management by golf course superintendent, support crew, club officials, etc. BMPs for turfgrass water conservation is complex, and when poor irrigation water quality is involved the costs and level of management complexity greatly increases—i.e., fertilization, leaching of salts, salt disposal/hydrological issues, complex irrigation systems and scheduling of irrigation.
> 9. Development of conservation and contingency plans. A formal BMPs document should be developed and agreed on by all club officials and members so that the golf course superintendent has support for any reasonable science-based measures to be taken. Also, a written plan may be required by regulatory agencies.
> 10. Proactively monitor and revise plans.
>
> **C. Assess benefits and costs of water conservation measures on all stakeholders.**
>
> Assessment of costs and benefits associated with development and implementation of a long-term BMPs water conservation plan is necessary not only for facility planning, but also to demonstrate to regulatory agencies and possible critics of golf courses that substantial effort and cost previously has been involved in water conservation by the facility.
>
> 1. Benefits
>   - Direct and indirect to the owner/manager and site customers
>   - Direct and indirect to other stakeholders, including water savings but also other benefits—economic, environmental, recreational, etc.
> 2. Costs
>   - Facilities costs for past and planned implementation of water conservation strategies—irrigation system changes, water storage, pumping, new maintenance equipment, water/soil treatments, course design alterations, water harvesting, storage
>   - Labor needs/costs
>   - Costs associated with changes in maintenance practices; different irrigation water sources (water treatment, soil treatment, storage, posting)
>   - Costs that may impact the community if water conservation strategies are implemented (especially mandated ones) such as revenue loss, job loss

Site Assessment and Planning

On complex turfgrass areas, such as golf courses with numerous microclimates, development of an effective water-use efficiency/conservation BMPs program is very complicated, time consuming, and often costly—in contrast to many other urban sites such as home lawns. The initial planning starts with identification of water conservation measures that already have been implemented by a golf course, including estimated costs of implementing these practices and possibly an estimation of the level of improvement in water-use efficiency on the site that arises from these practices, both individually and as a whole. This initial step helps clarify for the landscape management team and site owners exactly what is entailed in BMPs water conservation measures. Also, when the final document/program is shared with regulatory agencies, this information is valuable in pointing out that most landscape sites are not starting from zero in this arena, but have been implementing BMPs for many years at considerable cost and effort with little formal documentation. This information should be positioned in the front of the BMPs document developed for a specific site. A few examples of common water conservation strategies already in use on many recreational sites are the following:

- In many warm-season turfgrass areas, bermudagrass is the most widely used grass and it happens to be one of the most drought-resistant species.

- During severe water shortages, allowing selected turfgrass areas to go dormant and not receive any irrigation except survival of the grass cover to protect against soil erosion where needed.

- Water sources on a site such as a golf course may include stormwater harvesting of rainfall from the surrounding area and collection into irrigation ponds, or the use of reclaimed water as an alternative irrigation water source.

- Soil modification to improve water infiltration/percolation and deeper rooting, and on U.S. Golf Association golf greens, construction of a perched water table can be used enhance water conservation.

- Turfgrass cultivation programs and equipment to improve water infiltration/percolation and to enhance rooting.

- Higher-mowed areas with limited or no irrigation on a routine basis.

- Irrigation systems zoned to improve water distribution efficiency and aid in efficient scheduling.

- Irrigation scheduling programs based on local plant water requirements determined by a combination of turfgrass manager experience and on-site weather data.

- Educational training specific to water management for turfgrass managers and support staff. Community educational efforts have proved effective for the general public (CAST 2008).

Next, the purposes and scope of the initial site assessment phase should be determined. Site assessment is necessary to provide information to determine the best options (i.e., BMPs) for the specific landscape area. Site assessment and information collection often entail (1) determining the current water-use profile, (2) conducting an extensive irrigation/water systems audit, and (3) obtaining additional site infrastructure assessment information. That information includes an evaluation of alternative irrigation water sources; landscape design modifications; irrigation system design changes; microclimate soil/atmospheric/plant conditions affecting irrigation system design/zoning/scheduling; and drainage needs for leaching of salts or any surface/subsurface geo-hydrological considerations that may arise from use of any particular irrigation water source. Gathering information related to infrastructure changes often involves considerable time and costs. Thus, development of a BMPs water conservation plan may require more than a year on some sites, especially when alternative or multiple irrigation water sources must be identified, when the irrigation water is of initial poor or changing quality, when the irrigation distribution system is not efficient, and/or when major landscape design changes must be made. Multiple years also are normal for implementing required infrastructure changes.

Finally, future water needs should be determined, and an initial, realistic water-use efficiency/conservation goal should be identified. As implied by the process of gathering site-assessment information, plans may require flexible adjustment as new information becomes available because the entire ecosystem is dynamic and not static. But initially establishing a realistic water-use efficiency/conservation goal based on projected water needs is a necessary step. In instances where saline irrigation water is used, projected water needs must include an adequate leaching fraction to avoid soil degradation by salinization.

Identify, Evaluate, and Select Water Conservation Options

This is the stage where hard decisions must be made within the "Ten Core Water Conservation Strategies." Within each of these strategies, numerous options are available as noted in greater detail by Carrow, Duncan and Waltz (2005), Cathy (2003), the CUWCC (2005), GreenCO and Wright Water Engineers (2004), and the IA (2005). The choices are site-specific based on the water quantity requirements and conservation goals, expectation of the facility management and local governance, and actual resource requirements and availability. Essentially all major water conservation options can be classified under one of the following ten core water conservation strategies:

1. **Use of nonpotable water sources for irrigation—alternative water sources; water harvesting/reuse.** The decisions or choices associated with this strategy can become very costly or difficult, such as water quantity issues (multiple water sources, reliability over time, permitting, blending, storage, piping water to the location) and water quality issues (water treatment, soil amendments, changes in nutritional programs, leaching ability, salt disposal, effects on subsurface hydrology, drainage) (Duncan, R. R., R. N. Carrow, and M. Huck. Personal communication).

2. **Efficient irrigation system design and monitoring devices for implementing water conservation.** Items included in this strategy could be low-flow sprinklers in critical areas, adjustable heads, proper spacing of heads and nozzles, strategic placement of soil moisture and salinity sensors, as well as many other considerations. Upgrade or repair of any leakage areas, proper delivery system adjustment, and maintenance protocols also would be included in this category.

3. **Efficient irrigation system scheduling/operation.** Both irrigation system design and irrigation scheduling in the future will require much more site-specific information—i.e., a precision agriculture approach. Sensor technology integrated into a Global Positioning System/Geographical Information System approach will assist in development and interpretation of information for improved efficiency in irrigation distribution and scheduling.

4. **Development and selection of turfgrasses with respect to water uptake and utilization requirements in terms of quantity and quality.** Because lower-quality irrigation water may be used, many plants will require not only drought resistance but also multiple genetic-based stress tolerances, such as salinity, traffic, and cold and heat tolerance, across all turfgrass species used for permanent or overseeded grasses.

5. **Landscape design for water conservation.** Design for water harvesting; decreasing unnecessary acreage of highly maintained, closely mowed, irrigated turfgrass areas; avoiding excessive mounds or slopes; inclusion of nonirrigated turfgrass areas; and allowing for very limited or no irrigation on certain sites during water shortages.

6. **Altering practices to enhance water-use efficiency.** Some considerations are soil profile amendments, cultivation programs and equipment needs, mowing, fertilization, and chemigation. Maintenance of deep root systems is especially important to allow for deep and less frequent irrigation application; deep root systems favor improved capture and storage of rainfall to replace or delay irrigation events. Practices to enhance soil infiltration, percolation, and soil moisture retention are key options, as well as judicious use of wetting agents to enhance water infiltration and uniformity of percolation.

7. **Indoor water conservation measures in buildings, air conditioning units, pools, and other facilities associated with a landscape site.** Water conservation will not be a reality on some sites if it is confined to only the actual landscape area. Instead, it will be viewed as the responsibility of the turfgrass or landscape manager, and not as a policy or philosophy by the site owners, whether privately or publicly owned. Application of water conservation practices on a facility-wide basis—such as parks, large business grounds, sports complexes, or golf courses—should involve all facility owners/managers and site users.

8. **Education.** Complex issues require educated, science-based decision making. Planning for initial and continuing education on water conservation/management is essential for landscape managers, support crew, and facility officials with direct communication to state,

regional, and local water regulatory officials. BMPs for turfgrass water conservation are complex, and when poor irrigation water quality is involved, the level of infrastructure and maintenance costs and management complexity greatly increase. Fertilization, cultivation, leaching of salts, salt disposal/hydrological issues, complex irrigation systems, and scheduling of irrigation are just some of the complex issues involved.

9. **Development of formal conservation and contingency plans.** A formal BMPs document should be developed and agreed on by all facility officials so the landscape manager has support for any reasonable, science-based measures undertaken. Also, a written plan may be required by regulatory agencies. This should be an ongoing, flexible, and realistic plan subject to revision over time. Additionally, the components should be integrated into daily operation of the club or facility activities, implemented as routine practice, and subsequently documented for progress in achievement of the targeted goals. Previously, we noted that a rigid regulation approach to water-use efficiency/conservation (or any other environmental issue) is much less desirable for all stakeholders compared with a BMPs approach. A more positive regulatory approach is to foster BMPs for water conservation. For example, a governmental unit may require that managers of larger landscape areas develop and implement BMPs. Additionally, during a water shortage crisis, more rigid regulations are often necessary for all water users, but should be avoided as the long-term or primary means to deal with environmental issues. In the matter of water quantity, a state, region, watershed, or community may go incrementally into a series of increasingly restrictive water-use regulations during a prolonged water shortage. Normally, there are triggers for each step, such as a reservoir level, and all water users are affected by the restrictions.

10. **Monitor and revise plans.** Proactive monitoring is essential and may involve sensor technology on-site or sample acquisition and testing off-site. Regularly scheduled monitoring of specific conservation effectiveness, and of the overall BMPs plan, is essential for achieving goals and making effective adjustments. Flexibility in short- and long-term plan implementation is critical because climatic changes are major, uncontrolled variables.

Assess Benefits and Costs of Water Conservation Measures for All Stakeholders

Assessments of costs and benefits associated with developing and implementing a long-term BMPs water conservation plan are necessary not only for facility planning, but also for demonstrating to regulatory agencies and possible critics of perennial, urban landscapes that substantial efforts and costs in water conservation have been documented by the facility. Readers are encouraged to review the papers by Beard and Green (1994), Carrow (2006), CAST (2008), and Gibeault (2002) for information on economic, recreational, environmental, and other social benefits of turfgrasses to direct and indirect stakeholders. BMPs documents should define or at least list the benefits of the particular landscape facility, especially to indirect stakeholders who may not be aware of the benefits the turfgrass/landscape industry contributes to the local, regional, or state society.

BMPs for Water-Use Efficiency/Conservation on a Watershed or Community Basis

In addition to the components of a site-specific BMPs program, other practices can be used on a watershed or community basis to foster water-use efficiency/conservation. Some of these practices may be regulatory in nature whereas others are voluntary. An excellent example of a successful community-wide BMPs program for San Antonio, Texas (CAST 2008) is presented in Section 15 of this publication. Vickers (2001) and the U.S. Environmental Protection Agency (USEPA 1998) present good overviews of water conservation measures that may be used. Pricing for water conservation, consistent public outreach education efforts, and reasonable regulations to limit water waste are especially conservation-effective for sites without a professional turfgrass manager.

One aspect of turfgrass sites often not considered in relation to water-use efficiency/conservation is that turfgrasses can be allowed to go semi- or completely dormant. In fact, in most locations in the United States, both cool- and warm-season grasses naturally go dormant in the cold-season months. Perennial grasses also can be allowed to go dormant in water-shortage periods as part of a water conservation plan (Wade et al. 2003). In 2007 within San Antonio, lawn grasses for new home sites were required to be capable of surviving 60 days of drought (CAST 2008). Important aspects of drought-resistant dormant turfgrass include (1) irrigation is not needed; (2) pesticide and nutrient applications are not used during water-induced dormancy, yet the cover remains to prevent soil degradation by erosion, to limit sediment movement, and to foster rain infiltration when it occurs; and (3) dormant grass is not dead grass, so the groundcover can be regenerated when the water shortage is less severe.

Integration of BMPs

Stacking together several complex management issues is a challenge that will become more commonplace, especially on sites with a combination of poor irrigation water quality, water restrictions/conservation, and more salt-tolerant turfgrass and landscape species. Protection of water resources from pesticides, nutrients, and sediments, as outlined by the EPA (2003, 2005a) and the Florida Department of Environmental Protection (2002), is the first complex challenge. Second, increased emphasis on stormwater management in urban settings has resulted in more active attention to this issue, with many sites requiring a stormwater management plan (CASQA 2003). A third issue is cultural and irrigation practices for optimum water-use efficiency/conservation and turfgrass performance, which requires a systems or holistic BMPs approach with proactive monitoring and frequent adjustments in practices that influence water-use efficiency (Carrow, Duncan, and Waltz 2005; Carrow, Duncan, and Wienecke 2005a,b; Cathy 2003; CUWCC 2005; GreenCO and Wright Water Engineers 2004; IA 2005). A fourth complex management challenge arises from the quality of irrigation water. BMPs for salt-affected sites where the irrigation source is a major contributor of salt load are essential to avoid negative accumulation impacts on the entire ecosystem—soil, water, and plants (Carrow and Duncan 1998; FAO 2005; Oster 1994). Reclaimed water irrigation sources may or may not be high in total soluble salts, but generally contain higher levels of nutrients than domestic water sources (Bond 1998; Huck, Carrow, and Duncan 2000; Scott, Faruqui, and Raschid-Sally 2004; Stevens et al. 2004; Thomas et al. 1997). Proactive monitoring of soil, water, and plants should become more frequent in dynamic saline or reclaimed water situations to manage salt levels and nutrient status adequately. Poor irrigation water quality may necessitate a change in grass species or cultivar, which presents additional long-term maintenance adjustment challenges for the turfgrass manager, especially in terms of managing salt loading in soils and in budgeting for this dynamic continuum.

Therefore, when water conservation pressures increase to the point at which lower-quality irrigation waters are used, turfgrass management becomes more complex. As individual BMPs for water conservation, ecosystem salinity management, turfgrass nutritional programs, and new salt grass additions all interface—each complex in its own right—they face markedly increased challenges. Turfgrass managers of the future must become whole-systems (holistic) managers, with the ability to understand and apply multiple BMPs for site-specific water use, water quality, new grasses, fertilization, and other site-specific management aspects.

As more turfgrass sites use poorer water quality, turfgrass managers and facility owners must address the above challenges of salinization prevention, multiple water quality problems involving the hydrological cycle on a site, and the stacking of multiple, complex BMPs. The Council for Agricultural Science and Technology (CAST 2002) has summarized many of these environmental challenges within urban areas. Currently, the most comprehensive treatment of integrated environmental issues in the perennial, urban landscape has been by Audubon International (Dodson 2005).

In recent years, the EPA (2005b) has been promoting the Environmental Management Systems (EMS) approach to deal with multiple environmental concerns on a site, not just in agriculture, but across all entities that may have an environmental impact. The EPA (2005b) defines an EMS as "a set of processes and practices that enable an organization to decrease its environmental impacts and increase its operating efficiency. An EMS is a continual cycle of planning, implementing, reviewing, and improving the processes and actions that an organization undertakes to meet its business and environmental goals." This is a program in which plans developed to deal with environmental concerns are integrated into normal, daily operation of the organization at all management levels. Plans must be in accord with current environmental regulations, but the EMS is voluntary in nature.

Within the relatively near future, the authors anticipate that the integration of management protocols to address multiple environmental concerns, including the water quality and quantity issues addressed in this CAST Special Publication, will require an EMS approach on many sites. A component of the planning phase is to assess all potential environmental concerns on a site and then to develop and implement plans to minimize environmental impacts. Positive aspects of this approach for the turfgrass industry include the following:

- EMS is for all entities, public or private, that may have potential environmental impacts. Thus, it is not targeted toward a single industry.

- The EMS approach brings under one umbrella all environmental issues on a site. When a single issue is targeted by a group (e.g., water conservation) toward the turfgrass industry or a single facility, it is not uncommon for the only determination of success to be the decrease in water use without any consideration for economic/job or unintended environmental consequences. Within an EMS, all environmental issues are combined. Thus, potential adverse effects must be addressed. For example, the method to decrease water use may be to remove turfgrass acreage, but in an EMS approach the issues of soil degradation (wind and erosion loss, decreased organic addition to soils), human health effects from dust, and adverse effects of decreased grass surface on water infiltration, stormwater movement, and sediment movement must be addressed within the same EMS. Additionally, a basic premise of EMS is to consider "operation efficiency" or business impacts.

- EMS can be developed by stacking together the BMPs for each environmental issue of concern for the site. By using the BMPs model for each environmental concern on a site, the development of an EMS is simply an extension and integration of BMPs and not a whole new system or paradigm change.

Conclusion

The BMPs approach developed over the past 35 years by the EPA for protection of surface and subsurface waters from pesticides, nutrients, and sediment has a long track record for being successfully implemented because of several critical characteristics. It is science-based; incorporates all strategies in the ecosystem (holistic); embodies all stakeholders and their social, economic, and environmental concerns; values education and communication outreach; allows integration of new technologies and concepts; has been applied at the regulatory, watershed, community, and site-specific levels, as well as in educational realms; and maintains flexibility to adjust to new situations. Thus, this BMPs model is the template for dealing with other complex environmental issues.

The authors encourage adoption of the BMPs model with the previous characteristics for other water-related issues involving the turfgrass situations, such as water-use efficiency/conservation and irrigation water constituents. Adoption would have the following primary benefits:

- A basic, realistic approach to achieving water-use efficiency/conservation and management of irrigation water constituents will allow the turfgrass and landscape industries to go forward in a positive and unified manner to develop sound BMPs for these environmental issues.

- The BMPs model has all the characteristics necessary to resolve these complex environmental issues. Adoption of a BMP approach by various facets of the turfgrass industry for water issues would be an excellent environmental model and demonstrate a high degree of environmental stewardship.

- When confronted with pressures for rigid regulations that do not include the essential characteristics of the BMPs approach, those who have adopted and implemented BMPs programs would be able to show due diligence in these areas and to demonstrate their approach as being the best science and practical model to resolve complex environmental issues.

- Development of BMPs for each specific water-related problem would allow combining the BMPs into an EMS document and management style in the future.

- The BMPs model as a common approach will aid in focusing research, education, and extension needs to serve the turfgrass industry and society. For example, in addition to the traditional turfgrass science 4-year university programs, perhaps a future program would be the addition of an environmental turfgrass/landscape science option where the focus would be on whole ecosystems management and the ability of students to integrate knowledge into implementable BMPs and EMS management protocols.

Literature Cited

Australian Water Authority (AWA). 2000. *Australian and New Zealand Guidelines for Fresh and Marine Water Quality.* Paper No. 4, Chapt. 9. Primary Industries. Australian Water Authority, Artarmon, NSW, Australia, http://www.deh.gov.au/water/quality/nwqms/index.html (13 December 2006)

Ayers, R. S. and D. W. Westcot. 1994. *Water Quality for Agriculture. FAO Irrigation and Drainage* Paper 29, Rev. 1. Reprinted 1994. Food and Agricultural Organization. Rome, Italy, http://www.fao.org/DOCREP/003/T0234E/T0234E00.htm#TOC (13 December 2006)

Barton, L. and T. D. Colmer. 2005. Irrigation and fertilizer strategies for minimizing nitrogen and leaching from turfgrass. *Proceedings of the 4th International Crop Science Congress*, Brisbane, Australia, 26 September–1 October, www.cropscience.org.au (13 December 2006)

Beard, J. B. and R. L. Green. 1994. The role of turfgrasses in environmental protection and their benefits to humans. *J Environ Qual* 23:452–460.

Bond, W. J. 1998. Effluent irrigation—an environmental challenge for soil science. *Austral J Soil Res* 36:543–555.

California Stormwater Quality Association (CASQA). 2003. *Stormwater Best Management Practice Handbook: New Development and Redevelopment.* California Stormwater Quality Association, Menlo Park.

California Urban Water Conservation Council (CUWCC). 2005. Memorandum of understanding regarding urban water conservation in California. Amended 2004, www.cuwcc.org (23 January 2006)

Carrow, R. N. 2006. Can we maintain turf to customers' satisfaction with less water? *Agric Water Mgt* 80(1–3):117–131.

Carrow, R. N. and R. R. Duncan. 1998. *Salt-Affected Turfgrass Sites: Assessment and Management.* John Wiley & Sons, Hoboken, New Jersey. 185 pp.

Carrow, R. N., R. R. Duncan, and C. Waltz. 2005. Best Management Practices (BMPs) for turfgrass water conservation. *Golf Course Superintendents Association of America Seminar Manual,* www.georgiaturf.com (13 December 2006)

Carrow, R. N., R. R. Duncan, and D. Wienecke. 2005a. BMPs: Critical for the golf industry. *Golf Course Mgt* 73(6):81–86.

Carrow, R. N., R. R. Duncan, and D. Wienecke. 2005b. BMPs approach to water conservation on golf courses. *Golf Course Mgt* 73(7):73–76.

Cathy, H. M. 2003. Water right—conserving our water, preserving our environment. International Turf Producers Foundation, www.TurfGrassSod.org (23 January 2006)

Cohen, S., A. Surjeck, T. Durborow, and N. L. Barnes. 1999. Water quality impacts by golf courses. *J Environ Qual* 28:798–809.

Council for Agricultural Science and Technology (CAST). 2002. *Urban and Agricultural Communities: Opportunities for Common Ground.* Task Force Report 138. CAST, Ames, Iowa.

Council for Agricultural Science and Technology (CAST). 2008. *Water Quality and Quantity Issues for Turfgrasses in Urban Landscapes.* Special Publication 27. CAST, Ames, Iowa.

Dodson, R. G. 2005. *Sustainable Golf Courses—A Guide to Environmental Stewardship.* John Wiley & Sons, Hoboken, New Jersey. 267 pp.

Duncan, R. R. and R. N. Carrow. 2000. *Seashore Paspalum: The Environmental Turfgrass.* John Wiley & Sons, Hoboken, New Jersey. 281 pp.

Duncan, R. R., R. N. Carrow, and M. Huck. 2000. Effective use of seawater irrigation on turfgrass. *USGA Green Section Record* 38(1):11–17.

Environmental Institute for Golf (EIFG). 2006. Web portal with links to published documents relating to BMPs for water management, integrated plant management, wildlife and habit management, energy and waste management, and environmentally sound siting/design/construction of golf courses. Documents range from site-specific to state levels. EIFG, Lawrence, Kansas, http://www.eifg.org/ (13 December 2006)

Environmental Literacy Council (ELC). 2005. About us. What is environmental literacy? *About ELC page*, www.enviroliteracy.org (13 December 2006)

Florida Department of Environmental Protection (FDEP). 2002. *Best Management Practices for Protection of Water Resources in Florida*. Department of Environmental Protection, Orlando, Florida, http://www.dep.state.fl.us/central/Home/MeetingsTraining/FLGreen/FLGreenIndustries.htm (13 December 2006)

Food and Agricultural Organization (FAO). 2005. Global network on integrated soil management for sustainable use of salt-affected sites, http://www.fao.org/landandwater/agll/spush/degrad.htm (13 December 2006)

Gibeault, V. A. 2002. Turf protects the environment, benefits health. University of California at Riverside Turfgrass Research Advisory Committee Newsletter, December 2002. University of California, Riverside.

Gold, M. V. 1999. *Sustainable Agriculture: Definitions and Terms*. Special Reference Briefs Series No. SRB 99–02, Updates SRB 94–05. National Agricultural Library, Beltsville, Maryland, http://www.nal.usda.gov/afsic/AFSIC_pubs/srb9902.htm (13 December 2006)

GreenCO and Wright Water Engineers, Inc. 2004. *Green Industry Best Management Practices (BMPs) for the Conservation and Protection of Water Resources in Colorado*. 2nd ed. GreenCO and Wright Water Engineers, Inc., Denver, Colorado, www.greenco.org (23 January 2006)

Hanson, B., S. R. Grattan, and A. Fulton. 1999. *Agricultural Salinity and Drainage*. Division of Agriculture and Natural Resources Publication 3375. University of California, Davis.

Harivandi, M. A. 1991. *Effluent Water for Turfgrass Irrigation*. Leaflet 21500. Cooperative Extension, Division of Agriculture and Natural Resources, University of California, Oakland.

Harivandi, M. A. 1999. Interpreting turfgrass water test results. *California Turf Culture* 49(1–4): 1–6.

Huck, M., R. N. Carrow, and R. R. Duncan. 2000. Effluent water: Nightmare or dream come true? *USGA Green Section Record* 38(2):15–29.

Irrigation Association (IA). 2005. *Turf and Landscape Irrigation Best Management Practices*, September 2004 on-line publication, www.irrigation.org (23 January 2006)

Marcum, K. B. 2005. Use of saline and nonpotable water in the turfgrass industry: Constraints and developments. *Proceedings of the 4th International Crop Science Congress*, Brisbane, Australia, 26 September–1 October 2004, www.cropscience.org.au (13 December 2006)

Oster, J. D. 1994. Irrigation with poor quality water. *Agric Water Mgt* 25:271–297.

Pettygrove, G. S. and T. Asano. 1985. *Irrigation with Reclaimed Municipal Wastewater—A Guidance Manual*. Lewis Publishing, Chelsea, Michigan.

Rawson, J. M. 1995. *Congressional Research Service Report to Congress: Sustainable Agriculture*. CRC Report for Congress, 95-1062 ENRD. Congressional Research Service, Committee for the National Institute for the Environment, Washington, D.C., www.ncseonline.org/NLE/CRSreports/Agriculture/ag-14.cfm?&CFID=962773&CFTOKEN=76886153 (31 December 2006)

Rhoades, J. D., A. Kandiah, and A. M. Mashali. 1992. *The Use of Saline Waters for Crop Production*. FAO Irrigation and Drainage Paper #48. Food and Agricultural Organization, Rome, Italy.

Scott, C. A., N. I. Faruqui, and L. Raschid-Sally. 2004. *Wastewater Use in Irrigated Agriculture*. CABI Publishing, CAB International, Wallingford, Oxfordshire, U.K.

Snow, J. T. (ed.). 1994. *Wastewater Reuse for Golf Course Irrigation*. Lewis Publishing/CRC Press, Boca Raton, Florida.

Stevens, D., M. Unkovich, J. Kelly, and G. G. Ying. 2004. *Impacts on Soil, Groundwater and Surface Water from Continued Irrigation of Food and Turf Crops with Water Reclaimed from Sewage*. Australian Water Conservation and Reuse Research Program, CSIRO Land Water, Australia, www.clw.csiro.au/awcrrp/ (13 December 2006)

Thomas, J. R., J. Gomboso, J. E. Oliver, and V. A. Ritchie. 1997. *Wastewater Re-use, Stormwater Management, and National Water Reform Agenda*. CSIRO Land and Water Research Position Paper 1, Canberra, Australia.

U.S. Environmental Protection Agency (USEPA). 1998. *Water Conservation Plan Guideline*. EPA 832-D-98-001. U.S. EPA, Office of Water, Washington, D.C.

U.S. Environmental Protection Agency (USEPA). 2003. *National Management Measures to Control Nonpoint Source Pollution from Agriculture*. EPA 841-B-03-004. U.S. EPA, Office of Water, Washington, D.C.

U.S. Environmental Protection Agency (USEPA). 2004. *Guidelines for Water Reuse*. EPA/625/R-04/108. U.S. EPA, Office of Water, Washington, D.C.

U.S. Environmental Protection Agency (USEPA). 2005a. *National Management Measures to Control Nonpoint Source Pollution from Urban Areas*. EPA-841-B-05-004. U.S. EPA, Office of Water, Washington, D.C.

U.S. Environmental Protection Agency (USEPA). 2005b. *Environmental Management Systems (EMS)*. EPA website for environmental management systems information, http://www.epa.gov/ems/index.html (13 December 2006)

U.S. Environmental Protection Agency (USEPA). 2006. *Introduction to the Clean Water Act*, EPA website for water, http://www.epa.gov/watertrain/cwa/ (13 December 2006)

University of Florida (UFL). 2006. *Florida Green Industries: Best Management Practices for Protection of Water Resources in Florida*. Florida Department of Environmental Protection and University of Florida Pubs, June 2002, University of Florida, Gainesville, http://duval.ifas.ufl.edu/Agriculture/Commercial%20Horticulture/BMPs/bmp_manual_and_ifas_references.htm (13 December 2006)

University of Georgia (UGA). 2006. University of Georgia IPM Programs, Web portal for UGA Entomology Department integrated pest programs, Department of Entomology, Athens, Georgia, http://ipm.ent.uga.edu/ (13 December 2006)

Vickers, A. 2001. *Handbook of Water Use and Conservation*. Waterplow Press, Amherst, Massachusetts.

Wade, G. L., J. T. Midcap, K. D. Coder, G. Landry, A. W. Tyson, and N. Weatherly, Jr. 2003. *Xeriscape™ A Guide to Developing A Water-wise Landscape*. Revised from 1992 version. Cooperative publication between Georgia Water Wise Council, Marietta, Georgia, and the University of Georgia Cooperative Extension Service, Bulletin 1073, http://pubs.caes.uga.edu/casepubs/pubcd/B1073.htm (13 December 2006)

World Health Organization (WHO). 2005. *Guidelines for the Safe Use of Wastewater and Excreta in Agriculture and Aquaculture—1989 Guidelines*. Revised release. World Sanitation and Health (WSH), www.who.int/water_sanitation_health/wastewater (13 December 2006)

Yiasoumi, B., L. Evans, and L. Rogers. 2003. Farm water quality and treatment. *Agfact AC.2*, 9, www.agric.nsw.gov.au/reader/3825 (13 December 2006)

Appendix A: Workshop Participant List

Bruce Augustin
The Scotts Miracle-Gro Company
TEL: 937-644-7646
bruce.augustin@scotts.com

James Baker
Consultant
TEL: 515-268-1979
jlbaker@iastate.edu

James Beard
International Sports Turf
  Institute
TEL: 979-693-4066
FAX: 979-693-4878
isti@neo.tamu.edu

Greg Bell
Oklahoma State University
TEL: 405-744-6424
FAX: 405-744-6424
greg.bell@okstate.edu

Bruce Branham
University of Illinois
TEL: 217-333-7848
FAX: 217-244-3219
bbranham@uiuc.edu

Marc Campbell
Salt River Project
TEL: 602-236-2354
FAX: 602-236-2159
mccampbe@srpnet.com

Robert N. Carrow
University of Georgia
TEL: 770-228-7277
FAX: 770-412-4734
rcarrow@griffin.uga.edu

Jolie Dionne
RCGA
TEL: 514-705-3889
FAX: 905-845-7040
jdionne@rcga.org

Michael Dukes
University of Florida
TEL: 352-392-4092
FAX: 352-392-4092
mddukes@ufl.edu

Ron Duncan
Turf Ecosystems LLC
TEL: 678-481-0936
FAX: 210-568-6698
turfecosystems@yahoo.com

Kimberly Erusha
USGA
TEL: 908-234-2300
FAX: 908-781-1736
kerusha@usga.org

Douglas Fender
D. Fender & Associates
TEL: 847-381-7860
FAX: 847-381-4172
doug@dougfendertravel.com

Calvin Finch
San Antonio Water System
TEL: 210-233-3649
FAX: 210-233-5339
cfinch@saws.org

Mica Franklin
Aquatrols Corporation
TEL: 770-845-3392
mica.franklin@aquatrols.com

Frank Gasperini
RISE
TEL: 202-872-3843
fgasperini@pestfacts.org

Peter Gonzales
Bureau of Reclamation
TEL: 702-293-8610
FAX: 702-293-8774
anking@lcusbr.gov

Gary Grinnell
Las Vegas Valley Water District
TEL: 702-258-3909
FAX: 702-259-8299
Gary.Grinnell@lvvwd.com

Patrick Gross
USGA
TEL: 714-542-5766
FAX: 714-542-5777
pgross@usga.org

Midrar Haq
Grow Green International
TEL: 0092-300-2077983
midrarulhaq@yahoo.com

Ali Harivandi
University of California
TEL: 510-639-1271
FAX: 510-748-9644
maharivandi@ucdavis.edu

James Henry
Univ. of Cal. Coop. Ext.
TEL: 951-683-6491
FAX: 951-788-2615
mjhenry@ucdavis.edu

Dean Hicks
Hicks Interprises Inc.
TEL: 702-283-1916
FAX: 702-896-0863
deanhicks@cox.net

Brian Horgan
University of Minnesota
TEL: 612-624-0782
FAX: 612-624-4941
bphorgan@umn.edu

T. Kirk Hunter
Turfgrass Producers
  International
TEL: 847-649-5555
FAX: 847-649-5678
khunter@turfgrasssod.org

Richard Johnson
Penn State University
TEL: 814-865-8080
FAX: 814-863-8175
rhj3@psu.edu

Kevin King
USDA–ARS
TEL: 614-292-9806
FAX: 614-292-9448
Kevin.King@ars.usda.gov

Kelly Kopp
Utah State University
TEL: 435-797-1523
FAX: 435-797-3376
Kelly.kopp@usu.edu

Bernd Leinauer
New Mexico State University
TEL: 505-646-2546
FAX: 505-646-8085
leinauer@nmsu.edu

Gregory Leyes
ISK Biosciences Corp.
TEL: 440-357-4645
FAX: 440-357-4661
leyesg@iskbc.com

Steve Lohman
Denver Water
TEL: 303-628-5999
steve.lohman@denverwater.org

Rene Longoria
Silver Lakes Association
TEL: 760-245-1606
FAX: 760-246-7610
rslong103@aol.com

Eric Lorenz
Penn State University
TEL: 814-865-1074
esl1@psu.edu

Kenneth Marcum
Arizona State University–East
TEL: 480-727-1213
FAX: 480-727-1236
Kenneth.marcum@asu.edu

Joseph Massey
Mississippi State University
TEL: 662-325-4725
FAX: 662-325-0374
jmassey@pss.msstate.edu

Ed McCoy
Ohio State University
TEL: 330-263-3884
FAX: 330-263-3658
mccoy.13@osu.edu

Mike McCullough
Northern California Golf Association
TEL: 831-625-4653
FAX: 831-625-0150
mike@ncga.org

James Moore
USGA
TEL: 254-848-2202
FAX: 254-848-2606
jmoore@usga.org

Mary W. Olsen
University of Arizona
TEL: 520-626-2681
molsen@ag.arizona.edu

Punda Pai
Clark County Water Reclamation District
TEL: 702-639-5629
moxley@cleanwaterteam.com

Andrew Richardson
American Water Works Association
TEL: 602-275-5595
FAX: 602-267-1178
arichardson@greeley-hansen.com

Mike Richardson
University of Arkansas
TEL: 479-575-2860
FAX: 479-575-8619
mricha@uark.edu

Susan E. Rose
Colorado St. Univ. Coop. Ext.
TEL: 970-244-1841
FAX: 971-244-1700
susan.rose@mesacounty.com

Thomas Salaiz
University of Idaho
TEL: 208-397-4181
FAX: 208-397-4311
tsalaiz@uidaho.edu

Bob Shearman
University of Nebraska–Lincoln
TEL: 402-472-0022
FAX: 402-472-8650
rshearman1@unl.edu

James Snow
USGA
TEL: 908-781-1074
FAX: 908-781-1736
jsnow@usga.org

Laurie Trenholm
University of Florida
TEL: 352-538-0788
FAX: 352-392-1413
ltrenholm@ifas.ufl.edu

Adriana Ventimiglia
Clark County Water Reclamation District
TEL: 702-639-5629
moxley@cleanwaterteam.com

Hope Yu
City of Golden
TEL: 328-384-8184
FAX: 328-384-8161
hyu@cl.golden.co.us

Appendix B: Workshop Agenda

Water Quality and Quantity Issues for Turfgrasses in Urban Landscapes

Monday, January 23, 2006

1 p.m.	Session I. Opening Session: Urban Perennial Grasses in Times of a Water Crisis: Benefits and Concerns Douglas H. Fender	
2 p.m.	Session II. Regulatory Considerations Beth Hall	
3 p.m.	Break	
3:20 p.m.	Session III. Municipal Water Use Policies Andrew W. Richardson	
4:20 p.m.	Session IV. Water Quality Protection, Storm Water, and Contaminant Runoff Michael P. Kenna	

Tuesday, January 24, 2006

7 a.m.	Breakfast
8 a.m.	Session V. Soil Water Management Ed McCoy
9 a.m.	Session VI. Pesticide and Nutrient Fate—Leaching Bruce Branham
10 a.m.	Break
10:20 a.m.	Session VII. Pesticide and Nutrient Fate—Runoff Kevin W. King and J.C. Balogh
11:20 a.m.	Session VIII. Pesticide and Nutrient Modeling Stuart Z. Cohen, Qingli Ma, LaJan Barnes, and Scott Jackson
12:20 p.m.	Luncheon

1:15 p.m.	**Session IX. Species Aspects in Landscape Water Conservation** Dale A. Devitt and Robert L. Morris
2:15 p.m.	**Session X. Grass Water Use/Quantity Strategies, An Overview** Bingru Huang
3:15 p.m.	**Break**
3:35 p.m.	**Session XI. Turfgrass Cultural Practices for Water Conservation** Robert C. Shearman
4:35 p.m.	**Session XII. Irrigation Technology, Design, and Practices; ET-based and Moisture Sensing** David F. Zoldoske

Wednesday, January 25, 2006

7 a.m.	**Breakfast**
8 a.m.	**Session XIII. Recycled Water, Gray Water, and Salinity** M. Ali Harivandi, Kenneth B. Marcum, and Yaling Qian
9 a.m.	**Session XIV. Multiple Factor Considerations in Low-Precipitation Landscape Approaches** James B. Beard
10 a.m.	**Break**
10:20 a.m.	**Session XV. Practical Experiences in Urban Water Conservation Programs** Calvin Finch
11:20 a.m.	**Sessions XVI. Concluding Session. Best Management Practices: Holistic Systems Approach** Robert N. Carrow and Ronny R. Duncan

Water Quality & Quantity Issues for Turfgrasses in Urban Landscapes

Water Quality & Quantity Issues for Turfgrasses in Urban Landscapes